# NATO ASI Series

## Advanced Science Institutes Series

*A series presenting the results of activities sponsored by the NATO Science Committee, which aims at the dissemination of advanced scientific and technological knowledge, with a view to strengthening links between scientific communities.*

The Series is published by an international board of publishers in conjunction with the NATO Scientific Affairs Division

| | |
|---|---|
| A Life Sciences<br>B Physics | Plenum Publishing Corporation<br>London and New York |
| C Mathematical and Physical Sciences<br>D Behavioural and Social Sciences<br>E Applied Sciences | Kluwer Academic Publishers<br>Dordrecht, Boston and London |
| F Computer and Systems Sciences<br>G Ecological Sciences<br>H Cell Biology<br>I Global Environmental Change | Springer-Verlag<br>Berlin Heidelberg New York<br>London Paris Tokyo Hong Kong<br>Barcelona Budapest |

PARTNERSHIP SUB-SERIES

| | |
|---|---|
| 1. Disarmament Technologies | Kluwer Academic Publishers |
| 2. Environment | Springer-Verlag |
| 3. High Technology | Kluwer Academic Publishers |
| 4. Science and Technology Policy | Kluwer Academic Publishers |
| 5. Computer Networking | Kluwer Academic Publishers |

*The Partnership Sub-Series incorporates activities undertaken in collaboration with NATO's Cooperation Partners, the countries of the CIS and Central and Eastern Europe, in Priority Areas of concern to those countries.*

## NATO-PCO DATABASE

The electronic index to the NATO ASI Series provides full bibliographical references (with keywords and/or abstracts) to about 50000 contributions from international scientists published in all sections of the NATO ASI Series. Access to the NATO-PCO DATABASE compiled by the NATO Publication Coordination Office is possible in two ways:

- via online FILE 128 (NATO-PCO DATABASE) hosted by ESRIN, Via Galileo Galilei, I-00044 Frascati, Italy.

- via CD-ROM "NATO Science & Technology Disk" with user-friendly retrieval software in English, French and German (© WTV GmbH and DATAWARE Technologies Inc. 1992).

The CD-ROM can be ordered through any member of the Board of Publishers or through NATO-PCO, Overijse, Belgium.

2. Environment – Vol. 2

*The Partnership Sub-Series incorporates activities undertaken in collaboration with NATO's Cooperation Partners, the countries of the CIS and Central and Eastern Europe, in Priority Areas of concern to those countries.*

---

The volumes published as a result of these activities are:

Vol. 1: Clean-up of Former Soviet Military Installations. Edited by R. C. Herndon, J. E. Moerlins, J. M. Kuperberg, P. I. Richter, and I. L. Biczó. 1995

Vol. 2: Cleaner Technologies and Cleander Products for Sustainable Development. Edited by H. M. Freeman, Z. Puskas, and R. Olbina.

# Cleaner Technologies and Cleaner Products for Sustainable Devlopment

Edited by

Harry M. Freeman

U.S. Environmental Protection Agency
Cincinnati, OH 45268, USA

Zsuzsa Puskas

Engineering Bureau for Human Protection
& Modern Process Technology
Budapest, Hungary

Rada Olbina

Tufts University
Medford, MA 02155, USA

Springer

Published in cooperation with NATO Scientific Affairs Division

Proceedings of the NATO Advanced Research Workshop "Cleaner Technologies and Cleaner Products for Sustainable Development", held in Budapest, Hungary, September 12–15, 1994

ISBN-13: 978-3-642-79674-6 e-ISBN-13: 978-3-642-79672-2
DOI: 10.1007/978-3-642-79672-2

CIP data applied for



**Softcover reprint of the hardcover 1st edition 1995**

Typesetting: Camera-ready by authors/editors
SPIN: 10475249 31/3136 – 5 4 3 2 1 0 – Printed on acid-free paper

**DEDICATION**

In the memory of my mother,

**Desa Koštan-Olbina,**
a citizen of Sarajevo,

with love and thankfulness,

Rada Olbina

# PREFACE

As industrial developing and developed countries continue to struggle with issues related to how standards of living can be maintained and increased without reducing environmental quality, it is becoming increasingly clear that there must be developed cleaner production technologies and cleaner products that will result in waste elimination at the source. Otherwise, society is locked into a never ending cycle of developing and adopting better and more efficient treatment, disposal, and clean-up processes. This latter approach is just not cost effective.

Recognizing this fact, researchers around the world in industry, academia, and government are attempting to develop the needed more environmentally friendly products and the related better production technologies for producing these products. This work is often characterized as pollution prevention research, sustainable industrial development research, or industrial ecology research. Though much remains to be done, there have been to date impressive technological breakthroughs that have lead to the virtual elimination of many hazardous waste streams and the severe reduction of others.

On September 12-15, 1994 some 60 individuals from 22 NATO and Cooperation Countries gathered in Budapest in an Advanced Research Workshop, Cleaner Technologies and Cleaner Products for Sustainable Development, sponsored by the NATO Science Programme. The papers included in this book resulted from that meeting. This volume contains the proceedings of that workshop.

As countries around the world continue to make progress on economic and environmental fronts, the types of technologies, strategies, and training programs that are covered in this volume will come to be of increasing importance. It is our hope that the efforts of the workshop organizers, the presenters, and the publishers of this volume will contribute in some small way to achieving the goal of advancing the quality of life of individuals around the world both economically and environmentally.

Harry Freeman

Zsuzas Puskas

Rada Olbina

## CONTENTS

COMPARING CLEANER PRODUCTION EDUCATION PROGRAMS IN THE U.S. AND IN AUSTRIA . . . . . 1
*H.-J. Altenburger, J. Fresner, J. Jantschgi and H. Schnitzer*, Austria

CLEAN TECHNOLOGIES VERSUS END OF PIPE TECHNOLOGIES A CHANCE FOR ECONOMIES IN TRANSITION . . . . . 13
*M. Nussbaumer and H. Schnitzer*, Austria

SUSTAINABLE DEVELOPMENT & CLEANER PRODUCTION - HOW DO THEY FIT TOGETHER . . . . . 25
*H. Schnitzer*, Austria

AN OVERVIEW OF THE GREAT LAKES POLLUTION PREVENTION INITIATIVE - CANADA . . . . . 33
*T. Tseng*, Canada

CLEANER PRODUCTION IN THE CZECH REPUBLIC . . . . . 43
*V. Dobeš*, Czech Republic

LIFE CYCLE SCREENINGS OF TWO FOOD PRODUCTS . . . . . 53
*B.P. Weidema*, Denmark

ENVIRONMENTAL SITUATION IN ESTONIA AND POSSIBLE PATHWAYS FOR ITS IMPROVEMENT . . . . . 65
*R. Munter*, Estonia

CLEAN TECHNOLOGY RESEARCH IN GERMANY: EXAMPLES OF IMPLEMENTED PROJECTS . . . . . 81
*G.W. Panzner*, Germany

RESEARCH AND DEVELOPMENT IN CLEAN TECHNOLOGIES IN GERMANY . . . . . 99
*K. Schroeter*, Germany

HOW TO PREVENT NITROUS GAS EMISSION IN NITRIC ACID TREATMENT OF METALS . . . . . 127
*J. de Jonge*, Hungary

EMISSION REDUCTION OF CHLORINATED HYDROCARBONS BY A MINERAL-BASED METAL-OXIDE CATALYST . . . . . 133
*S. Kántor and Z. Adonyi*, Hungary

WASTE REDUCTION STRATEGIES IN THE CHEMICAL INDUSTRY . . . . . . . . . . . 141
*P. Mizsey and Z. Fonyo*, Hungary

TO THE REDUCTION OF ENVIRONMENTAL POLLUTION DURING
REPRODUCTIONS USED BY DEADS AND WASTES IN HUNGARY . . . . . . . . . . . 153
*J. Mujzer*, Hungary

ENVIRONMENTAL EDUCATION AND SOME EXAMPLES OF TEACHING
FOR SUSTAINABLE DEVELOPMENT IN CHEMICAL ENGINEERING
AT THE TECHNICAL UNIVERSITY BUDAPEST . . . . . . . . . . . . . . . . . . . . . . . . 157
*G. Széchy and I. Szébenyi*, Hungary

CEMENT INDUSTRY - IS IT CLEAN? . . . . . . . . . . . . . . . . . . . . . . . . . . . . . 165
*F. Tamás*, Hungary

LITHUANIAN EXPERIENCE IN DISSEMINATING P2
IDEAS AND TECHNIQUES IN INDUSTRY . . . . . . . . . . . . . . . . . . . . . . . . . . . 179
*J. Staniskis and A. Sileika*, Lithuania

THE DUTCH ECODESIGN PROGRAM . . . . . . . . . . . . . . . . . . . . . . . . . . . . . . 185
*A. Zweers and H. te Riele*, Netherlands

INTRODUCTION OF MODERN METHODS FOR DEGREASING
OF METAL PARTS . . . . . . . . . . . . . . . . . . . . . . . . . . . . . . . . . . . . . . . . . . 191
*G. Kjaerheim*, Norway

OBSTACLES IN APPLICATION OF CLEANER PRODUCTION
IN THE POLISH INDUSTRY . . . . . . . . . . . . . . . . . . . . . . . . . . . . . . . . . . . . 207
*A. Doniec*, Poland

THE PHARE ENVIRONMENTAL PROGRAMME IN POLAND . . . . . . . . . . . . . . . . 213
*A. Mierzwinski, L. Dygas-Ciolkowska and P.L. O'Brien*, Poland

THE EVALUATION OF TEXTILE AIDS AND CHEMICALS
USED BY THE TEXTILE INDUSTRY IN ORDER TO
REDUCE ENVIRONMENTAL IMPACT . . . . . . . . . . . . . . . . . . . . . . . . . . . . . . 225
*J. Rouba*, Poland

AN OVERVIEW OF POLLUTION PREVENTION IN PORTUGAL . . . . . . . . . . . . . . 243
*A. Gonçalves, V. Limpo and F. Bartolomeu*, Portugal

CONSIDERATIONS ON SUSTAINABLE DEVELOPMENT AND
ITS PERSPECTIVE IN ROMANIA . . . . . . . . . . . . . . . . . . . . . . . . . . . . . . . . . 251
*C.A.L. Negulescu and V. Harceag*, Romania

MAJOR CONSTRAINTS TO SUCCESSFUL IMPLEMENTATION OF CLEANER TECHNOLOGIES TO SLOVAK INDUSTRY . . . . . . . . . . . . . . . . . . 259
*M. Chodák*, Slovakia

WASTE MINIMIZATION IN A DYE HOUSE . . . . . . . . . . . . . . . . . . . . . . . . . . . 271
*J. Petek and P. Glavič*, Slovenia

CLEANER ENERGY PRODUCTION TECHNOLOGIES FOR FOSSIL FUELS . . . . . . . . . . . . . . . . . . . . . . . . . . . . . . . . . . . . . . . . . 279
*A.T. Atimtay and A. Aksoy*, Turkey

PREVENTION OF AIR POLLUTION PROBLEMS IN THERMAL POWER PLANTS IN TURKEY . . . . . . . . . . . . . . . . . . . . . . . . . . 291
*A.T. Atimtay and G. Özenirler*, Turkey

A LIFE CYCLE STUDY FOR COMMON CONSTRUCTION MATERIALS . . . . . . . . . . . . . . . . . . . . . . . . . . . . . . . . . . . . . . . . . . 303
*A. Müezzinoğlu and H. Toprak*, Turkey

POLLUTION PREVENTION STUDIES IN COTTON TEXTILE INDUSTRY IN TURKEY . . . . . . . . . . . . . . . . . . . . . . . . . . . . . . 331
*F. Sengul and N. Azbar*, Turkey

UTILISATION OF COPPER SLAGS FOR CHEMICALS PRODUCTION . . . . . . . . . . . . . . . . . . . . . . . . . . . . . . . . . . . . . . . . 351
*H. Sesigür, E. Açma, O. Addemir, A. Tekin and E. Ekinci, Turkey*

AN APPROACH FOR REDUCING AOX IN THE BLEACHERY EFFLUENTS OF A TURKISH PULP AND PAPER PLANT . . . . . . . . . . . . . . . . . . . 359
*Ü. Yetiş, A. Selçuk and C.F. Gökçay, Turkey*

CLEAN SYNTHESIS AS ENVIRONMENTALLY FRIENDLY CHEMISTRY . . . . . . . . . . . . . . . . . . . . . . . . . . . . . . . . . . . . . . . . . 371
*C.J. Suckling*, UK

EDUCATIONAL MATERIALS FOR WASTE MINIMIZATION AND POLLUTION PREVENTION EDUCATION AND TRAINING . . . . . . . . . . . . . . 399
*D.Y. Boon*, U.S.A.

ALL ABOUT POLLUTION PREVENTION . . . . . . . . . . . . . . . . . . . . . . . . . . . 417
*H.M. Freeman*, U.S.A.

GREEN TECHNOLOGY BY DESIGN: A NEW PARADIGM FOR ENGINEERING EDUCATION FOR SUSTAINABLE DEVELOPMENT . . . . . . . . . . . 443
*C. Overby*, U.S.A.

EVOLVING CONCEPTS IN LIFE CYCLE ANALYSES . . . . . . . . . . . . . . . . . . . . . 455
*M.R. Overcash*, U.S.A.

PREVENTION OF INDUSTRIAL PROCESS ACCIDENTS FOR
SUSTAINABLE DEVELOPMENT . . . . . . . . . . . . . . . . . . . . . . . . . . . . . . . . . 471
*A.S. Markowski*, Poland

INDEX . . . . . . . . . . . . . . . . . . . . . . . . . . . . . . . . . . . . . . . . . . . . . . . . . 483

# Comparing Cleaner Production Education Programs in the U.S. and in Austria

H.-J. Altenburger, J. Fresner, J. Jantschgi, and H. Schnitzer
Stenum GmbH
Sparbersbachgasse 22/2
8010 Graz
Austria

Abstract: *This paper describes the results of an analysis of US-American cleaner production education programs. They are reviewed for their transferability to Austria. Concepts for training programs for students in technical and business courses, permit writers and the management of companies will be developed taking into account the specific characteristics of industry in Austria.*

Stenum consists of a working group at the Institute of Chemical Engineering of Graz University of Technology and an engineering company. Stenum is active in three fields: education, research, and consultancy in cleaner production. People of Stenum are involved in teaching in chemical engineering courses at university giving lectures in cleaner production, in seminars (e.g. for the chamber of commerce and similar institutions) and conduct research work concerned with methodological assessment of the environmental performance of companies, balancing materials and energy in industrial processes, improving the efficiency of processes and on how to approach industry towards sustainability.

Stenum has conducted about 30 case studies in cleaner production together with companies of the machine building, pulp and paper, textile, food processing, wood and furniture, printed board circuit manufacturing and plating industry in Austria. The projects are funded both by the industry and the public sector, that is the City of

NATO ASI Series, Partnership Sub-Series, 2. Environment – Vol. 2
Cleaner Technologies and Cleaner Products
for Sustainable Development
Edited by H. M. Freeman, Z. Puskas, and R. Olbina

Graz, the federal state of Styria, the Austrian Departments of the Environment, Youth and Family and of Science and the Innovation and Technology Funds.

The results of these projects /Heitzinger, 1991; Sage, 1993; Fresner, 1994/ show, that hundreds of options for reducing waste and emissions at the source exist, scores of measures can be taken, ranging from good housekeeping, which amounted for over 30% of all measures, changing raw materials, closing internal cycles (e. g. for water) to improving process technology (e. g. by using better control systems and adopting new technologies) to research projects where there is no technology available to solve existing problems. And it could be shown that there was an economic benefit to many of the measures taken. These results resemble those from the Dutch Prisma Project and the Landskrona project /Huisingh, 1991/.

However, there are many barriers to the wide adoption of the preventive approach to solve existing waste and emissions problems:

- There is a lack in knowledge and in confidence, that cleaner production is an approach, that works and yields economic benefit
- There is a lack in knowledge on existing technology
- There is a lack in experience with these technologies
- There is a lack in knowledge of the true costs of waste and emissions (disposal or treatment costs and costs of the raw materials, plus indirect costs (which can range from permitting to future liabilities for environmental risks)
- There is a lack of committment of the people involved
- There is a lack in leadership of the management in the companies

Most Austrian companies are small or medium sized (up to 500 employees), they show a broad variation in sectors, products and technologies. In small companies there are usually limited resources for research and development, and only little staff to deal with environmental aspects of production. Usually, the one in charge of the production also has to deal with security, health, permitting, and the environment. So his time for considering environmental effects of the production processes usually is scarce.

These results again resemble the ones from other reports. From interviews with companies involved in cleaner production projects, we concluded that one of the most important reasons for the deficiencies is insufficient education. Therefore we have focused this study on successful education approaches, methods, and materials. The facts for the preparation of this paper were taken from a recent study trip to 10 relevant US institutions.

**The situation in the US**

In the US, states like California or Massachusetts are most advanced in their environmental policy and in their support to industry in meeting pollution minimization goals. The federal government has recently also put stress on environmental policy. The US strive for international leadership in environmental protection by "harnessing the power of the marketplace in the service of the environment", as President Bush stated in 1989 /President of the United States, undated/. There is no choice between a strong economy and a healthy environment. The US have to work to achieve both while sacrificing neither: Economic development and environmental protection go hand in hand.

The administration developed a National Strategy for Environmental Quality, based on six goals:

- Harnessing the power of the marketplace
- Managing natural resources as responsible stewards
- Promoting creative partnerships
- Developing cooperative international solutions
- Preventing pollution before it starts
- Enforcing environmental laws firmly and fairly

And, as it seems to us, the strategy has worked: The budgets for pollution prevention efforts have grown considerably and programs like the EPA's 33/50 and programs for clean energy technologies have proved highly successful /EPA, 1993/.

The national environmental policy act (NEPA) is the nation's environmental magna charta. With passage of NEPA in 1969 the US adopted a national environmental policy. The NEPA established the council on environmental quality (CEQ) to advice the president and assist federal agencies for compliance and the act mandated procedural requirements to fulfill its substantive goals.

One of the eight priority efforts of this council was environmental education. Others include environmental data acquisition, international affairs, environmental awards, etc.

To coordinate environmental education efforts, EPA was required to establish an office of environmental education. This office shall guarantee good organization and consistency among the US environmental education efforts. It emphasized in 1992 that environmental educators have a dual change:

1. Instill an environmental ethic in America's young people that will prepare them to deal responsibly with the environment throughout their lives and
2. Raise the environmental awareness of adults as informed consumers in the global shift toward sustainable development and pollution prevention.

The goals set were:
(1) Expand communication with environmental experts
(2) Develop partnerships to increase effectiveness and to maximize use of limited resources
(3) Increase the number and skills of teachers who infuse environmental issues into their curricula

(4) Encourage and support grassroots efforts

(5) Promote careers in the environmental field

(6) Provide educators and the public with access to appropriate environmental education materials

(7) Educate the general public to make informed decisions

(8) Reach beyond national boundaries

Some examples for current efforts to achieve these goals are:

Goal 1: A strong body of research and literature exists along with an experienced cadre of environmental educators. Federal agencies are establishing communication and advisory networks with educators to ensure that the nation has full benefit of their expertise. Advisory councils to Congress and the EPA have been installed to make recommendations and assessing the quality of environmental education. A Pollution Prevention Advisory Group shall recommend the EPA on developing pollution prevention educational materials for students and on teacher training.

Goal 2: federal agencies try to engage all stakeholders to form new coalitions, respond to local needs and involve groups, that traditionally are not involved with environmental education, such as industry, religious organizations and senior citizen associations.

Goal 3: federal agencies are supporting the development of training programs and materials to ensure that all teachers, including in nonscientific fields, have the opportunity to receive training and materials that will allow the infusion of environmental issues into their curricula. In 1992 the EPA awarded a 1.6 million US$ grant to a consortium of academic institutions, corporations and nonprofit organizations headed by the University of Michigan to develop and operate a national environmental education and training program.

A main topic on this issue is also to establish partnerships with youth organizations, business, health organizations, state and government, academia and also senior citizen groups.

Goal 4: local initiatives are supported; they range from community cleanup to developing computer games. EPA selects projects for their innovation and applicability in other communities.

Goal 5: As the nation needs environmental scientists, engineers and other professionals capable of developing preventive solutions to environmental problems, the EPA promotes environmental careers in environmental degree programs through a number of programs, such as:

- Internship and fellowship
- President's environmental youth awards
- National environmental education awards
- President's environment and conservation challenge awards
- National network for environmental management studies

Goal 6: A wealth of environmental education materials exists throughout the US. However finding appropriate good material can be difficult. To provide easy access to the best materials available, EPA is developing an interactive computer system for environmental education materials produced by federal agencies and other organizations They plan to have it available shortly. The quarterly periodical "Earth notes" provides a forum for innovative ideas from educators and for the exchange of teaching ideas concerning environmental education in the elementary grades.

Goal 7: Federal agencies are improving their ability to communicate complex scientific and technical information to the public. To prepare citizens to make informed decisions on the environment, projects as the following are sponsored: Development of a media strategy with the help of private partners and the media, including press, television, and radio. Other awareness programs include programs of the US coast guard and a program by the US postal service: They distribute a brochure to explain, what each person can do at work or at home to help the mail and mother earth. The brochure contains tips as the design of recyclable mailings (use of waterbased adhesives, water based inks, uncoated paper), dealing with suppliers that use environmentally sound products, papers, and packaging, printing on both sides of the paper, reuse and recycle packaging materials, target mailings, and reduction of undeliverable mail by keeping addresses current. Poster are used throughout the postal lobbies countrywide to promote the three R's: Reduce, Reuse, Recycle.

In California, there is a "Pollution Prevention Week" every year, when the governor delivers a speech on pollution prevention, TV and radio spots on pollution prevention are broadcasted, local programs are introduced to the public, company visits are organized, etc. to heighten the awareness of the people.

Goal 8: Environmental risks transcend political boundaries. Therefore international information exchanges and joint training ventures have become an environmental education priority, and federal agencies are supporting initiatives around the world. These programs include sport fishing education, trilateral cooperation of the US with Canada and Mexico, Fulbright Fellowships, sponsorship for institutions in Hungary, Puerto Rico and the Ukraine.

## The situation in Austria

In Austria, a national environmental plan is being discussed following an initiative of the Department of the Environment, Youth and Family. The discussion has been conducted for two years now involving authorities, unions, the academia, non-governmental organizations and the industry. A national environmental plan is scheduled to be approved in spring 1995. Whereas e.g. in general efficient use of resources and energy will be demanded as well as switching to renewable energies to reach sustainability, up to now no reference is made explicitly to education and training.

The awareness of the general public has significantly risen by recent laws requiring separation of waste to paper, glass, metals, biodegradables and plastics and by collecting dues for packaging materials.

Separate courses in cleaner production up to our knowledge are incorporated mainly in engineering curricula in chemical engineering in Graz and in environmental engineering in Leoben. Also business economics curricula in Linz and at the University of economics in Vienna contain lectures in cleaner production. Other initiatives include lectures in cleaner production in courses in training programs for environmental managers, e. g. held by the trade chambers.

Activities to implement waste minimization in companies are up to our knowledge mainly conducted by us and our cooperation partners. There are programs in place, e.g. in Upper Austria now to train other consultants in the methods used for analyzing companies, defining weak points in the use of materials and energy and eliminating them.

Up to now, these efforts are not coordinated in a formal way.

## Conclusions

Cleaner production and design for the environment, which means taking cleaner production into consideration as early as possible in the product design phase involve everyday awareness and complex cooperation. All the actors in the field of industrial production (consumers, business people, technical people) must develop an understanding of environmental issues.

As it is crucial for a successful waste minimization program in any company to obtain the support of the top management and have a formal committment to environmental goals in the company, it would be important for a nationwide education program in cleaner production to have a relevant statement by the national government to recognize the importance of environmental education to the success of a national policy striving for a sustainable development of our industry. This would make it easier to get committment by the relevant individuals and institutions, coordinate the efforts and allocate resources to the programs.

Target groups who will have to be addressed include

- the public in general, as a general environmental consciousness has to develop on an informed basis
- students at university level, as they will be the future decision makers, designers and operators as well as students in high school
- management of the companies, as they have to develop strategies and allocate resources for a sustainable industrial development
- permit writers of the authorities, who have to give companies a fair chance to try pollution prevention approaches
- consultants, who have to assemble a body of knowledge in business strategies and technologies to support the companies

Key goals of each education program then would be:

- building up responsibility
- teamwork
- developing a basic understanding of ecology
- teaching systematic and multi media approaches to pollution prevention
- practicing common sense
- giving actual examples of how pollution prevention works

- hands-on-experience
- teaching where to obtain relevant (technical, legal, toxicological etc.) information from, and how to interpret and apply it
- teaching total cost analysis: learn how to put a dollar on it
- providing incentives: award systems

The main focus of these programs will have to vary according to the task groups although they will basically address the same key elements. In general, to our opinion, existing channels for the distribution of information should be used. Thus, in the long run, pollution prevention should be included to all relevant curricula, ranging from primary school to university courses. The aim of the programs is not to create an additional flood of paper, but to get the relevant messages through: So what is needed is a pool for information, tools and materials, out of which teachers and trainers can take self explaining units for their courses. The people at the University of Michigan call these "compendia": They are currently developing such compendia for business, law, and chemical engineering /Keoleian, 1994/. They illustrate the connection of the relevant field to environmental problems, discuss educational aims and offer materials to use in business, law, and engineering courses. Comparable resource books should be compiled for Austria also.

For the education of the public education at an early stage, e. g. in primary schools is very important. Early education is also very important to develop a culture of pollution prevention among the people who run businesses, design processes and operate them as well. And most of the key elements are not only concerned with acquiring technological knowledge but with acquiring skills like analyzing the way processes are run, interdisciplinary groupwork, and building up general understanding and awareness.

Special seminars should be offered to the target groups (business, law, and technical people) tailored to fit their needs to spread pollution prevention efficiently. As these seminars have to account for a highly complex and broad area, they should feature open problems and teamwork. By "open problems" we mean problems, which do not necessarily have clear, well defined solutions. Finding strategies to address problems, setting priorities, developing cooperations should be the main fields addressed to help develop creative approaches to problem solving.

Key elements to use are:

- roll playing
- simulation
- games
- computer tools: games, cdroms, spreadsheets, touchscreens
- videos
- group work
- discussions with experts
- site visits, hands-on experiences
- real world examples and cases

Models for similar courses are available /Aqua, 1994, Luskin, 1994, Savage, 1994/.

**Recommendations**

So we feel, that the important steps to take to have a broad knowledge of cleaner production are:

1. Establish a national policy of cleaner production
2. Check, in which courses in K12 to university pollution prevention topics can be integrated (e. g. strategic product design and costing in business courses, liabilities in laws, reaction synthesis and use of solvents in thermal chemical engineering)
3. Create a pool for tools to teach cleaner production and provide the relevant teachers access to that pool (cases, videos, spreadsheets, readers, ...)

To help small and medium sized companies immediately, special seminars should be held, where people learn to approach their problems actively and in a creative way. These seminars could last for some three days and address the top to middle management. They should be focused on open ended cases, interaction and discussion. Product exhibitions and site visits should be included. Wherever possible, actual examples from the companies should be prepared and used.

**References**

Aqua, E. personal communication

EPA Pollution prevention accomplishments: 1993, EPA-100-R-94-002, Office of the Administrator

Fresner, J., 'Avoiding wastes and emissions in industry: experiences from Austria', J. Cleaner Prod. 1994, Volume 2 Number 1, 43

Heitzinger, P., Schnitzer, H., Nussbaumer, M., Regatschnigg, H., Thiel, W., Grabher, A., Heitzinger, M. and Novak, K. 'Ökoprofit', Institute of Chemical Engineering, Graz University of Technology, 1991

Huisingh, D. and Baas, L. W., Eur. Water Pollut. Control 1991, 1(1), 24

Keoleian, G. personal communication

Luskin, J. personal communication

President of the United States, U.S. Actions for a better enviroment - a stustained commitment, undated brochure

Sage, J. Thesis, Institute of Chemical Engineering, Graz University of Technology, 1993

Savage, D. personal communication

# Clean Technologies versus End of Pipe Technologies
## A Chance for Economies in Transition

Martin Nußbaumer, Hans Schnitzer
STENUM GmbH
Sparbersbachgasse 22/2
8010 Graz
AUSTRIA

**Abstract:** *This paper deals with a project for waste minimisation in Eastern European countries commissioned by the Austrian Federal Ministry for Science and Research and elaborated by the working group STENUM at the Graz University of Technology. Out of the experiences of the project some evaluations will be made regarding future chances from End of Pipe Technologies towards Clean Technologies in the countries in transition. First results of two case studies in a textile company in Slovenia and a food processing company in Hungary are very encouraging and will lead to further efforts in the direction of Cleaner Production in these countries.*

## Background

In the year 1993 the working group STENUM was commissioned by the Austrian Federal Ministry for Science and Research with a project for waste minimisation in Eastern European countries. The idea was to carry out some case studies with the PREPARE-Method developed in a EUREKA/EUROENVIRON project. PREPARE (**P**reventive **E**nvironmental **P**rotection **A**pproaches in **E**urope) was initiated by the Dutch Ministry of Economies and is based on the PRISMA project for waste minimisation carried out in ten medium and small size enterprises in the Amsterdam and Rotterdam region.

NATO ASI Series, Partnership Sub-Series, 2. Environment – Vol. 2
Cleaner Technologies and Cleaner Products
for Sustainable Development
Edited by H. M. Freeman, Z. Puskas, and R. Olbina

It was our task to find some interested partners in Slovenia and Hungary and after that to find some interested companies in Hungary and Slovenia.

**Tab. 1: Project partners**

| |
|---|
| × University of Maribor - Department of Chemical Engineering; Prof. Glavic |
| × University of Veszprem - Department for Information Technology and Process Automatisation |
| × Hungarian Academy of Science |
| × Engineering Bureau HPMPT Ltd Budapest |
| × STENUM-Graz University of Technology/Austria |

**The main goals of this project are:**

1. To translate the PREPARE-manual into Hungarian and Slovenian language as well as preparing basic material for information and promotion
2. To carry out case studies which should
   - × minimise the environmental impact of the analysed companies
   - × strengthen their economic performance
   - × reduce input of raw materials and energy
3. To evaluate the results regarding several aspects. One of these aspects is the evaluation of the social, economic and political circumstances for the implementation of integrated Cleaner Production in comparison with End of Pipe Technologies in economies in transition.

In order to achieve the third goal we spreaded a questionnaire among 130 experts in Western and Eastern Europe asking for information in the following fields:

1. Statistical information about food and textile industry
2. Environmental regulations
3. Economic preconditions
4. Experiences made in environmental projects

Our consulting partners tried to find and select companies in branches which are important for the respective country. They found partners in the textile (Slovenia) and the food industry (Hungary). The textile industry for example is the second most important branch at the Slovenian industry with 14% of all workers employed in industry and a 10% share in total Slovene exports value.

## Sustainable Development

In the year 1987 the World Commission on Environment and Development published its report „Our Common Future". This work commissioned by the General Assembly of the United Nations was expected *„to*

- *propose long-term environmental strategies for achieving sustainable development by the year 2000*
- *recommend ways concern for the environment may be translated into greater co-operation among developing countries and between countries at different stages of economic and social development and lead to the achievement of common and mutually supportive objectives that take account of the interrelationships between people, resources environment and development*
- *to consider ways and means by which the international community can deal more effectively with environmental concerns and*
- *to help define shared perceptions of long-term environmental issues and the appropriate efforts needed to deal successfully with the problems of protecting and enhancing the environment, a long-term agenda for action during the coming decades, and aspirational goals for the world community"* [The World Commission on Environment ad Development; Our Common Future; page IX; Oxford University Press 1987]

According to this goals the Commission worked out a concept for sustainable development. The essence is that this *„development is a process of change in which*

- *the exploitation of resources*
- *the direction of investments*
- *the orientation of technological development and*
- *institutional change*

*are all in harmony and enhance both current and future potential to meet human needs and aspirations"* [The World Commission on Environment ad Development; Our Common Future; page 46; Oxford University Press 1987]

There are a lot of efforts all over the world to reach these very demanding goals. One of these activities is the introduction of Clean Technologies instead of End of Pipe Technologies which were used till now in general. A comparison which emphasises the differences between Clean Technologies (CT or CP for Cleaner Production) and End of Pipe (EoP) Technologies is proposed in a draft paper by the Erasmus University of Rotterdam (D. Huisingh) and is shown in the following table:

**Tab. 2: Comparison CP-EoP**

| CP | EoP |
|---|---|
| **Handling Pollutants** | |
| × Pollution prevention through measures at the source | × Pollutants controlled by filters and treatment methods |
| × CP is a continuos work to achieve higher standards | × Only goal is to fulfil standards set by the authority |
| **Environmental control** | |
| × CP is integrated part of product and process development | × EoP starts when problems arise |
| × Waste is considered to potential resources | × Activities cost the company time and extra money |
| × CP includes technical and non technical approaches | × Only technology oriented |
| **Responsibility** | |
| × All people trough the company are responsible | × Improvements addressed by environ-mental experts |
| **Focus** | |
| × CP is a continuous evaluation and innovation | × EoP solves only one strict problem |
| × Goal is to meet customers needs with minimal impact on health and environment | × Quality is defined as meeting customers requirements |

A comparison between the goals of CP on one hand and EoP on the other hand shows that CP is the more likely way into the direction of a sustainable development. The consideration of technical as well as non-technical approaches, the reduction of material and energy input as well as the consideration of the whole system (company) in CP means a reorientation of technological development and also a change in institutions and attitudes. These are only a few examples out of a great variety which would break up the frame of this paper. The basic goal of the paper is to give some good arguments why CP is a chance for economies in transition.

## Ways to Clean Production in Eastern Europe

According to the ideas of Clean Technology as integrated environmental protection which solves problems at the source, a lot of successful case studies have been carried out in Western Europe. The main goals of the performed case studies were to minimise the environmental impact and to strengthen the economic performance of the analysed companies.

Since the quiet revolutions in the late eighties and early nineties in Central and Eastern Europe (CEE), a lot of efforts were made to adopt the Western system of free market economy. It was thought that the transition from central planned economy to free market economy would automatically lead to environmental benefits. Beside this the representatives of these countries also started to consider Clean Production as a step out of the economic and ecological disaster. *„Bronislaw Kaminski, Poland's Minister of Environment said in autumn 1990: The process of reversing [the] neglect of environmental protection in Western countries continued for many decades and has been crowned with success thanks to radical reform of the economic structure and investment in clean production technology. This same mode of thinking guides our intentions.“* [R. Manser (1993); The squandered dividend-The free market and the environment in Eastern Europe; Earthscan Publications Ltd-London; page 70].

Literature shows that the countries in transition already started to anchor the ideals of sustainable development in their environmental protection actions. Two examples -one from Hungary and one from Slovenia- are showing the efforts in this field.

**Environmental Law Act No. II; Hungary 1976**

This act expresses the right of every citizen [... to live in an environment worthy of man ... ] and then it states that

[... in course of economic activities the aspects of environmental protection should be inforced *in accordance with the capabilities of the economy* ...]

The restriction in this sentence is one of the most serious problems in Eastern Europe. There is a strict way the politicians follow in these countries. First the economy has to be restructured and afterwards there can be a discussion about changing into the direction of sustainable development. A lot of experts agree in the fact that this way will cause the same mistakes as the Western European economies made several years before.

Another example, the Environmental Protection Act of the Republic of Slovenia published in 1993 is more likely to follow the principles of CP. Only a few examples of basic principles should illustrate this.

**Environmental Protection Act - Republic of Slovenia 1993**

- Principle of Prevention
- Principle of Liability
- Principle of Restitution for Environmental Strain
- Principle of Incentives
- Principle of Public Access and Dissemination of Information

These principles are well expressed, but as in all other countries in Europe, the implementation of these ideas might not succeed in near future. In some of the CEE countries for example the laws for environmental protection were as least as strict as the laws in Western Europe but were neglected in most cases.

First project results from our project PREPARE-Eastern Europe (shown in table 3) as well as experiences from other projects like a Norwegian/Polish and a Slovakian/Norwegian project on CP are available now.

**Tab. 3: Project results-Calculated reduction**

| | Calculated reduction | Percentage from today's amount |
|---|---|---|
| Waste Water | 153.000 $m^3/a$ | 21% |
| Condensate | 3.060 t/a | 10% |
| Dyes | 7,3 t/a | 24% |
| Textile Auxiliaries and Chemicals | 73,8 t/a | 14% |
| Waste Heat | 1.928.000 MJ/a | 3% |

The main goal of the initial phase of the project PREPARE-Eastern Europe was to find some options with low investment costs and high reduction potential in order to show people the effectiveness of preventive environmental protection and to motivate the employees for further efforts. Some practical improvements in the production have already been made.

## Occurred Problems

Beside the very motivating benefits mentioned above also few problems occurred.

- ***Uncertainty regarding the future:*** The most important problem was the uncertainty regarding the future of the companies. The companies are now in the situation of privatisation and therefore unwilling to start new projects before the circumstances can be made clear.
- ***Uncertainty regarding future laws:*** This was also a restriction during the initial phase of the project. Most of the CEE countries have now democratically elected governments which started to review and renew the laws for environmental protection.

- ***Lack of faith in the new idea:*** Most of the usual environmental technologies as filters or waste treatment (incinerators) and waste water treatment are well known in the CEE countries though they are not established everywhere. CP the new approach of handling environmental problems is not considered as a tool to solve the rised problems.
- ***Lack of academic acknowledgement:*** Regarding to the traditional assessment of University researchers, industrial CP projects are not acknowledged as real science.
- ***Lack of experience:*** Organising (getting money from authorities) and handling of these projects turned out to be very difficult.

## Ideas to overcome problems

Most of the problems mentioned can be can be solved using only few easy measures which are shown in the following list.

- ***Training of people in CEE countries:*** This means on one hand direct training on the job (exemplary case studies) and on the other hand
- ***Support with information material and methods:*** Translation of manuals and guidelines for the successful implementation of CP in industry as well as information material to anchor the idea in the minds of representatives of authorities. This support means also
- ***Open access to data bases*** with case studies and technologies for waste minimisation.
- Information also includes motivation of people by ***stressing the self interest of Clean Production***. While End of Pipe Technology eliminates pollutants, CP reduces the mass flow at the source using preventive methods. In many cases it leads to savings of money and often also time.

## Chances for economies in transition

Beside these very easy measures for implementation of Clean Technologies in CEE countries there have to be substantial changes in the way of thinking. Mrs. Iza Kruszewska from Greenpeace International believed that the region of the CEE countries was by 1993 on the way to become [ ... Europe's dumping ground for obsolete technologies, withdrawn products and waste ... ] and this has to be avoided.

There are several good ideas of experts all around the world to avoid the failures of the industrialised countries by changing into a sustainable development on an earlier stage.

- ***Changes in environmental laws (promoting CP)***

  Some experts in the OECD area suggested that the lowest release levels attained by a company in a given sector -assuming these release levels fulfil existing laws- should become the new standard for all companies in this field after a reasonable period (suggested 5-10 years). Existing facilities not meeting the standard within the proposed period would have to cease operation. This means that companies have no intrinsic right to contaminate the environment.

- ***Economic instruments***

  The introduction of economic instruments like

  a) charges/taxes on emissions

  b) charges/taxes on products

  c) deposit refund systems

  d) tradeable permits

  e) enforcement incentives

  would be a possible way to promote CP. Beside this restrictive measures there could also be supporting measures like subsidies or tax provisions for ecological aware companies. The implementation of the polluter pays principle -according to the principle of strict liability- is also considered to be an impulse for companies to implement CT.

- ***Voluntary agreements***

  Beside all the regulations set up by authorities, voluntary agreements and ECO-labelling programs are a good opportunity to promote CP. The 33/50 program (reduction of defined toxic chemicals up to 33% until 1992 and reduction up to 50% until 1995) of the United States EPA is a good example for the effectivnes of voluntary agreements.

  In Europe the EC set up the order Nr. 1836/93 for the voluntary installation of eco-management systems in companies. This order is expected to motivate companies to improve their awareness in environmental fields like the ISO 9000 in field of quality management.

The influence of political, economic and social circumstances on the application of these proposals on the current branches in Slovenia an Hungary will be part of the final report of the project PREPARE Eastern Europe. The report also should contain new proposals for improved preconditions for implementing CP in the CEE countries A thesis based on the results of this report is also in preparation.

## REFERENCES

HORVATH V. (1994) Overview of the legal and economic control of environmental protection; Review Environment, Nature Conservation, Building and Regional Policy in Hungary Budapest 1994

HUISINGH D. (1994) A comparison between Pollution Control and Cleaner Production attitudes and approaches for achieving improved environmental quality and sustainability. Unpublished

MANSER R. (1993) The squandered dividend-The free market and the environment in Eastern Europe. Earthscan Publications Ltd., London

MINISTRY of the Environment and Regional Planning (1993) The Environmental Protection Act, Republic of Slovenia 1993

NIKOLIC M. et. al. Restructuring of Slovene Textile Industry-Strenghts, Weaknesses, Risks and Chances; Tekstilec-letnik 36-7-8/93, page 229

NUSSBAUMER M. et. al. (1994) Preventive integrated environmental protection in Eastern Europe. Paper presented at the 1st European Roundtable on Cleaner Production Programs, Graz Austria-October 1994

OECD (1994) Managing the environment-The role of economic instruments, Paris 1994

OECD (1992) Technology and Environment-Government policy options to encourage cleaner production and products in the 1990s, Paris 1992

PETEK J., GLAVIC P. Waste minimisation in a dye house; Paper at the NATO meeting in Budapest; September 1994

SCHNITZER H., NUSSBAUMER M. (1994) Co-operation in Cleaner Production between Austria, Hungary and Slovenia. Paper presented at an UNEP-Meeting in Kaunas/Lithuenia, September 1994

WORLD COMMISSION ON ENVIRONMENT AND DEVELOPMENT (1987) Our common future, Oxford University Press, Oxford

# Sustainable Development & Cleaner Production -How Do They Fit Together

Hans Schnitzer
Institute for Process Engineering
Graz University of Technology
Inffeldgasse 25
A-8010 Graz
Austria

## 1) What is the nature of our environmental problems?

For thousands of years, man's development based on renewable forms of materials and energy. Their use and conversion has always played an important role for the inhabitants of the planet, and apart from a period of negligible length - relative to evolutionary and historical time scales - the renewable sources have been the only ones accessible to mankind (Sorensen 1979). Food and (fire-) wood - the basic requirements for well-being, mobility and comfort - are converted and stored solar energy. When man began to harness animals, his demand for materials and energy increased rapidly. Starting with human settlements, the society developed specialisation and division of labour. Politicians, priests, artists, craftsmen and soldiers did not any more care for their food and energy by themselves. Trade demanded an infrastructure, and so goods and energy had to be provided commercially.

Up to two centuries ago, everything used has been based on biomass or minerals. Each farmer used a remarkable fraction of his farmland for energy supply: to feed horses and oxen or to grow wood as construction material and energy source. Rather late in history, the technical use of solar energies not based on biomass began. Sailing boats for transportation on sea and wind mills for pumping and grinding may have been the first and only applications for hundreds of years.

First industries in Europe have been located on sites, where energy has been available: along the rivers and near the large forests. Hydro power and char coal was essential for the blacksmith, wood was heating material for the sugar refinery. It

NATO ASI Series, Partnership Sub-Series, 2. Environment – Vol. 2
Cleaner Technologies and Cleaner Products
for Sustainable Development
Edited by H. M. Freeman, Z. Puskas, and R. Olbina

was easier to transport and store ore and metals than to bring energy to the mines. Many industrial centres of today still are located in such regions, where hydro power or wood is available. Close to these industries, lack of fire wood and deforested mountains were one of the first essential "environmental" problems. Forestry was far from being "sustainable" that time. Mining of coal helped to overcome this "environmental crisis" about two hundred years ago. Through the beginning utilisation of crude oil, not only a new source for energy was found, but also a stuff that could be used to produce materials of unknown variety and fantastic characteristics.

But new problems arose. Emissions from industries and human settlements into air, water and soil caused serious problems with animals, humans and nature in general. The answers have been very simple in the beginning: higher stacks and longer pipes. But *dilution is no solution to pollution* and so mankind soon had to look for new strategies. *End of Pipe Technologies* (EoP) seemed to be rather effective in the next step. Waste water treatment plants, waste incineration and flue gas cleaning helped to reduce imissions of many components, but on the other hand cause high costs and require further input of chemicals and energy.

Today environmental problems are of a new dimension again. Thinning of the ozone layer, global warming and the nitrification of the ground water in many agricultural regions cannot be fighted with EoP-Technologies effectively anymore. A change in the pattern of production, consumption and dispersion is necessary now.

## 2) Sustainable Development

The complex nature of the environmental problems of today in conjunction with social problems (unemployment, north-south conflict, etc.) forced the development of a new approach. What is required now, is an answer that includes resource conservation, environmental protection, social compatibility and the responsibility for future generations. The World Commission on Environment and Development (1987) developed the following definition:

*Sustainable development is development that meets the needs of the present without compromising the ability of future generations to meet their own needs. It contains within it two key concepts:*

*the concept of 'needs', in particular the essential needs of the worlds poor, to which overriding priority should be given; and*

*the idea of limitations imposed by the state of technology and social organisation on the environment's ability to meet present and future needs.*

Apart from a general agreement to this definition, there is no consensus about how to act practically in the different fields of human activities: agriculture, production, products, traffic, tourism, energy supply, etc.

One of the central postulates of Sustainability is the reduction of the material flows of any kind down to a level of about 10% of the present. This mainly regards to all streams coming from or going into the biosphere, while streams within the antroposhere could stay in any size in principal. Of course, there is an inter-linkage between internal and external streams and the energy needed, so that a reduction of the flows in these three fields has to go hand in hand.

Up to now, environmental protection did not contribute to Sustainability, since the classical EoP strategies rather added mass flows by the installation of further equipment demanding more energy and more chemicals. Cleaner Production on the other hand could be able to reduce environmental hazards and make a step towards Sustainability at the same time, if applied in a^right way.

## 3) Cleaner Production

Cleaner Production (CP) concentrates at the minimisation of the generation of wastes and emissions rather than to eliminate them when they have been produced. There are several attempts to define Cleaner Production. One of the best definition has been set up by the United Nations Environmental Program UNEP:

Cleaner Production is the conceptual and procedural approach to production that demands that all phases of the life-cycle of a product or of a process should be addressed with the objective of prevention or minimisation of short and long-term risks to humans and to the environment.

Cleaner Production is one of the fundamental strategies within a sustainable industrial development. Since in any economic and social system products have to be manufactured, the way how we organise our factories and enterprises is essential to the development.

Cleaner Production means that the best technology available is applied for

- product modifications
- changes in process operation and technology
- changes in raw- and input materials

The "best" technology will always be depending on the environment. Environment in this case does not only mean nature and climate, but also the social contexts like education, social life, environmental awareness, type and size of industry, taxes, regulations and available local resources.

## 4) Sustainable Development - Postulate 1: Reduce all Mass Flow Rates in Size

De-materialisation of the mass flows between the environment and the antroposphere by the factor of 10 is one of the essential postulates within a sustainable development. Cleaner Production can contribute to that in many aspects:

- increase the efficiency of material use, thus reducing the need for virgin raw materials; strategies to do so are manifold as training of the personnel, increased automation, new equipment, etc.
- closing cycles within the production processes to reduce the mass flow rates into the environment (water, solvents, packaging, etc.)
- use of recycled material from other production processes or from the consumers to replace virgin raw materials
- at the same time, "wastes" from the production should possibly be offered as raw material for other companies (industrial symbiosis)
- energy consumption should be held at a minimum, at the same time using biogeneous and renewable resources as fuels as far as possible
- biodegradable wastes are to be brought back to the biosphere in such a way, that there is no harm to any form of life

We can see from this list that several aspects of Cleaner Production do perfectly fit into the requirements of Sustainability. This is mainly true regarding auxiliary materials and utilities. On the other hand, Cleaner Production does not help to reduce the enormous mass flows caused by the products.

5) SD - Postulate 2: Increase the Service Intensity of Products

For our comfort, we ask for a certain amount of services. In general, those are provided through goods. But beside goods, we need many other parameters to meet the needs and provide the comfort (fig. 1).

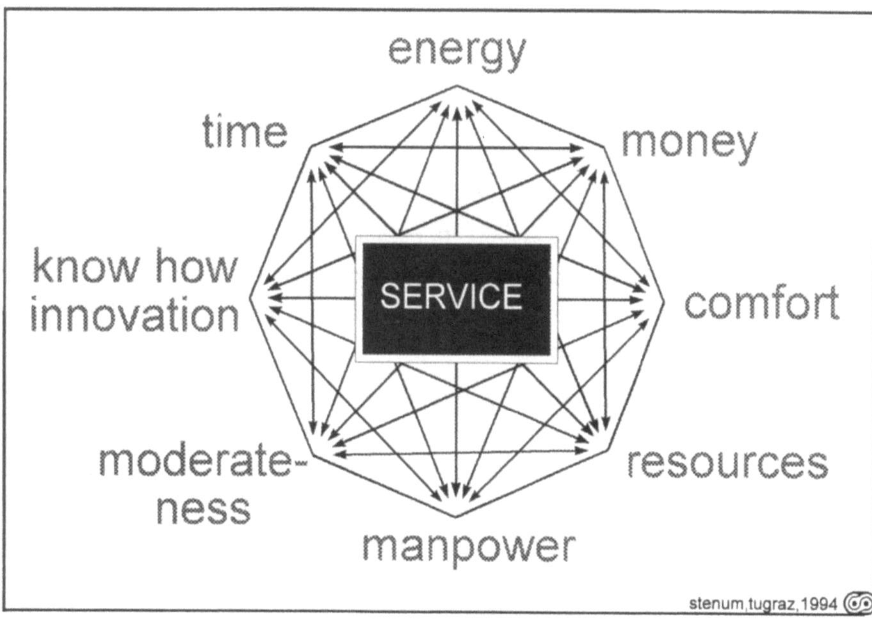


Fig. 1; Parameters required to meet the needs and provide services

Typical services asked for, are comfort / well-being, mobility and information. Our way of life usually meets these needs by providing goods as heating and air conditioning devices, cars and TV-sets. Simultaneously the utilisation of these goods requires energy and further material during the whole time of consumption. Sustainability demands a reduction of the amount of goods in exchange to more intelligent solutions: the *dematerialisation* of services through the replacement of mass through ideas. One has to be very carefully not to replace materials by energy as long as the energy supply system itself is not sustainable. Regarding energy, Sustainability means a fully solar based system, as far as we know now.

A further aspect here is the durability of the products. Enlarging the life-time of products, including the possibility of repair, reconstruction and reuse will help to cut down the amount of materials used. At present there is hardly any attempt within Cleaner Production to make steps in this direction.

## 6) SD-Postulate 2: Replace Scarce Materials by such Ones that are Plenty Available

One of the basic postulates of Sustainability is it, not to compromise the future generations ability to meet their needs. For production this means that the utilisation of scarce materials should be minimised. In general we can assume that anything which is scarce on the market is expensive at the same time, so that economic considerations will lead to a replacement of such materials through others - less scarce and therefor less expensive ones. But the future generation is not present on the market place, and consequently cannot influence the price. It must be one of the guidelines of Cleaner Production therefor to replace scarce materials before the price indicates the economic priority.

In times of transition to Sustainability, it is of course important to put priorities to scarce and toxic materials when we start with dematerialisation. But many researchers believe that there is no absolute criterion on what is toxic or scarce, so that a general reduction should be strived for. Many examples can be given from recent years, where "inert" components as halocarbons and $CO_2$ turned out as environmental problems through new theories.

## 7) SD-Postulate 3: Prefer Renewal Resources and Biotechnology

Since any non-renewable raw material will be scarce once upon a time, it will then be necessary to switch over to renewables or to recycled matter. Of course, this postulate goes hand in hand with the previous one, and the substitution has to follow the aspects of scarcity and toxicity.

But this is not the only reason to use renewal resources of energy and materials. We can assume, that nature can handle any substance that has been produced by biological or biotechnical processes through decomposition and recycling into natural cycles. There are exceptions to that, mainly mineral ones as coral reefs, but this can be taken for true regarding organic components.

## 8) SD-Postulate 4: Keep the Renewing Rate of Renewables Constant

Seen in a long term, renewables have to provide most of the goods to cover the needs mankind has: food, energy, building materials, clothing, etc. It is therefore of greatest importance not to diminish the rate, how these materials are reproduced by nature. There is danger in various aspects as soil, climate, clean water and biodiversity. Production is not permitted not diminish the capability of nature to evolve and produce.

There are many ways how industrial and private activities reduce the capability of nature to produce the different kinds of services we ask for: food, materials, energy, clean water, fresh air, a nice landscape, etc. One of the most serious risk is the change of he climate. But there are much more than this, as acid rain, heavy metals in the soil, lowering the level of ground water, use of land for other purposes, etc.

This postulate is closely connected with the first one, which generally claims the reduction of he mass flow rates.

## 9) Conclusion

It is not clear in any detail, how Sustainability looks like in the different aspects of human life. But we know very often, where Sustainability is violated through our activities. Cleaner Production as such does not inevitably lead to a sustainable development, but there are many aspects of correspondence.

One of the most serious aspects of the responsibility of our generation to future is hardly covered by Cleaner Production at present: the lack of sufficient jobs. Since man is one of the main weak points in many production processes, very often a replacement of an employee by a electronical control device can reduce losses and emissions. Cleaner Production supports the tendency towards automation in production, thus setting free people from the jobs. There is a need to search for new strategies, that help to overcome this problem effectively. On the other hand, Cleaner Production provides a chance for new developments:

- The utilisation of biogeneous resources offers a change for agriculture to produce new material for industry, thus getting the chance to create new jobs.
- Cleaner Production is the of the prerequisites for the compatibility of industry with agriculture and tourism, thus providing a chance for industry to survive outside the large industrial areas in regions where the environmental consciousness is high.
- recycling wastes and by-products is much more labour-intensive than the utilisation of virgin resources in general. Recycling and utilisation of by-products can create new jobs therefor.

Concluding what has been said, CLEANER PRODUCTION is an essential stone in the mosaic of Sustainability. But it is only a partial utopia. Many more aspects have to be included, as products, life stile, energy, transport, traffic, etc.

## 10) References

Fischer K., Schot J. (edt) (1993); Environmental Strategies for Industry. Island Press, Washington

Goldemberg J. (1988), et al.; Energy for a Sustainable World. John Wiley Eastern, New Delhi,

Sorensen B. (1979); Renewable Energy. Academic Press, London

World Commission on Environment and Development (1987); Our Common Future, Oxford University Press, Oxford

# AN OVERVIEW OF THE GREAT LAKES POLLUTION PREVENTION INITIATIVE - CANADA

Thomas Tseng, Ph.D., Manager
Pollution Prevention & Abatement Division
Environmental Protection
Ontario Region
Environment Canada
25 St. Clair Avenue East, 7th Floor
Toronto, Ontario
M4T 1M2 CANADA
Tel: (416) 973-1087
Fax: (416) 973-7509

## BACKGROUND

In March 1991 Environment Canada announced the Great Lakes/St. Lawrence Pollution Prevention Initiative. It is designed to produce verifiable reductions/elimination in the use, generation and release of persistent toxic substances; and to promote pollution prevention, as opposed to pollution control, as the preferred approach to environmental protection. The initiative is funded by Canada's Green Plan with an annual budget of $2 million for 7 years. Under this initiative, we have:

NATO ASI Series, Partnership Sub-Series, 2. Environment – Vol. 2
Cleaner Technologies and Cleaner Products
for Sustainable Development
Edited by H. M. Freeman, Z. Puskas, and R. Olbina

created several projects with industries and municipalities to develop and implement pollution prevention plans; and

established the Great Lakes Pollution Prevention Centre, working at arm's length, to provide both the public and private sectors with training and information services.

These activities have been complimented by the Canadian Council of Ministers of the Environment's "National Commitment to Pollution Prevention" which has confirmed minimizing or avoiding the creation of pollutants and wastes can be more effective in protecting the environment than treating them or cleaning them up after they have been created; and the pollution prevention approach is needed to secure a safe and healthy environment and a sound and prosperous economy.

In 1993, Environment Canada has undertaken a re-organization and established a Pollution Prevention Directorate at headquarters to integrate prevention into existing national programs. Within this Directorate, the National Office of Pollution Prevention co-ordinates departmental, national and international pollution prevention activities.

The Great Lakes Pollution Prevention Initiative (GLPPI) was originally housed in Ontario Region's Great Lakes Office. In November 1993, the Initiative was moved into the Environmental Protection Branch to facilitate integration with existing protection programs.

## STRATEGY

The Great Lakes, with one fifth of the world's fresh water, are a valuable resource for Canada and the United States. The main concern for the Great Lakes is persistent toxic substances. In July 1994, the governments of Canada and Ontario signed a major agreement detailing their co-operative activities to restore, protect and conserve the Great Lakes Ecosystem.

While this Canada - Ontario Agreement (COA) focuses considerable resources on the clean up of past pollution and the rehabilitation of degraded areas, it also makes pollution prevention a priority. To achieve the COA targets of significant interim reductions in the releases of persistent, bio-accumulative and toxic substances, the Environmental Protection Branch - Ontario Region is:

- integrating pollution prevention into existing environmental protection programs;
- developing and maintaining pollution prevention projects with industries and communities;
- demonstrating green technologies and clean processes; and
- strengthening actions by industries, communities and individuals through training, technical assistance and information services, mainly provided by the Great Lakes Pollution Prevention Centre (GLPPC).

**1994/5 FOCUS**

In 1994/95 resources are committed to achieve results on more than 50 projects covering commercial chemicals, hazardous waste management, federal facilities, industries and municipalities and the Lake Superior basin. Figure 1 provides an overview of these projects which fall under two main categories: program integration and voluntary agreements. In addition, Figure 1 shows a number of institutions which work with the Environmental Protection Branch closely on green technologies and clean processes, and on training, information and communications. The Province of Ontario is a major partner in many projects, and several initiatives also involve co-operative efforts with industry, other basin jurisdictions and the United States. The advancement of several projects' objectives is enhanced by key linkages with community groups, technical assistance agencies, training organizations and information service providers. Figure 1 also shows linkages with other major Environment Canada programs such as the National Pollutant Release Inventory and the Environmental Citizenship Program.

Table 1 provides a summary of voluntary pollution prevention agreements that have been initiated or being developed with industries, municipalities and federal facilities. In general, each of these agreements commits the participating industries and facilities to: (1) establish a list of target substances for reduction and elimination, the list is unusually drawn from information on persistent toxic substances; (2) develop an inventory of chemicals/substanes used, generated or released by the participant to identify opportunities for reducing/eliminating the targeted substances; and (3) develop and implement pollution prevention plans to achieve reduction and elimination. A steering committee comprised of government (federal and Ontario) and industry representatives meet regularly to establish project directions and prepare annual status reports on agreement implementation and pollution prevention project achievements.

Table 2 provides a summary of projects being undertaken/funded by the Environmental Protection Branch - Ontario Region in 1994/95 to achieve pollution prevention results. Again, emphases are placed on reduction/elimination of toxic substances. Most of these projects are implemented by existing Environmental Protection program staff who have been provided with pollution prevention training recently. They are now tasked to work on prevention projects to achieve program integration and to establish linkages with Environment Canada's other priority initiatives such as: the National Pollutant Release Inventory (NPRI), the Strategic Options Process (SOP) and the Accelerated Reduction/Elimination of Toxics (ARET) initiatives.

## SELECTED PROJECT UPDATES

**Federal Facilities**

For federal facilities operating in Ontario a pollution prevention strategy has been developed, focusing on hazardous waste minimization, reduction and elimination of persistent toxic substances, ozone depleting substances, and spill prevention. Delivery of demonstration projects, workshops and fact sheets, and implementation of a comprehensive program at CFB Trenton (8 Wing) are underway.

**Regional Municipality of Hamilton-Wentworth**

The Regional Municipality of Hamilton-Wentworth is developing a Comprehensive Municipal Pollution Prevention Management Plan. This plan aims to integrate preventive environmental solutions into existing municipal structures and develop a model template for other municipalities by 1997. The project consists of five components:

- inventory of by-laws, policies and programs;
- identification of prevention opportunities;
- implementation of pollution prevention plans;
- production of a generic pollution prevention template; and
- evaluation.

The inventory phase is well underway, identifying resources used, wastes generated, audit of bylaws, and the impact of municipal activities on the environment.

**Dry Cleaning Industry**

The Ontario Fabricare Association, Korean Dry Cleaners Association, Environment Canada and the Ontario Ministry of Environment and Energy are signing a three-year agreement to reduce the use of non-aqueous solvents in the fabricare industry, with particular focus on the possible elimination of chlorinated solvents.

As part of this agreement, Green Clean, Toronto's first clothes cleaning depot to demonstrate aqueous alternatives to chemical-based dry cleaning, opened in June 1994. The depot is testing technical viability and consumer acceptance of wet cleaning. Data will be used to evaluate the potential for solvent reduction, expand wet cleaning to other sites in Ontario and train cleaners.

**Automotive Manufacturing**

Initiated in 1992, the Automotive Manufacturing Pollution Prevention Project was the first pollution prevention agreement in the Ontario Great Lakes basin. Fifteen case studies included in the second progress report (April 1994) demonstrate 2.24 million kilograms of targeted substances have been reduced or eliminated from the manufacture of the automobile at the Ontario facilities of Chrysler, Ford and General Motors. Participating parties have agreed to extend the project to September 1995, and will prepare two more progress reports on further implementation of pollution prevention plans.

**Automotive Parts Manufacturing**

This project is being undertaken by the Automotive Parts Manufacturers' Association and six member companies, Environment Canada and the Ontario Ministry of Environment and Energy. In addition to the development of site-specific pollution prevention plans at six facilities, the project includes training and technical assistance. Research is in progress on clean technology for cleaning and degreasing, chromium releases, surface coatings and metal working fluids.

**Metal Finishing**

Four industry associations and the federal and provincial governments are developing and implementing site-specific pollution prevention plans. Methods and tools being used include training, testing of a planning manual by four facilities and the development and delivery of a technical assistance program. Pollution prevention opportunities will be discussed at sector workshops.

**Printing and Graphics**

This initiative involves printing and graphic art companies, equipment and chemical suppliers, and a newspapers association. A joint industry and government task force is finalizing the project agreement and action plans which will emphasize preventive practices at small and medium-sized printers and encourage pollution prevention planning and implementation.

**Lake Superior**

In 1991 the governments of Canada, the United States, Michigan, Wisconsin, Minnesota and Ontario agreed to an action plan to protect and restore the Lake Superior ecosystem through pollution prevention, regulatory measures and remedial programs. Canadian projects for 1994/95 include (1) customized pollution prevention training for key sectors in September 1994, (2) industrial demonstration projects, (3) public awareness campaigns for household and agricultural hazardous wastes, and (4) comprehensive pollution prevention plan for the City of Thunder Bay.

Figure 1

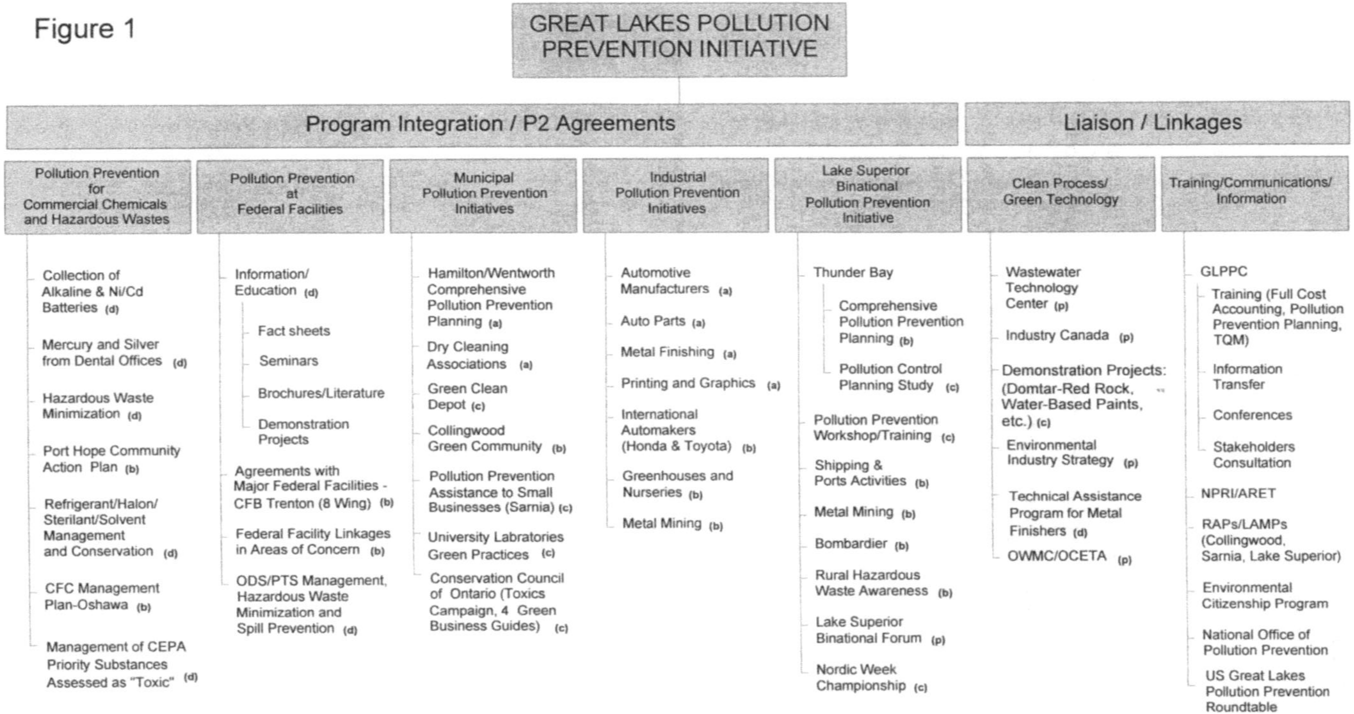


**(a)** : pollution prevention project initiated by industries/municipalities
**(b)** : pollution prevention project being developed with industries/municipalities
**(c)** : project funded by Environment Canada
**(d)** : project implemented by Environmental Protection - Ontario Region
**(p)** : partners

| | |
|---|---|
| ARET | - Accelerated Reduction/Elimination of Toxics |
| CEPA | - Canadian Environmental Protection Act |
| CFC | - Chlorofluorocarbons |
| GLPPC | - Great Lakes Pollution Prevention Centre |
| LAMPs | - Lakewide Management Plans |
| NPRI | - National Pollutant Release Inventory |
| OCETA | - Ontario Centre for Environmental Technology Advancement |
| ODS | - Ozone Depleting Substances |
| OWMC | - Ontario Waste Management Corporation |
| PTS | - Persistent Toxic Substances |
| RAPs | - Remedial Action Plans |
| TQM | - Total Quality Management |

September, 1994

Table 1: **VOLUNTARY P2 AGREEMENTS**

Purpose: to produce verifiable reductions/elimination of persistent toxic substances used, generated or released by the participating industries/communities

| | INDUSTRIAL | MUNICIPAL | FEDERAL FACILITIES |
|---|---|---|---|
| Agreements Initiated | Automotive Manufacturers | Hamilton - Wentworth | CFB Trenton * |
| | Auto Parts | Dry Cleaning | |
| | Metal Finishing | | |
| | Printing & Graphics | | |
| | Chemical Producers ¥ | | |
| Agreements Being Developed | Ontario Mining Association | Collingwood | |
| | Honda & Toyota | Thunder Bay | |
| | Greenhouses Association | Port Hope * | |

* Partnership forged equivalent to an agreement
¥ Observer

Table 2: **PROGRAM INTEGRATION / LINKAGES**

| Commercial Chemicals/ Hazardous Wastes | Federal Facilities | Municipal & Industrial | Clean Process & Green Technology |
|---|---|---|---|
| Management of PSL Substances assessed as "Toxic" | Information, Training, Factsheets, Brochures, Seminars | Green Clean Depot | Domtar Red Rock Demonstration Project (chlorine free) |
| Collection of Alkaline & Ni/Cd Batteries | ODS/PSL Toxics Management & Conservation | Conservation Council of Ontario: Toxics Campaign | Technical Assistance Program for Metal Finishers |
| Mercury & Silver from Dental Offices | Hazardous Wastes Minimization | P2 Assistance to Small Businesses (Sarnia) | Nutrient Recycling at Greenhouses |
| Refrigerant/Halon/ Sterilant Management/ Conservation | Spill Prevention | University Laboratories Green Practices | $NO_x$/VOC Codes Training for 1300 Ontario Dry Cleaners |
| CFC Management Plan for City of Oshawa | | Nordic Week Championship | |

# Cleaner Production in the Czech Republic

Vladimír Dobeš
The Czech Cleaner Production Center
Politických vězňů 13
110 00 Praha 1
The Czech Republic

KEYWORDS/ABSTRACT
/cleaner production/ waste minimisation/ Czech Republic/ environmental legislation/ sustainable development/ case studies/ VOC/ Czech-Norwegian Cleaner Production Project/ interactive training/ incentives/ barriers/ policy

This paper focuses on the experience from the first cleaner production activities in the Czech Republic which are initiated by the Norwegian Society of Chartered Engineers (NSCE), the World Environment Center (WEC) and the Czech Cleaner Production Center. The main identified driving forces and constraints consequent to these activities are screened and a possible future strategy is suggested.

## INTRODUCTION

At present Czech industrial enterprises find themselves in a period of transfer to a market economy and they have to face the heavy burden of environmental pollution caused by their production, the shortage of investments and the unsustainability of non-effective technologies. This situation creates an appeal for the implementation of cleaner production.

NATO ASI Series, Partnership Sub-Series, 2. Environment – Vol. 2
Cleaner Technologies and Cleaner Products
for Sustainable Development
Edited by H. M. Freeman, Z. Puskas, and R. Olbina

The Czech Republic has a set of new environmental legislation developed in the last 4 years following the blueprints of the EC standards. Sustainable development is laid down by Czech law in the Environment Protection Act. No. 17/92. This Act § 17 and the Vaste Act No. 238/91 § 3 state the duty to prevent the generation of pollution by reduction at the source of origin. However, waste minimisation techniques are not used systematically except the first activities in the field of cleaner production.

## THE FIRST ACTIVITIES

The first activities in the field of Cleaner Production in the Czech Republic are initiated and manged from abroad. The main organizations which are bringing the cleaner production approach into the Czech Republic are the Norwegian Society of Chartered Engineers (NSCE) and the Vorld Environment Center (VEC). One case study was launched by the University of Amsterdam.

The first cleaner production education programme for university students exists at the University of Economics in Prague within the frame of the Tempus programme.

### VEC

The first cleaner production (waste minimisation) case study in Czechia was successfully implemented by VEC in the big chemical enterprise Chemopetrol Litvínov in 1992/93. This project was launched in co-operation with the Czech Environment Management Center (CEMC) - the first Czech non-govrnmental and non-profit organisation which incorporated waste minimisation in its program. The case study was focused on VOC loses in the production line of the petrochemical plant.

**Some results:**

* the VOC emission was reduced by over 940 t/y (86% reduction)
* this represents savings of almost 4 millions Czech crowns (Kc) per year.

VEC has done further great step to commit top managment of other chemical plants after this demonstration. The directors of 8 biggest chemical plants in Czechia were invited to visit USA to see how cleaner production works in the country of its origin. One-week waste minimisation course for the managers, plant engineers and local consultants followed this trip in May.

Waste minimisation projects are being started in factories now and they are conducted by the trained in-plant engineers and local consultants. The VEC programme is financed by US A.I.D.

NSCE

Since 1992 the Czech-Norwegian Cleaner Production Project is being implemented within the Czech-Norwegian agreement on cooperation on the protection of the environment (signed by ministers of environment). NSCE participates on the project implementation for the Norwegian part.

The aim of the Czech/Norwegian Cleaner Production Project is to provide Czech enterprises with methodology for implementation of cleaner production. To reach this aim, the same model as earlier in Poland - the interactive training programme in Cleaner production - was implemented in the Czech Republic.

Two interactive training programmes on cleaner production were concluded successfully under the professional management of

NSCE at the Faculty of Technology, Zlín (FT Zlín). The interactive training programmes were performed by Norwegian lecturers and advisors and were composed of two basic parts:

1) classroom sessions providing teoretical introduction to market economy and cleaner production methodology
2) the practical application of the teoretical knowledge in the industrial enterprise implemented to a specific production line where the cleaner production case study is developed.

The training programmes were executed in the form of 240 hours postgraduate study and the successful participants are awarded the Professional Development Certificate in Environmental Management and Cleaner Production in Industry. The participants were professionals from enterprises, consultancy firms and government agencies.

The PDC certificate was awarded to 64 participants and 21 case studies were developed within the frame of this training courses.
The graduates of the course founded **the Association of Managers of Cleaner Production (AMCP)**, using their skills as lecturers and consultants.

**Some results:**

NSCE case studies developed within the frame of two interactive training programmes brought

* identification of possiblities for a significant reduction of the industrial pollution at examined production lines (some of good-housekeeping measures were already implemented within the frame of the case study)
* economic savings from good-housekeeping measures of more than 3,8 mil Kc and the potential for economic savings when implementing remaining recommended cleaner production measures 86,3 mil. Kc/year, while pay-back period for investments vary from 1 month up to 7,5 years). Some more information on the second training programme is given in the ANNEX.

This educational programme was assessed by e.g. EC DG XI (or sooner by OECD in Poland) as one of the most successful ones in the CEE countries. The interactive training course bringing together professionals from different industrial branches seems to be the most effective for the transfer of the cleaner production concept.

Follow-up actions in participating industries differs very much from company to company.

The ultimate goal of the Czech-Norwegian Cleaner Production Project is the founding of **the Czech Cleaner Production Center (CPC)**. CPC was formaly established in April 1994. CPC is taking the responsibility to continue in the activities started by Norwegians and to widen them. From September 1994 organises new interactive training courses situated already at two training centers (FT Zlín and University of Chemistry and Technology Prague). 15 participating companies agreed to recover a proportion of training and consultancy costs if they benefit from cleaner production measures.

## CONCLUSIONS

We can make folowing conclusions consequent to the first activities implemented in the field of cleaner production in Czechia:

POSITIVE

1) **Driving forces for the voluntary implementation of cleaner production by enterprises already exist:**

* **The development of market conditions** puts pressure on the present production to be more effective and pushes an increase of the production quality. An improvement of the competetive edge of a company is in every case the main driving force which includes the following ones.

* **Legislative provisions** concerning the reduction of the burden on the environment (restrictive regulation using limits, fees and penalties).

* **The increasing prices** of the production inputs, the increasing costs of the waste treatment and the costs of the pollution control support more effective use of resources.

2) **The implementation of cleaner production concept in Czech companies shows that it is possible to reduce significantly the burden to the environment caused by industrial pollution and to lower production costs at the same time** (see ANNEX).

3) **The potential for cleaner production in the Czech industry is higher than in the industrially most developed countries,** especially due to the extensive possibilities for a good-housekeeping improvement. Some end-of-pipe technologies have not been implemented yet.

4) **The creativity of the Czech technicians is very high and the technical knowledge concerning some cleaner production options already exists.**

NEGATIVE (existing barriers)

1) **Inertia causing resistence to changes in the approach towards production on the level of enterprise and government as well.** Company's usual approach to environmental problems is to wait for a regulation and than use end-of-pipe solutions.

2) **Old management system in enterprises.** Unclear ownership, unclear responsibility, lack of clear enterprise strategy.

3) **Lack of information on cleaner production.** Especially on waste minimisation method and its technical and organisational possibilities.

SUGGESTED STRATEGY

**The final aims of dissemination of cleaner production are:**

1) Cleaner production philosophy is a natural part of everybody's daily practice (he or she does not call it "cleaner production" any more).
2) Cleaner production information is avalable on the right place at the right time.
3) The structure supporting the implementation of cleaner production has been established.

**To reach these aims we can use the following tools:**

1) **Training and education**

2) **Encouraging examples:**
   - **case studies on production lines** - developed by:
     - approaching people on all levels in a company,
     - good evaluation of benefits so as to provide incentives to implement feasible measures,
     - creating funds for implementation of measures
     - evaluation of results (monitoring system, audit before and after to get clear results)
   - **demonstrations in a whole enterprise** - a new management system has to be introduced
   - **demonstrations in a chain of enterprises** (enterprises

exchanging goods and services are communicating with the aim to implement cleaner production approach on the life cycle of initial inputs).

3) **Government policy**: Indirect regulation, economic incentives, space for voluntary activities of industries, public disclosure of information about pollution generated by enterprises

4) **Cleaner production services** (focal points offering consultancy and training esp. for small and medium size industries).

5) **Information**: Communication with all actors influencing the use of resources with the objective to let the cleaner production philosophy penetrate into their daily activities. Establishment of the national database on cleaner production options, cleaner production manuals etc.

ANNEX

# Czech - Norwegian Cleaner Production Project

## II training programme 1993/1994

# CASE STUDIES - INFORMATIVE SUMMARY

## I. ECONOMICAL BENEFITS

| CASE STUDY No | 1 | 2 | 3 | 4 | 5 | 6 | 7 | 8 | 9 | 10 | 11 | total |
|---|---|---|---|---|---|---|---|---|---|---|---|---|
| A [thous. Kč] | 25 | 250 | 750 | 318 | 74 | ? | <90 | - | 210 | 516 | ? | ca 2230 |
| B [mil. Kč] | 4,9 | 4,5 | 33 | - | 3,2 | 17,4 | 0,9 | - | 0,4 | 7 | ? | ca 71 |
| C | + | + | - | + | + | + | + | + | ? | + | + | |

**A** = economic benefits from the cleaner production measures which were being implemented during the work on the case study (mainly "good housekeeping" measures without investments)
**B** = potential for economic benefits (from the investment needing measures with pay-back period from 1 month to 7 years)
**C** = commitment of the company management to implement investment needing measures

approx. exchange rate for the Czech krown (Kč): 1 USD = 28 Kč
1 XEU = 34 Kč

## II. CASE STUDIES AND THEIR POTENTIAL ENVIRONMENTAL BENEFITS (EB)

1) * **rubber industry**
- minimization of rubber waste
**EB: cca 75 t/y of rubber waste avoided (33%) and 86 MWh/y saved**

2) * **metalurgy**
- waste minimization in the production of the steel cold-pilgered tubes
**EB: decreasing of the amount of oiled wastes by 44t/y (44%)**

3) * **chemical industry**
- waste minimization in the production of titanium dioxide
**EB: 2 500 t/y of waste sulphuric acid avoided**

4) * **chemical industry**
   - minimization of waste produced by heavy oil gasification unit

   **EB: 47,8 t/y of hazardous waste avoided**

5) * **food industry**
   - reduction of water consumption and minimization of blood content in waste water

   **EB: water consumption reduced by 12 159 $m^3$/y (8%)**
   **blood content in waste water reduced by 20 t/y (50%)**
   **reduction of fat contents in waste water by 20 t/y (20%)**

6) * **chemical industry**
   - minimization of VOC emissions in the crude oil refinery

   **EB: VOC emissions reduced by ca 1 900 t/y**

7) * **chemical industry**
   - reduction of raw material losses in amonium chloride prod.

   **EB: raw material consumption reduced by ca 30%**
   **air emission of $NH_4Cl$ reduced**

8) * **metal scrap processing**
   - waste minimization on the crashing line of car wreckages

   **EB: reduction of metal waste by 350t/y**

9) * **household goods prod.**
   - waste minimization in the teflon coating

   **EB: reduction of inhalation of organic compounds from teflon dispersion**
   **consumption of teflon dispersion reduced by ca 1 800 l/y**

10) * **metallurgy**
   - minimization of the waste foundry sand

   **EB: 8 900 t/y of waste foundry sand avoided**

11) * **energy production**
   - reduction of ash generation in the heat generating plant

   **EB: ash generation reduced by 50%**
   **decrease of emissions: $SO_2$ by 13,5 kt/y**
   **$NO_x$ by 1,8 kt/y**
   **$CS_2$ by 0,5 kt/y**
   **$H_2S$ by 0,3 kt/y**

---

*This summary was prepared by the Norwegian Society of Chartered Engineers and the Czech Cleaner Production Center.*

*Source of information: Participants of the second cleaner production interactive training programme 1993/94 and the final reports from the case studies, FT VUT Zlín, April 1994.*

*For more information or contact addresses ask the Czech Cleaner Production Center, Politických vězňů 13, 110 00 Praha 1, the Czech Republic.*

# Life cycle screenings of two food products

Bo Pedersen Weidema
Krüger Consult AS
Gladsaxevej 363
DK-2860 Soeborg
Denmark

**Abstract**

*Life cycle screening, defined as the initial data collection and interpretation of the environmental effects of a product system, is applied to two food products. Examples of the results are given and some methodological suggestions are given to simplify the procedures related to product definition, geographical origin and co-product allocation. It is suggested that life cycle studies of food products may be more easily performed and their results become more comparable if industrial conventions can be agreed to on these aspects.*

## Introduction to life cycle screening

Assessing the overall environmental impacts of a product (product life cycle assessment) is becoming an important tool in industrial product development as well as for environmental labelling. In the food sector, however, the use of life cycle assessments is still limited. Therefore, the Danish Academy of Technical Sciences has initiated a project to develop a method for life cycle assessments to be used within in the food sector and to demonstrate its practical application through case studies. So far, life cycle screenings of two food products have been carried out: one on a vegetable product (rye bread) and one on an animal product (cured ham). The full results of these screenings are published in a report by the Academy (Weidema et al. 1994).

A life cycle screening is the first data collection (inventory analysis) and initial interpretation (impact assessment) of the environmental effects of a product system. In the

NATO ASI Series, Partnership Sub-Series, 2. Environment – Vol. 2
Cleaner Technologies and Cleaner Products
for Sustainable Development
Edited by H. M. Freeman, Z. Puskas, and R. Olbina

screening phase, emphasis is on obtaining data for all processes and environmental effects in the product system, but *not* on obtaining very precise or detailed data. The main purpose is to obtain enough data to be able to judge whether a given process or parameter is important or not, and thus to give directions for more in-depth studies. To avoid ignoring anything important, worst case estimates are used when actual or average data are not available.

Whether a process is insignificant or not can only be determined when the overall result and the contribution of each process is known, i.e. when the data has already been collected. Thus, the following iterative procedure must be applied (for each of the investigated environmental parameters unless their relative importance is already determined in advance):

Initially, it is estimated which process(es) are most important for the result. In case of a *comparative study*, what is to be identified is the largest contributions to the *difference* between the investigated product systems. In case of a *non-comparative study* what is to be identified is the largest contribution to the *uncertainty* of the result.

Secondly, the similar contributions of all other processes in the system are estimated.

Finally, the processes to be *included* in the study are determined by listing them in order of decreasing importance until the contributions from the remaining part of the life cycle is below the identified difference (comparative studies) or uncertainty (non-comparative studies). This remaining part may then be excluded from the more detailed analysis.

It follows from the above description that in a non-comparative study, more and more processes are to be included as the uncertainty decreases with improved data quality. Thus, a non-comparative study is in theory a never-ending task. Obviously, in any practical study the procedure will be terminated by the impossibility of infinitely improving the data quality.

**The distinctive characteristics of food products**

Food products differ from most other (industrial) products in that:

* food production affects the environment in ways which are untypical for other products, especially a substantial use of surface area, diffuse emissions, emissions directly to soil and influence on animal welfare.

* the spatial and temporal variation in the production is often very large, both economically,physically and with regard to environmental effect,
* food fulfils a multitude of needs besides the basic supply of energy for human body functions,
* consumers have a very strong opinion on the quality of food products,
* food products are very short lived, giving special significance to loss of product,
* biological products are more difficult to give an unambiguous technical characterisation,
* agricultural production is difficult to partition in relation to individual products.

These distinctive characteristics and their possible practical and methodological consequences for application of the life cycle concept was the background for the project. Especially, the project has investigated the consequences for:

* obtaining data, both in general and for geographically specific conditions,
* defining the product to be compared, and
* allocating the environmental effects to the individual co-products of multi-product processes.

**The investigated products**

Two products - a vegetable product and an animal product - were chosen for the screenings:

1) A sliced rye bread in 0.9 kg packages,
2) A cured ham product packed in 0.95 kg pieces with maple syrup.

The system definition for the example products can shortly be summarised as follows:

* It has been attempted to obtain enterprise specific data. When this has not been possible, averages have been used for the most specific geographical area possible. Worst-case estimates have been used when actual or average data were not available.
* The technological level is present average (approximately 1992).
* The life cycle and the environmental parameters are not limited in advance of the screening, although initial emphasis has been placed on establishing the mass and energy flow.

Figure 1 shows the overall flowchart for the rye bread. Energy and water has not been included in the figure since these are inputs to nearly all processes.

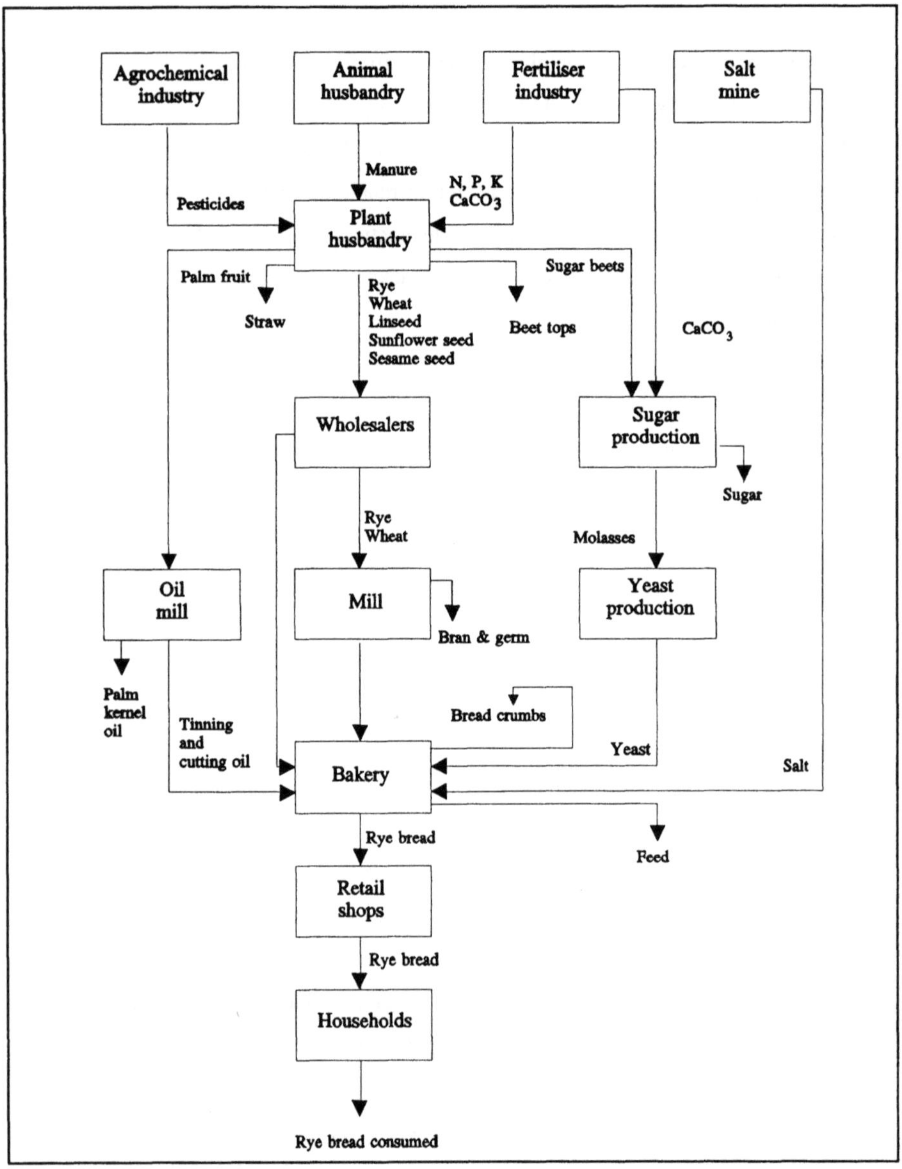


*Figure 1. Flow chart for rye bread*

**Results of the screenings**

The result of the initial data collection on the consumption of primary fuels in the life cycle of the rye bread is shown in table 1.

| *Process in rye bread life cycle* | *Product* | *Fuels (MJ)* | *%* | |
|---|---|---|---|---|
| Agrochem. industry | Pesticides | 0.066 | 0.8 | |
| Animal husbandry | Manure | 0.360 | 4.5 | |
| N-fertiliser production | N-fertiliser | 0.591 | 7.4 | |
| P-fertiliser production | P-fertiliser | 0.105 | 1.3 | |
| K-fertiliser production | K-fertiliser | 0.035 | 0.4 | * |
| Chalk mine | $CaCO_3$ | 0.016 | 0.2 | * |
| Field operations | Crops | 0.605 | 7.5 | |
| Crop drying | Grain | 0.158 | 2.0 | |
| *Total crops excl. seeds* | *Crops* | *1.936* | *24.1* | |
| Seed production | Seeds | 0.097 | 1.2 | |
| *Total crops incl. seeds* | *Crops* | *2.033* | *25.3* | |
| Salt mine | Sodium chloride | 0.006 | 0.1 | * |
| Oil mill | Oils | 0.005 | 0.1 | * |
| Grain mill | Flour | 0.194 | 2.4 | |
| Sugar factory | Molasses | 0.076 | 0.9 | * |
| Yeast factory | Yeast | 0.019 | 0.2 | * |
| Bakery | Rye bread | 2.847 | 35.4 | |
| Packaging industry | LDPE packaging | 0.319 | 4.0 | |
| Packaging industry | HDPE packaging | 0.023 | 0.3 | |
| Packaging industry | PS packaging | 0.021 | 0.3 | * |
| Packaging industry | Paper packaging | 0.628 | 7.8 | |
| Water supply | Water | 0.005 | 0.1 | ** |
| Waste water treatment | Treated water | 0.000 | 0.0 | ** |
| Detergent industry | Cleaning agents | 0.000 | 0.0 | ** |
| Capital goods | Machinery etc. | 1.062 | 13.2 | |
| Transport | | 0.798 | 9.9 | |
| *Total per kg rye bread* | | *8.037* | *100* | |

*Table 1. Primary fuel consumption per kg rye bread sold*

The largest contributor to energy consumption is the bread factory. Even though the uncertainty on this consumption is relatively low (10% or 0.28 MJ) it may still be used as an indicator of the uncertainty of the overall result. Thus, processes contributing *in total* with less than 0.28 MJ can be disregarded. For food products in general, it is concluded that under Danish conditions processes marked with two stars in table 1 (cleaning agents, water supply and waste water treatment) can be excluded *as far as energy is concerned*. The same calculations would have to be made for the other environmental parameters in order to exclude these processes entirely. In the specific case it is suggested further to exclude processes marked with one star (production of potassium fertiliser and lime, all non-grain/seed ingredients of the bread, the closing clip for the package as well as the secondary packaging).

The similar results for the cured ham product showed that the main energy consumption was related to agriculture.

It seems likely that possibilities for improvements in energy efficiency are largest in the processes with the largest energy consumption. Thus, the results of the example screenings call immediately for a more detailed investigation of the energy use in the bakery while for the cured ham a closer analysis of the agricultural production is warranted.

As a general conclusion, the screenings revealed that the production and maintenance of machinery for agriculture requires a relatively high proportion of the overall energy consumption compared to what is typical for industrial products. More than 10% of the entire primary energy consumed in the life cycles of the investigated food products could be related to capital goods in agriculture.

For other environmental parameters than energy (such as emission of nutrients, emission of toxic substances and effects on animal welfare) a qualitative assessment was sufficient to identify the agricultural production as the most important part of the life cycles.

## Methodological suggestions I: A simplified procedure for product definition

The environmental effects of the product system must all be related to a certain quantity of the investigated products. This quantity is called the functional unit of product, since the actual quantities compared must be related to the function of the product. This means that if the different alternatives included in the investigation have different product properties (such as texture, keeping quality, nutritional value) to an extent that influences consumer use and preferences, the quantities to be compared must be adjusted as to fulfil comparable functions.

As an example, rye bread may be compared per kg or per slice. Whether to choose a slice or a certain weight as the basis for comparison must be determined from consumer behaviour: A slight change in the weight per slice may not mean a change in the number of slices consumed, but at a certain point the weights are too different for two slices to be regarded as comparable leading to a change in the consumed amount of slices. The relationship is non-continuos. The question may be further complicated by different sizes and shapes of the slices and by the relation between this and the complementary foodstuff which is placed on the slices.

As can be seen from this example, an unambiguous product definition may be difficult. Both because consumers have very different needs, opinions and behaviour in relation to the different properties of food products, and because actual investigations at the consumer level are very time consuming compared to industrial investigations.

For a life cycle screening it may be necessary to use a more simple procedure to determine what is seen as comparable. It is suggested to look at the way the product is presented to the consumer. In the case of a rye bread the sales unit is determined by weight and not by slices, while in the case of some vegetables the sales unit is determined in "pieces".

Based on this "sales-unit-method" the functional unit of product in our investigation was determined as a weight (1 kg) of rye bread *sold*. However, it may be argued that it would be more correct to use the weight *consumed*. To this end, it was attempted to determine the waste of bread in the households by comparing production figures with information from household waste analyses and diet investigations. This approach did not prove workable since information was not sufficiently detailed or gave unrealistic estimates. A more reliable estimate (2.7% of the sold rye bread to be wasted) was achieved by combining information on the distribution system, the consumption pattern and laboratory tests of the bakery.

## Methodological suggestions II: Simplified procedures for geographical origin

When investigating a specific product from a specific enterprise (e.g. for product declarations or product improvements) it is natural to attempt to collect data from the specific enterprises in the life cycle. However, raw materials typically come from a large number of farms for which specific information is not available. Agricultural products can typically be traced back only to country of origin, not to specific producers. Therefore, data on fodder, fertiliser and pesticide consumption and other farm practices can only be obtained as national averages.

Sometimes, it may not even be possible to trace the national origin of an agricultural product as suppliers change due to market fluctuations or as different lots are mixed to obtain a specific composition or quality. In such situations one may use the national import statistics to identify the average origin of the imported products. A complication for this approach is that quite a number of products are re-exported and it is difficult to identify this in the trade statistics. Even when tracing products back to a specific enterprise (e.g. a mill) it may be impossible to ascertain its country of origin (it may actually be imported from another country by this enterprise). Therefore, it is suggested to use a simplified procedure in which the true geographical origin is approximated by the main exporting country for the product group in question (easily identifiable in the international trade statistics). Other possible procedures would be worst-case/best-case estimates and more precise estimates such as the average origin of exports (equally identifiable though the trade statistics) or average origin of imports (from the import statistics corrected for re-export).

## Methodological suggestions III: Simplified procedures for co-product allocation

For processes which produce more than one product (outputs with a positive economic value as opposed to waste), a procedure is needed to allocate the environmental effects of the process and its upstream processes between the products (unless all products are used inside the investigated product system).

A quick worst-case approach (relevant in a life cycle screening) would be to allocate all environmental effects to the main product (the product which is used in the analysed product system). If the accumulated environmental effect of the process is identified as significant when following this procedure, i.e. if its exclusion would affect the result, the more refined analysis is applied to obtain more correct values:

By a causal analysis it may be possible to split up processes which are initially regarded as multi-product processes into two separate processes each with one single product thus rendering co-product allocation unnecessary. A causal analysis could also be extended to include the substitution effects of the by-products (those co-products which are used for purposes outside the scope of the investigation). The part of the environmental effects to be carried by the by-products is then calculated as the environmental effect avoided as the by-products replace other products. Friedrich et al. (1993) use this method on production of rape oil where the by-products rapeseed (colza) cake and meal is assumed to be used to substitute soy meal for fodder and the by-product glycerine is assumed to substitute artificial glycerine.

However, not all allocation problems can be solved by analysing the technical causalities. Instead, a social causality may be used: the economic value of the products. This economic value is the reason for the existence of the process and thus for its environmental effect. Economists have generally identified the gross sales value as the most appropriate factor for cost allocations when a technical causality cannot be identified (Huppes 1992). In some instances, the co-products will have different finishing costs (the costs incurred from the split-off point where the two products are separately identifiable and until they reach the market where the sales prices are determined). In such instances, the gross sales values will not correctly reflect the relative value of the products. Instead, the net realisable value should be used, i.e. the shadow value of the products at the split-off point calculated as the sales value *minus* the finishing costs (Huppes 1992). For example, a hectare of barley yields 3310 kg grain and 1590 kg straw. The sales price of these products are DKK 4189 and DKK 652 respectively giving a total value of DKK 4841 per ha. Thus, using the gross sales values method, the environmental effects of barley growing (including upstream processes) should be allocated with 4189/4841 (87%) to the grain and 652/4841 (13%) to the straw. However, a more detailed analysis reveal that *at the split-off point* (i.e. when the straw is still lying in the field after the passing of the combine) the price of straw and grain is different from the gross sales price. The correct price of the straw (the shadow price at the split-off point) can be calculated as the sales price *plus* the saved cost of the alternative post-harvest treatment of the straw (DKK 281) *minus* the costs of baling (DKK 525), resulting in a shadow price of DKK 408 per ha. The correct price of the grain can be calculated as the sales price *minus* the drying costs (174 DKK). Using these values, the allocation ratio becomes 4015/4423 (91%) to the grain and 408/4423 (9%) to the straw.

For the milling industry and similar processing industries, quite a number of the by-products used for fodder are split off at an early stage in processing. It may be quite difficult to establish the actual finishing costs and shadow prices of these fodder by-products. In fact, it can be argued that the entire processing is a finishing cost of the main product (e.g. the wheat flour or the rape oil) since the processing is done with the aim of producing this main product and adds no value to the by-product (the fodder value of wheat bran or rape cake is not increased by the processing as such). Following this argument, the environmental effects allocated to the fodder by-product should be based on its price relative to the price of the *incoming* raw material rather than to the price of the co-products.

A farm is a complicated technical-biological system in which many production factors recycle internally, thus involving a large number of multi-product processes. In the screenings, a number of simplifying procedures were used as approximations of the true functioning of some of the multi-product processes in agriculture:

Application of fertiliser was allocated according to the nutrient requirement of the crops, but the rotational value of the crops (the reduced need for fertiliser and crop protection for the following crops) was disregarded, thus suggesting that the advantage of rotation may be regarded as a "free good".

The environmental effects of farming are often diffuse, which further complicates the allocation of the environmental effects to the individual products. For example, the nitrogen leaching should partly be allocated to artificial fertiliser and partly to animal manure. In the screenings, the average leaching for different crops was compared to the leaching from the same crops at recommended fertilisation level with artificial fertiliser, assuming the difference to be caused by the application of animal manure.

Animal manure is a typical marginal by-product (i.e. it is sometimes a waste and sometimes a valuable product) which further complicates the co-product allocation, since environmental effects are not allocated to wastes but only to co-products. In the screenings, the amount of animal manure to be regarded as a co-product was determined as the difference between the recommended fertilisation level and the amount of artificial fertiliser actually applied, assuming maximum efficiency in utilisation of this amount.

**Suggestions for further work**

Identifying processes which may be excluded (as described under "results of the screenings" above) involves some time-consuming data collection and calculations which per definition are of no significance to the further study. Therefore, to limit the work load and cost of future studies, it is recommended that conclusions of this kind are generalised into industrial standards for the specific product groups, so that certain processes may always be excluded from life cycle studies of these products without further justification. The expected positive implication of such simplifications is a general increased use of life cycle studies and their results, which should outweigh the possible negative implication that such general exclusions may lead to wrong decisions in a few specific cases.

The simplified procedure suggested for product definition (described above) can only be used as a rough guide. To enable a more qualified comparison of different life cycle studies of the same product categories it is recommended to establish a list of aspects which should be included as a standard reference when defining the functional unit for each product type. Such a standard reference list cannot be derived solely from technical arguments but must be developed as a convention by the involved industry.

When co-product allocation is done by economic allocation factors (as suggested above), the allocation factors may not be stable in time and place, which complicates the comparison of different life cycle studies on the same product. It is recommended to overcome the problem of fluctuating allocation factors by establishing industrial conventions prescribing for each product group the use of specific physical product properties chosen to reflect as closely as possible the economic value of the co-products.

The justification for the described simplified procedures for multi-product processes in agriculture is based purely on theoretical reasoning and the consequences for the result has not been investigated. Similarly, the errors introduced by the described simplified procedures on geographical origin have not been estimated, due to the large amount of calculations a full error analysis would involve. For future studies, it is recommended that assessments are made of the implications of the errors introduced by the above mentioned simplifications and possible alternative procedures, e.g. by the use of empirically based models of the agricultural systems.

Through such assessments it may be possible in future to reach a consensus (e.g. in the form of an industrial standard) on the situations in which it is acceptable to apply such simplified procedures or worst-case assumptions.

It has been suggested that the most correct reflection of the actual environmental effects would be obtained if the environmental effects of the marginal production facilities (the production facilities which would be taken into use or out of use when the total market demand respectively increases or decreases) was used in the investigation. However, this method was not been used in the screenings, since the identification of the marginal production facilities demands quite extensive information on market trends and fluctuations and the responses of the producers. Compared to industrial products and energy products, the application of the marginal method to agricultural products is complicated by unpredictable natural fluctuations in production, large number of decision makers, long planning periods required for production changes, and large amount of possible alternative uses of land and production facilities. Further studies are recommended to determine the extent to which it is possible to identify the marginal production facilities in relation to agricultural products and thus to predict the actual environmental effects of implementing the results of life cycle assessments.

**References**

Weidema BP, Pedersen RL, Drivsholm TS (1994). Life cycle screening of food products - two examples and some methodological proposals. Lyngby: Danish Academy of Technical Sciences (in preparation).

Friedrich A, Glante F, Schlüter C, Golz C, Nöh I, Reinhard G, Höpfner U, Satorius R, Benndorf R, Blümel H, Schärer B, Rodt S (1993). Ökologische Bilanz von Rapsöl bzw. Rapsölmethylester als Ersatz von Dieselkraftstoff. Berlin: UmweltBundesAmt (Texte 4/93).

Huppes G (1992). Allocating impacts of multiple economic processes in LCA. Pp. 57-70 in SETAC-Europe: Life-cycle assessment. Brussels: SETAC-Europe. (Report from a workshop in Leiden 1991.12.2-3.).

# Environmental Situation in Estonia and Possible Pathways for its Improvement

Rein Munter
Department of Chemical Engineering
Tallinn Technical University
5 Ehitajate Street
Tallinn EE-0026
Estonia

## Introduction

The Republic of Estonia is a member of the United Nations since September 17, 1991. After the re-establishment of the state the Estonian government has faced enormous efforts which need to be taken in order to rehabilitate nature and build up economy based on sparing use of natural resources. The nearest goal of environmental policy in Estonia has been carrying out the land reform in such a way that nature reserves established in our territory would be maintained and that biosphere would not suffer from the reforms. Another environmental protection priority has been the assessment, localization and liquidation of environmental pollution caused by the residence of the former Soviet Army within our territory (National Report of Estonia, 1992). 1993 was a year of restructuring of the Ministry of the Environment. Control over natural resources is now subjected to one institution. This is a prerequisite of sustainable development.

## General Information

The territory of the Republic of Estonia is 452,150 $km^2$ (plus 2000 $km^2$ currently

NATO ASI Series, Partnership Sub-Series, 2. Environment – Vol. 2
Cleaner Technologies and Cleaner Products
for Sustainable Development
Edited by H. M. Freeman, Z. Puskas, and R. Olbina

occupied by Russia). Population comprised 1,574,955 inhabitants (1990), about 70% of them citizens. Average density of population is about 35 persons per square km. Ethnic composition: Estonians 61.5% (963,269), Russians 30.3%, Ukrainians 3.1%, Byelorussians 1.8%, other nationalities 3.3%. Our capital Tallinn has 484, 400 inhabitants (45% Estonians). There are 33 towns in Estonia. Larger towns are Tartu (115,400), Narva (82,300), Kohtla-Järve ( 76,800) and Pärnu (54,200).

Mean annual precipitation in Estonia is 500 to 700 mm. There are over 1400 lakes in Estonia (6.1% of the territory). The most larger is Peipsi (shared with Russia). There are over 1500 islands and islets in Estonia ( 9.2% of the territory). The larger ones are Saaremaa and Hiiumaa. Coastline is 3780 km long. Agricultural land comprises 32.5% of the territory and forests 40.1%.

The first nature protection law in Estonia was approved already in 1935. 47 nature reserves were established before 1940. Nowadays we have 1 National Park (Lahemaa, 64,400 ha), 1 West-Estonian Islands' Biosphere Reserve, 5 State Nature Reserves, 1 Hydrological Reserve, 14 Landscape Reserves and different locally protected areas (National Report of Estonia, 1992).

## Natural Resources and Economy

Although not huge in area, Estonia is relatively rich in different natural resources, both mineral and biological (Table 1) (Estonian Environment, 1993):

Table 1. Mineral Resources

| Resource | Unit | Explored | Mined in 1993 |
|---|---|---|---|
| Oil Shale | mil tons | 3800 | 15 |
| Phosphorite | mil tons | 260 | 0 |
| Peat | mil tons | 565 | 0.3 |
| Limestone, dolomite | mil cu | 300 | 0.6 |
| Sand and gravel | mil cu | 180 | 0.27 |
| Clay | mil cu | 50 | 0.08 |
| Sea mud | mil tons | 4 | 0.002 |

The world's largest exploited deposits of oil shale in North-East of Estonia form the basis for power production and chemical industry. The production volume of oil shale was highest in 1980, when 31.3 mil tons of oil shale were mined. *Oil shale production is one of the main polluters of the environment in Estonia.* Underground mines cause the "sinking" of the surface after stopping of mining activities, which leads to formation of swamps and forests destroying. Another considerable impact of mining is caused by oil shale solid waste (semi-coke) piled into artificial ash mounds which pollute both air and water with toxic phenols. On January 1, 1994, 8,600 ha of the area disturbed by oil shale mining (10,560 ha in total) had been reclaimed. To improve the quality of the reclamation works, a reclamation project has been ordered from Richards, Moorehead and Laing (UK).

Phosphorite deposits are the largest in Europe. Dictyonema argillite (alum-shale), one of the layers above the phosphorite, contains valuable micro-elements (V, Mo, U, Th, Re etc.). The production of phosphorite was stopped in the beginning of 1991 due to the pressure of public opinion. The out-of-date soviet technology and methods used caused serious damage to the environment. The real impact of the mining activities will only be determined in the future as the natural balance in the areas of quarries has been changed and the migration of hazardous compounds, including radioactive substances, can cause considerable danger to ground- and surface water. Air pollution caused by self-ignition of waste landfills of dictyonema has also been a very serious problem. Economic calculations made recently, showed that the price of produced in Estonia phosphorite would probably be too high to be competetive in the world market.

The demand for natural building materials also continued to decline in 1993. Accordingly, mining decreased. For example, the output of technological limestone for lime production has dropped dramatically: 5,000 cu m were mined (6% of the output in 1992). The production of construction sand decreased as well. It was caused by the fact that former major producers sold sand mined and stored during the previous years.

The output of peat was only 46% of that of 1992. Peat production has decreased 3.5 times from the level of the 1980s.

Estonia is quite rich in renewable natural resources. Forests cover over 40% of the territory. The area of the woodlands has decreased by 5,200 ha (0.3%) from the level of 1992. There were 10,179 private farms with total area of 252,200 ha in Estonia in 1993. The average area of the farm was 25 ha, which included 43.2% of fields, 33.8% of woodlands and 16.1% of grasslands ( Estonian Environment, 1993). 87 native and over 500 introduced tree and bush species are recorded in the dendroflora of Estonia. Damage caused by the acid rains has been detected at over 3200 ha of forests. New forests' pests and diseases have appeared.

The structure of economy which was developed in the mid-sixties has almost preserved its main proportions until now. However, some changes have happened. Industry has the leading position in employment (over 40% together with construction), with light (textile, clothing, footwear) and food industry providing roughly half of the industrial output. Engineering production includes instruments, oil and gas processing equipment etc. Power production and chemical industry based to the local oil shale resources are of importance. The privatization process in industry is still going on.

## Environmental Problems

### Air

Atmosphere, being the most unstable of all geospheres, unites the environmental problems of all countries because of its natural laws. The fatal effect of this was demonstrated in April , 1986, when the radioactive isotopes released into atmosphere due to accident in the Chernobyl Power Plant (Ukraine) were detected in Central Sweden within 60 hours. In 80% of the Estonian territory radioactivity measurements have been made which proved that the pollution load with Cesium-137 was the highest in North-East Estonia where the average load on the surface was 0.3 to 0.5 Ci/km2 (National Report of Estonia, 1992). Since the dominating winds in Estonia are south-westerly and southerly, the "imported" air pollution originates from Central and Eastern Europe (e.g. Poland and the former East Germany), the "export" in its turn is a considerable pollution source for North-Western Russia, Finland and Sweden.

For some air pollutants, Estonia has been the worst polluter in the region of the Gulf of Finland. Hundreds of thousands of tons of sulfur dioxide, nitrogen oxides and solid particles have been annually emitted from stationary sources within its territory.

Approximately 75% of the main pollutants have been emitted by the Baltic and

Estonian Thermal Power Plants (in North-East Estonia, at the Russian border) which rank among the 10 biggest sources of air pollution in Europe. The power plants are heated with local oil shale of low calority and high ash content (65%) and use out-of-date gas cleaning equipment , desulfurization equipment is still lacking. Since the chimneys of the power plants are relatively high (150-200 m), the radius of the polluted area is considerable. In 1990 611,000 tons of sulfur dioxide, nitrogen oxides and solid particles were emitted from stationary sources in Estonia including 302,000 tons of solid pollutants, 208,000 tons of sulfur dioxide and 22,000 tons of nitrogen oxides (National Report of Estonia, 1992). In 1993 639 enterprises reported the volume of air pollutants emitted by them into atmosphere. 379,800 tons of pollutants (62% of the amount in 1990) were emitted during the year, including 189,000 tons of solids, 145,000 tons of sulfur dioxide and 12,000 tons of nitrogen oxides (Figure 1) (National Report of Estonia, 1992).

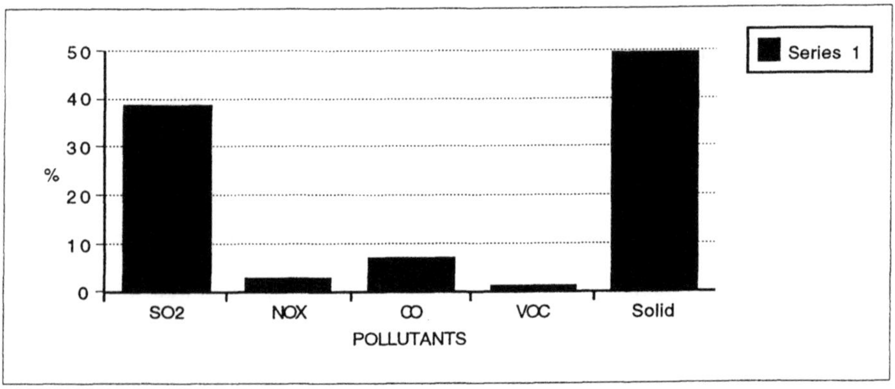


Figure 1. Distribution of the individual air pollutants emitted from stationary sources in 1993 (%)

The main pollutants represented the following branches: energetics (61.3%), building materials industry (19.4%) and oil shale processing (3.5%) (Figure 2) (Estonian Environment, 1994)

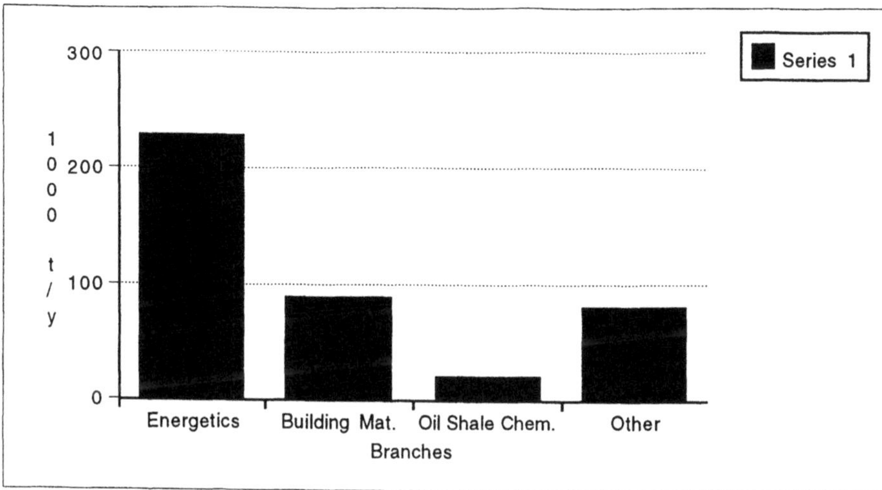


Figure 2. The biggest air polluters in Estonia (thousands tons/year of gaseous and solid wastes)

The decrease in the volume of pollutants (99,400 tons less than in 1992) is caused by the drop in the output volumes at the Baltic and Estonian Thermal Power Plants, as well at the Kunda Cement Plant. The dust emitted into atmosphere annually mostly originates from the chimneys of the power plants and contains oxides of alkali and and alkaline-earth metals, due to which the pH of precipitation is alkaline in the vicinity of the plants. In addition to this, the fly ash of oil shale contains heavy metals, including the toxic ones, in relatively high concentrations: for example, some 50 tons of lead, 30 tons of mercury, 30 tons of zinc and 20 tons of copper are emitted annually (National Report of Estonia, 1992). Some 90% of pollutants remain within 30 km from the pollution sources via dry and wet deposition. Besides inorganic compounds, the stationary pollution sources in North-East of Estonia in 1992 emitted more than 8000 tons of organogenic substances, including over 5000 tons of hydrocarbons, some

2000 tons of bensene, 1000 tons of toluene and other toxic compounds.

According to the need for sustainable development the energy concept for the next 20 to 25 years includes economizing in energy consumption, redistribution of power production and exploitation of alternative energy sources, especially windpower. For example, the part of oil shale in the energy balance will probably be reduced from 52% to 20% and the part of natural gas will be increased from 13% up to 40% in 2020 (Figure 3). Correspondingly will be decreased sulfur dioxide emissions. The better technology for the power plants flue gases treatment must be introduced ( wet scrubbers using the alkaline suspension of the fly ash are perspective).

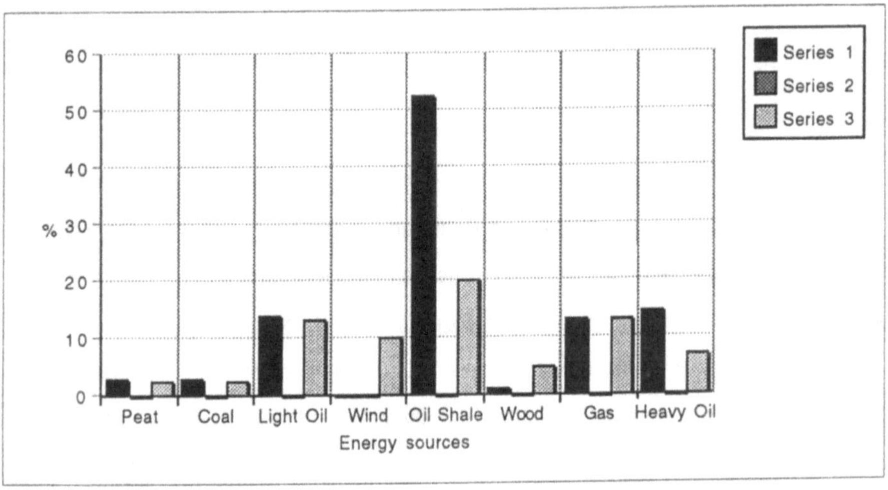


Figure 3. Energy consumption in Estonia in 1990 and a proposed sustainable alternative for 2020

Increasing in natural gas consumption is planned on the basis of the natural gas from the Norwegian Sea. Estonian government has made a proposal to build the gas pipeline from the Norwegian Sea through Sweden, Finland, Estonia, Latvia, Lithuania up to Poland and Central Europe countries. At the present time the feasibility study of the project is being carried out. This project is very important not only from the point of view of atmosphere protection but , first, from the point of view of Estonian national security.

Water

The main share of pollution originating from the Estonian waters is carried into the Gulf of Finland. In 1990 many towns and settlements lacked sewage treatment facilities. Besides, no biological-chemical sewage treatment plants were in operation. Therefore biological pollution equivalent to 49 thousand tons of total BOD was carried into the Baltic Sea, in addition to some 322 tons oil products and 150 tons of total phenols (Table 2) (Estonian Environment, 1994).

Table 2. Pollution load into the Baltic Sea, tons/year

| Parameter | 1990 | 1991 | 1992 | 1993 |
|---|---|---|---|---|
| Total BOD | 49,000 | 44,200 | 23,500 | 14,640 |
| Oil products | 322 | 286 | 154 | 127 |
| Sulphates | 122,000 | 109,000 | 102,000 | 107,553 |
| Chlorides | 13,000 | 16,000 | 14,600 | 12,827 |
| Total phosphorus | 792 | 909 | 673 | 445 |
| Total nitrogen | 7530 | 8230 | 5640 | 4241 |
| Fats | 443 | 169 | 146 | 160 |
| Phenols | 153 | 100 | 82 | 23 |

The inflow of nutrients (nitrogen and phosphorus) is especially important for the coastal waters where eutrophication processes are increasingly intensive. Estonian

Republic is a member of the Baltic Sea Convention since 1992. According to this, our country also adopted responsibility for implementation of the joint decisions. One of the main decisions was passed in 1988, forseeing reduction of the sea pollution load by 50% by 1995, especially as regards heavy metals, organic compounds (phenols) and nutrients (National Report of Estonia, 1992). Up to 10% of the oil shale mined in Estonia (about 1.5 mil tons annually) is treated chemically at the temperature 500-550 C to get oil. The solid waste, semi-coke, is disposed in an ash dump. Wastewater (leachate) from the ash dump is derived from atmospheric precipitation which has accumulated in areas of semi-coke storage. The pH of the leachate is 10-12 and total phenols content is in the range 500 to 700 mg/L. The results of the pilot plant tests of the Finnish company "Vesi-Hydro" showed that under the condition, if the relative amount of this effluent in the total wastewater stream does not exceed 7%, all these wastewaters can be treated biologically. However, to meet the established by HELCOM a very strict MPC for total phenols in the treated effluent (0.5 mg/L), an effective posttreatment method for the biologically treated effluents is needed. Ozonation and advanced oxidation processes (different combinations of ozone, UV and hydrogen peroxide) have been shown to be very effective for the posttreatment of phenolic effluents (Kallas et al., 1992). Some facts are encouraging, too. The numbers of total BOD, phosphorus, nitrogen and other pollutants load in Table 2 are decreasing. The concentration of toxic pollutants such as PCB, DDT, oil and oil products, which reached its peak in 1970s, is steadily decreasing in the Baltic Sea.

Due to the depression, for the first time in 30 years, no municipal wastewater treatment plants were put into operation in 1992, but 1993 has been very successful: the first stage of the biological treatment block in Tallinn was completed. Mechanical and chemical treatment plant with the capacity of 370,000 cu m/d was completed already in 1984. After years of standstill, the construction was continued in 1993. EEK 80 mil was invested in the construction, EEK 24 mil of it being aid from Finland. The second stage of the construction which is estimated to cost up to EEK 286 mil, has to be completed by 1998 according to international agreements. The project will probably be funded with a loan from EBRD. Pollution load into the Gulf of Finland decreased also because wastewater from Rakvere was led through a treatment plant. Planning and construction of wastewater treatment plants was started or continued also in Tartu, Türi, Valga, Haapsalu, Pärnu and other towns (Estonian Environment, 1994).

The quality of Estonian drinking water did not undergo major changes in 1993. However, there were fluctuations in many towns and counties. Water quality has somewhat worsened in Tallinn and Hiiumaa in respect of many characteristics (general analyses, water sources, chemical and bacteriological quality). In 1995 in Tallinn a significant improvement in drinking water quality is expected after putting into operation the ozonation station. The plant will be equipped with 2 ozone generators from the French firm "Trailigaz". The construction of the ozonation plant is estimated at EEK 142 mil., which will guarantee ozonation of 200,000 cu m/d.

Groundwater suitable for drinking is one of the most important natural resources in Estonia. In 1991, the total consumption was 175 mil cu m (480,000 cu m per capita) , covering 2/3 of the demand for drinking water. In rural settlements and most of the towns groundwater is the main source of water supply. Only in Tallinn and Narva treated surface water gives a considerable share of consumption (260,000 cu m in total) as the local groundwater reserves are unsufficient. Due to anthropogenic load the state of groundwater in Estonia has considerably worsened in the last decades. Nitrate pollution of the upper groundwater layers caused by agricultural runoff of fertilizers and careless management of stores has been detected within nearly a half of the Estonian territory. In 1992-1993 the groundwater quality has improved mostly due to the decreased use of mineral fertilizers. The military aircraft bases of the USSR have led thousands of tons of the plane fuel into Estonian surface waters, soil and groundwater. Groundwater pollution due to fuel has been a problem for years. Fuel pollution is most extensive in Tapa, where there is up to 5 m layer of pure fuel in drillwells. In 1993, Maves Ltd. started research and cleaning works here. The works are supervised by the Danish firm "Hedeselskabet" and financed mainly by the Danish Environmental Agency. The extent of the fuel pollution is shown in Table 3 ( Estonian Environment, 1994).

Table 3. Extent of fuel pollution in groundwater and surface

| Airfield | Polluted groundwater layer | | Polluted soil layer | |
|---|---|---|---|---|
| | area (ha) | depth (m) | area (ha) | depth (m) |
| Tapa | 1600 | 25 | 2 | 1 |
| Haapsalu | 7.3 | 10 | 1.5 | 1 |
| Ämari | 237 | 20 | 24 | 1.5 |
| Rakvere | 50 | 20 | 0.2 | 2 |
| Pärnu | | | 12.1 | 2 |
| Tartu | | | 16.2 | <4 |

The extent of groundwater pollution depends on the hydrogeological conditions of the airfield. Fuel layer concentrates on groundwater in rock crevices. The thick fuel layer does not decompose, as the decomposing bacteria live only on the layer between the groundwater and the fuel. To eliminate the pollution, the fuel has to be pumped out. Cleaning the polluted areas and groundwater is estimated to cost tens of millions of USD.

## Solid Waste

1993 was the first year after the Waste Law and minor legal acts on waste had taken effect. 1993 was also the first year of summarizing data about waste generation in Estonian enterprises (in 1992-1993). In 1993, 14.5 mil tons of waste were generated, which is somewhat less than in 1992. Landfills, ash mounds etc. contained over 425 mil tons of waste by the end of the year (Estonian Environment, 1994). The bulk of waste is generated during the process of oil shale mining, oil shale energetics and chemical treatment in North-East Estonia. *Waste materials of oil shale mining and processing cover thousands of hectares, there are waste heaps with relative heights exceeding 100 m.* Those terricones contain a number of compounds easily washed out with atmospheric precipitation (as it was shown above), including toxic heavy metals and phenols. *Due to the high content of minerals in oil shale (60-65%), solid waste generation in unavoidable.* Unfortunately, the present technological level does not allow the utilization of the huge amount of oil shale semi-coke and ash.

1,546,000 tons or 10.6% of the solid waste generated in enterprises was recycled in 1993, mostly oil shale gangue and ashes. Traditional recycling materials, such as glass, paper, textiles etc., were not extensively gathered due to the insufficient recycling facilities in Estonia and the collapse of the previously existed link between Estonia collectors and recycled product tradesmen and the Eastern neighbors.

In the field of hazardous waste, the joint project with Danish specialists concerning the treatment of such waste in Estonia was continued. The most dangerous radioactive waste deposit is situated in Sillamäe (NE Estonia). *The local military industrial complex who has earlier (1950s) been dealing with concentration of Uranium, has dumped in a tailing which lies on the coast of the Gulf of Finland 1200 tons of Uranium and 750 tons of Thorium. Radioactively polluted area covers over 100 ha, having impact on the health of local inhabitants.* The washed-out toxic compounds can reach both the groundwater and the sea. Risk assessment needs to be carried out and a safe solution found for the conservation of the tailing. Swedish and Finnish experts are working on it.

Military Environment

The army of the former USSR had over 500 objects with a total area of 85,000 ha at its disposal in Estonia, i.e. 1.8% of the Estonian territory. Only in Tallinn, there were 174 Russian military units who were using 872 ha of land.

The largest area used by the Russian army was Aegviidu artillery range (over 33,000 ha). Major areas were also occupied by military airfields at Tapa (771 ha), Tartu (682 ha), Pärnu (731 ha), Ämari (930 ha) and Haapsalu (799 ha).

Provisional estimates of the damages caused to the environment by the former USSR army were begun in 1991. In 1993, inventories of the damages caused by military units continued. All these works were funded from the State budget. Estonian environmentalists were trained in Germany. *Preliminary estimate show that the damages to our environment amount at least to EEK 55 bn.*

Nearly all territories taken over from the military units of the former USSR are polluted with oil products, chemicals, ruins and domestic waste. The most hazardous military objects are the airfields, especially the fuel tanks. Fuel from the territories of the airfields has immersed in the soil and polluted most of the groundwater ( see more detailed information presented above).

After the removal of the core fuel from the two Russian nuclear reactors up to the end of September, 1995 the direct danger of radioactive pollution will disappear, but the radioactive waste pond of the former military factory at Sillamäe is still a remarkable source of danger.

A major jeopardy both to the environment and people are the offence and training chemicals left behind by the Russian army. So, for example, 20 tons of napalm was left unguarded in Männiku and Lääne-Virumaa district. Large quantities of ammunition

and explosives are still stored on Naissaare Island and at Astangu near Tallinn.
The former USSR army managed large areas of Estonian forest (approx. 65,000 ha). Large areas have been destroyed in fires caused by shooting or bombing practice. On Pakri Island, for example, nearly 1,000 ha of land has been disturbed by bomb craters, at Utsala artillery range the bombing area amounts to 500 ha. In addition, large quantities of unexploded mines, shells and plane bombs have been found at the ranges (Estonian Environment, 1994).
Estonia is ready and willing to participate in wide-ranging international cooperation for dealing with the most critical environmental problems, in spite of our current economic difficulties originating from recent past.

## References

Estonian Environment 1993 (1994). Ministry of the Environment of Estonia. Tallinn. 94 p.

Kallas J., Pikkov L., Viiroja A., Kamenev S., Joarand H., Tali E., Munter R., Palosaari S. (1992) Treatment technology of wastewater containing phenols and phenolic compounds. Research Paper N26. Lappeenranta University of Technology. Lappeenranta. Finland. 40 p.

National Report of Estonia to UNCED (1992). Ministry of the Environment. Tallinn.40 p.

# Clean Technology Research in Germany: Examples of Implemented Projects

Günter W. Panzner
Projektträger Umweltschutztechnik des BMFT
Deutsche Forschungsanstalt für Luft- und Raumfahrt e.V. (DLR)
Südstraße 125
D-53175 Bonn
Germany

## Introduction

The German Federal Ministry for Research and Technology (BMFT) has funded innovative R&D projects in the area of clean technologies and clean products since 1975. Until 1993, 295 projects were performed in this area, at the cost of around DM 265 million in sponsorship grants, within the framework of the BMFT programme "Environmental Research and Technology". Further details on clean technology research in Germany have been given in another paper (Schroeter 1994).

With respect to those production processes and industry branches most responsible for environmental pollution, funds have mainly been allocated to the industrial sectors shown in figure 1.

Some examples of implemented projects are presented in this paper. Further projects supported by the BMFT have been described in the literature (Project Management Division Environmental Protection Technology, Projektträger Umweltschutztechnik).

NATO ASI Series, Partnership Sub-Series, 2. Environment – Vol. 2
Cleaner Technologies and Cleaner Products
for Sustainable Development
Edited by H. M. Freeman, Z. Puskas, and R. Olbina

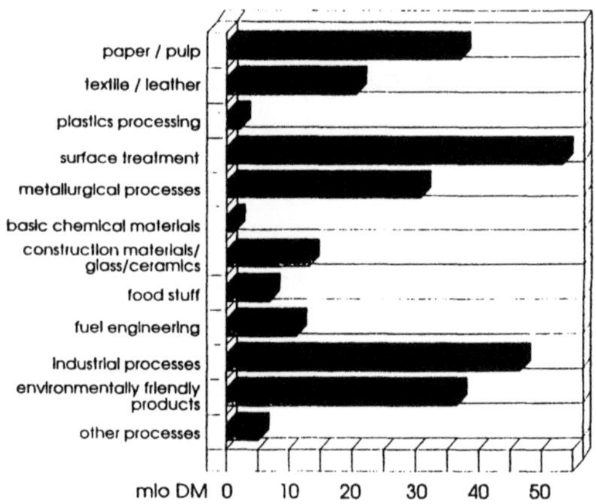


Fig. 1: Allocation of funds 1975 - 1993

**Surface Treatment: Nickel and Chromium Electroplating**

Large amounts of water are required in electroplating installations for a variety of processes and rinse steps because the workpieces must be cleaned after each treatment step. The result is that pollutants from the process baths accumulate in the waste water and the solutions must regularly be replaced.

In purification plants the current state of the art enables the contaminant load in the waste water to be substantially reduced. The substances are converted into metal-containing special waste, or what is known as electroplating sludge. In this way, 10-15 kg of drained sludge from 1 kg of metal can be expected. Around 80,000 tons of electroplating sludge are obtained annually. The complex composition of the sludge leads to considerable problems in recycling in that energy and chemicals, which themselves again contaminate the environment, would need to be used to seperate the components.

Improvements in the process technologies employed in the surface treatment of metals, therefore, offer the most effective means of preventing a pollution of the enviroment through waste water and residues.

Within the framework of BMFT sponsorship the prominent aim of research was to develop a technique that would completely avoid waste water, simultaneously ensuring a coating of excellent quality.

A waste water free and low waste technology was developed in the course of several successive projects. In a first step, an effective reduction of environmental overload was achieved by means of the improved rinsing technology (multistep dip-spray-rinse cascade), regeneration systems at all process steps and an interlinking with the multistep waste water purification processes. In comparison to a conventional electroplating istallation, water consumption and, correspondingly, waste water volume could be reduced considerably.

By upgrading of conventional technology, this meant a reduction from approx. 7 $m^3/h$ to approx. 0.3 $m^3/h$. At the same time, the loss of chemicals could also be reduced to a minimum and the service life of the solutions could be prolonged. The heavy-metals load in the waste water was reduced to 1.4%, which falls considerably below current threshold limits.

In a second step, new technologies were necessary for the complete elimination of waste water. Among others, a new adsorption technique for the separation of organic degradation products was developed, using special adsorber polymers, complete nickel and chromium recycling by membrane destillation was implemented, and a high-temperature destruction of cyanide from the electrolytic degreasing system was established.

These improvements resulted in a waste water free and low waste demonstration plant operating as a commercial electroplating enterprise (Galvano Dresden). Only residual waste waters, e.g. from spillage or leakage, are to be treated leaving very low amounts of salt residues to be deposited (ca. 3 kg/d instead of 100 kg/d for a conventional process)

**Surface Treatment: Paint and Protective Coating Agents**

A considerable pollution of air results from the use of solvent-containing paint and other surface protection systems. In this area, around 340,000 tons of solvents such as aliphatic and aromatic hydrocarbons, glycols, ketones and alcohols are used annually in Germany.

The variety of products to be painted and the very different tasks to be perfomed by these coating systems have already spawned a wide range of processing techniques. So far, research objectives have been to examine the development potential for new or improved spraying techniques and new coating systems.

Within the framework of a project, investigations were conducted into the reduction of solvent emissions during the spray painting of plastics to a technically unavoidable minimum. Possible ways of achieving this were identified on the basis of an analysis of the various process steps.

Results revealed that solvent emissions could be lowered through the implementation of various measures, such as:

- using circulated air in the painting chambers, whereby only a small portion of the air in the chamber is replaced by fresh air (reduction of solvent emissions by at least 80%),
- combining water-based paint in the primary coating process with a solvent-containing topcoat finish or with a solvent-containing base and topcoat finish (reduction of emissions by approx. 30%),
- using electrostatic spray painting, by which coating is performed by means of an electric field between the spraying device and the workpiece, depending on the coating system used and the workpiece geometry (reduction by approx. 30-40%),
- using automatic spray jet adjustment, subject to the workpiece geometry (reduction by approx. 10-15%).

The goal of another project was to develop environmentally sound waxing systems and techniques for the protective coating of automobiles. After they have been assembled and before they leave the plant, new vehicles receive two types of protective coating: the outside of the car is coated for transport protection, as is the motor against corrosion (85% gasoline and 15% wax). In this way, approx. 4 million kg of organic solvents and 1.1 million kg of wax are emitted in western Germany as special waste requiring disposal in the production of approx. 4.1 million automobiles.

The main goals of the completed project were:

- the development and testing of aqueous protective agents,
- the development of plant and process technologies,
- large-scale experiments and process optimisation,
- the development of the recyclability of residue materials.

The material developed for the outer protective coating consists of microwaxes (paraffins) dispersed in water with a solids content of around 30%, and has good protection characteristics. Investigations show that coatings can be removed without problems after storage times of up to 9 months.

The water-wax application method has been successfully tested in practice at Volkswagen in Wolfsburg (cf. fig. 2).

Fig. 2: Installation for the protective coating of assembled vehicles (Photo: Volkswagenwerk Wolfsburg)

The overspray waxes are largely recycled, while the water is returned to the water circuit. In addition to VW in Wolfsburg and Mosel this environmentally friendly method is also being used by Opel.

**Industrial Processes: Substitution of Halogenated Hydrocarbons**

Many halogenated hydrocarbons (HHCs) and in particular many chlorofluorocarbons (CFCs) are used in several fields of application on account of their excellent technical characteristics, such as stability, non-toxicity, incombustibility, high boiling point and especially their specific solvency for organic contaminants.

For this reason, around 300,000 tons of HHCs were still used in 1989 in the Federal Republic of Germany, most of which was released into the atmosphere. Included in this total volume were approx. 90,000 tons of CFCs.

Only later was it recognised that HHCs and especially CFCs are partly responsible for the depletion of the ozone layer and anthropogenous global warming. Because various HHCs (including perchloroethene) were also shown to have carcinogenic effects, the special sponsorship area "Reduction of Halogenated Hydrocarbon Emissions" was prepared and released on June 20, 1989.

About 30 projects have been sponsered within the three principal fields of application up to now: solvents and cleaning agents, plastic foaming, and refrigeration and air-conditioning technology.

## Solvents and Cleaning Agents

Following the soldering process, CFCs are used to wash off fluxing agents remaining on printed circuit boards. The BMFT has funded the development of water-washable solder paste that can be soldered under inert gas in the reflow soldering process. Residues from the solder paste are washed off in a machine developed specifically for this purpose. Furthermore, the development of flux systems low in solids has been promoted for use in the wave soldering process, which would eliminate the necessity of subsequent washing.

In manufacturing mechanical and optical precision components from metal, plastic or glass, cleaning of the workpieces is necessary in the course of different process steps carried out. While in the past CFCs and chlorinated hydrocarbons (CHCs) proved to be technically the best cleaning medium, alternative methods must now be developed which can achieve at least the same standards of quality.

For instance, in the grinding, polishing and coating of high quality optical components, grinding or polishing residues must be removed as best as possible from the glass surface following each process step. The main alternative for use of CFCs and CHCs should be aqueous cleaning processes, paying special attention to problems associated with drying and corrosion. Field testing of a newly developed CHC-free cleaning process for the highest quality of optical glass has been performed successfully.

Within the framework of the on-going sponsorship, the CFCs and CHCs used in textile cleaning processes are also to be replaced by aromatic-free hydrocarbon mixtures as cleaning medium.

In degrasing metal components, e.g. in the automobile industry, aqueous cleaning of components utilizing closed water circulation is an attractive alternative. As a result of a project funded, an aqueous cleaning installation is in regular operation at the Mercedes Benz AG factory in Düsseldorf (cf. fig. 3). By employment of 4 more such installations, this plant now operates CHC-free.

Fig. 3: Wet cleaning installation for steering gear cases; overall view of the pilot installation (Photo: DÜRR GmbH).

## Plastic Foaming

One of the largest fields of application for CFCs is the foaming of plastics. In this process, CFCs are used as foaming agents and also to separate the product from the molds.

The target of the project, however, is not the whole field of application for plastic foaming but only those sectors where considerable research and development potential is necessary for a changeover to halogen-free processing methods. These include polyurethane integral foam and insulating rigid foam, as well as polystyrene bulking foam. In other applications, especially those using polyurethane flexible foam, the state of the art already permits a changeover to CFC-free systems.

Since 1992 two medium-sized manufacturers have successfully switched to producing insulating rigid foam and integral rigid foam using CFC-free processes. On the basis of the research results, CFCs are now no longer used for foaming or separating in the manufacture of insulating containers and casings, for instance for computers and energy-absorbing components in automobiles.

In other applications, intermediate results look very promising for a changeover to water- or pentane-based systems in the manufacture of insulating panels for refrigerated vehicles. A new panel design permits a new wall and ceiling system to be used without stability-related problems with halogen-free polyurethane foam (cf. fig. 4). At present, 60 refrigerated trailers are already in regular use.

Fig. 4: Application of foam using a lowered mixing head (Photo: Schmitz-Anhänger Fahrzeugbau)

## Refrigeration and Air-Conditioning Technology

In refrigeration and air-conditioning technology, CFCs are used as refrigerants in a variety of ways in compression-type refrigeration installations for various ranges of temperatures, capacities and applications.

The main alternatives, besides ammonia, for compression-type refrigeration installations are chlorine-free, partially halogenated fluorocarbons, such as R 134 a, R 152 a, R 32 etc. Research has been performed concerning the thermodynamic characteristics, heat transfer, and oil, combustion and flow behaviour of these alternative refrigerants. The results of these advanced investigations are considered largely responsible for the current introduction of R 134 a onto the industrial market.

## Metallurgical Processes: Lead Smelting

The smelting of sulphidic lead concentrates by the classic two-step roasting-reduction process was responsible for considerable environmental overload. Emissions per ton of lead produced in this way amounted to approx. 40 kg $SO_2$ and 0.4 kg lead (dust).

Based on knowledge collected from an experimental installation employing the QSL process and working discontinuously, a pilot plant with a production capacity of 30,000 tons of lead annually was constructed and put into operation to demonstrate the technical operability of the process.

The concept of this technology, named after Professors Quenau, Schumann and the company, Lurgi, is based on the smelting of lead concentrates in a single closed reactor using a one-step process.

The reactor is a brick-lined, horizontal cylinder with bottom-blowing nozzles (cf. fig. 5).

The concentrates, fluxes and recycled flue-dust are moistened, pelletised and fed into the reactor. In the oxidating zone, oxygen is blown in to oxidise the sulphur while a fraction of the lead is already reduced to primary lead, at this stage. In the reduction zone, residual reduction is completed by blowing in pulverised coal. Both the slag and lead are continuously tapped from the reactor.

The $SO_2$ in the waste gas is separated from the flue-dust in a hot-gas electrofilter and processed into sulphuric acid. The separated flue-dust is refed into the reactor.

The elimination of various sources of pollutants resulted in a considerable reduction in emissions. With the QSL process lead emissions could be reduced by 80% and $SO_2$ emissions by 95%. At the same time, the metal content in the slag was lowered to below 3% and specific energy consumption reduced by 50%. Operational costs decreased by 10-15% compared to the conventional process.

The first large-scale installation began operation in 1991 with an annual capacity of 75,000 tons of lead and replaced the existing high-emission sintering and shaft furnace installations. A second QSL installation has been in operation at Korea Zinc in Onsan, Korea since 1992. In both installations, in addition to lead concentrates large amounts of secondary materials such as residues recovered from zinc processing and other lead-containing wastes were also further processed. A third lead works employing the QSL process will begin operating in the course of 1994 in Baiyin, China.

Fig. 5: QSL installation at the lead works Bleihütte Berzelius Stolberg GmbH in Stolberg (Photo: Werkfoto Lurgi/Stelljes)

**Textile and Leather Industry: Textile Finishing**

Auxiliary agents (sizing agents) are used for processing yarn in the textile industry. These are used to seal the yarn for protection against the mechanical stress of the weaving process, thus enabling high weaving speeds. These auxiliaries must be removed from the fabric in order to enable the further processing of the yarn (bleaching, dyeing, printing).

Usually the sizing agents are not water-soluble. For the removal from the fabric, they are first treated by enzymes and can then be washed off, thus ending up in the waste water. Sizing agents are responsible for up to 80% of the total COD load in the waste water of the textile finishing industry and are partly not biodegradable. This results in an annual COD load of 22,000 to 26,000 tons in Germany.

The goal of several research projects was to develop a process which can simultaneously clean sizing waste water and recover the sizing agent. Furthermore, the separated water could also be reused.

Ultrafiltration was shown to be suitable for the recovery of sizing agents (cf. fig. 6). Concentrates obtained in this way could be reused without impairing the weaving process. This newly developed technology enables a recovery of 80-85% of the sizing agent (based on the amount of agent used).

Fig. 6: Ultrafiltration installation with membranes in pipe modules (Photo: Gesellschaft für Trennverfahren mbH, Bodelshausen)

The high value of recycled sizing agent, water and thermal energy enables the ultrafiltration installation to pay off within 1 to 1.5 years, depending on the size of the plant. Ultrafiltration-based recycling of sizing agents has meanwhile been, in many cases, successfully tested in practice: in Germany, mainly on synthetic sizing agents and, also under BMFT sponsorship, in Egypt on locally-produced sizing agents modified by partner scientists from the National Research Center, Cairo.

A disadvantage of this process is the limitation to polyvinyl alcohol and other water-soluble sizing agents (e.g.carboxymethylcellulose). Solvent-soluble sizing agents and mixed agents cannot not be separated using this process. Moreover, the process is only economically feasible with sufficiently large quantities of output.

Thus a further project deals with the development and application of biodegradable and removable sizing agents. In this way, the conditions for a substantial reduction of waste water load within the entire textile industry are to be established.

**Foodstuff Production: Beverage Industry**

In some areas of beverage filtration (beer, juice and wine), kieselgur could not be replaced by other filtration techniques until now. Filtration was carried out in several steps thus requiring large amounts of filter materials and filtering agents. This led to the situation where, in the sector of beverage filtration alone, 120,000 to 150,000 tons of wet kieselgur had to be disposed of annually in Germany, in part as special waste. In addition to requiring depositing space in landfills, kieselgur contaminated by product residues is high-odorous and results in biologically polluted seepage water.

The use of cross-flow microfiltration technology reduces the amount of filtering auxiliaries required, especially kieselgur. The special design of the modules permits a compact construction of the cross-flow installation (cf. fig. 7). Therefore, only very small amounts of rinsing water and lye are necessary for cleaning, thus assuring an economic operation that meets the demands of the beverage industry.

Fig. 7: Cross-flow installation for the clarification of liquids, in particular beverages (Photo: Seitz-Filter-Werke)

The membrane material used remains completely inert against the filtered products. Even after many months of storage the filtered products do not show any signs of negative changes. Machines operating with this technology at production scale are meanwhile being employed by various companies, e.g. in wine filtration.

By the way, wines filtered by cross-flow microfiltration have already been prize-winners in wine competitions. Wine tasters can distinguish, from visual and sensory inspection, if wine from the same vineyard has been filtered by conventional or by cross-flow technology. And usually they decide that wine treated by cross-flow filtration tastes better. In this way, cleaner technologies can not only reduce environmental pollution, but can also contribute to a higher quality of life in another way.

## References

Project Management Division Environmental Protection Technology (ed) (1993) Environmental Protection Technology, Selected projects supported by the ministry for research and technolgy 1975 - 1992

Projektträger Umweltschutztechnik (ed) (1991) Umweltschutztechnik, Band 1, Fördermaßnahmen des Bundesministers für Forschung und Technologie 1975 - 1990

Projektträger Umweltschutztechnik (ed) (1993) Umweltschutztechnik, Band 3, Fördermaßnahmen des Bundesministers für Forschung und Technologie 1990 - 1992

Projektträger Umweltschutztechnik (ed) (1993) Umweltschutztechnik, laufende F+E-Vorhaben 1993, Berichte zu Fördermaßnahmen des Bundesministers für Forschung und Technologie

Schroeter K. (1994) Introduction to clean technology research in Germany. Ibid.

# Research and Development in Clean Technologies in Germany

Dr. Klaus Schroeter
Federal Ministry for Research and Technology
Office of Environmental Technologies

Mister Chairman, Ladies and Gentlemen!

This is my first time in Budapest and this is the first time I have participated in an event of the NATO Science Program. Budapest is wonderful. This workshop is perfectly organized. Our Hungarian hosts do a very fine job. Thank you very much. I am honored to attend this workshop. I am glad and grateful to be in Budapest.

Now I would like to deal with the environmentally oriented R&D policy and activities in Germany.

In the beginning of the seventies dying forests and extremly polluted rivers and lakes made the need for a healthy environment more and more urgent (Figure 1). A healthy environment became a real public demand. "Waldsterben" became an international word.

The response of the government was the formulation of an environmental policy with many new laws and regulations. The government also started to financially support R&D in environmental technologies. Therefore, in the beginning, the main R&D areas were focussed on

NATO ASI Series, Partnership Sub-Series, 2. Environment – Vol. 2
Cleaner Technologies and Cleaner Products
for Sustainable Development
Edited by H. M. Freeman, Z. Puskas, and R. Olbina

- water technology, that means technologies to provide clean drinking water and technologies to purify waste water from municipals as well as from industrial plants.
- An equally important area was research and development for cleaner air for instance to remove $SO_2$ and NOx or dust out of flue gases from power stations, industrial boilers and furnaces, households and cars.
- Later on, the list of technologies for cleaner air was extended by the task "low emissions in products and manufactoring processes" and phasing out or replacing CFC's in accordance with the Montreal Protocol to protect the ozone layer of our planet.

In the eighties the list of environmental technologies to be supported by governmental money was extended by Abandoned Hazardous Sites, Landfillings and Polluted Industrial Sites and most recently Abandoned Military Sites e. g. how to remove and treat TNT and substances for chemical warfare out of the soil. The problem of Landfillings and Polluted Industrial Sites in turn drew the attention to Waste Management with the subtopics Thermal Treatment, Conditioning, Recycling and Prevention of Wastes.

The latest topic added to the list of supported technologies is the "Production-Integrated Environmental Protection" or in short "Clean Technologies".

The environmental policy and the developed technologies of the seventies were quite successfull (Figure 2, 3). Monitoring measurements of air and water quality show significant reductions of pollutants, like

$SO_2$, NOx and dust in the air. The water quality of rivers and lakes improved significantly.

Unfortunately, all these improvements on the one side mean a greater environmental burden on the other. Cleaner air, reductions of emissions in production processes led for instance to more filters and collected dust, which have to be disposed of. This waste can be very problematic because it contains heavy metals, organic chemicals, of which quite a few can cause cancer. All the pollutants removed from waste water led to huge amounts of sludges from the purification process. To give you an impression of the amounts of wastes generated per year in Germany I would like to show the next transparency (Figure 4), which compares the Cathedral of Cologne with the volume of annual wastes.

The next transparency (Figure 5) shows the different sources of this waste. 20 - 30 % of the waste - mainly the more hazardous waste - is generated by the producing or manufacturing processes in industry. In comparison to the past more waste is being accumulated because industry has increasinly not been allowed to blow its emissions into the air or to "enrich" the water with many substances wich are unwanted side effects or side products in the production process. Laws and regulations limiting the permitted "releases" are for instance the "Federal Immissions Control Act" and the "Federal Water Act". The laws and regulations normally set standards which must not be exceeded but do not lay down which technology should be applied to meet these standards. Because these new regulations have been forced on an existing production system, which has all the machines and all the production facilities already in place, incorporating many, many years of experience, this system favours "end-of-pipe technologies". With their assistance the environmental standards can be met simply by the use of "add-on measures". Therefore, even current industrial environmental investments (Figure 6) are mainly concentrated on add-on technologies or as we call it "secondary technologies". Comparatively small are the investments in primary technologies or clean technologies such as process-integrated or production-integrated technologies protecting the environment.

The Pro's and Con's of end-of-pipe technologies are shown by the next figure (Figure 7). End-of-pipe technologies are highly efficient in removing specific pollutants to a level below the required standards. They are reliable because the huge rest of the production infrastructure is unchanged. The additional costs, that means the additional financial risks, are clearly calculable.

On the other hand, end-of-pipe technologies mean a shift of pollutants from air and water to more waste which has to be disposed of posing a potential burden to soil and groundwater (Figure 8). If secondary measures don't work or if they are switched off the former polluting conditions are re-established. They cause additional costs for investment and operation and very often they become excessivley expensive, when new regulations require higher standards.

Therefore it seems reasonable trying to find technical approaches where end-of-pipe technologies can be replaced by integrated solutions which avoid the unwanted emissions altogether or at least to a great extend (Figure 9, 10, 11).

Is that possible?

Looking back on 295 R&D projects supported with 265 Million DM by the Federal Ministry for Research and Technology between 1975 and 1993 with the objective to develop "low-emission processes and products", the answer is yes. It should be possible in many cases. The following figure shows the allocation of the total financial support of 265 million DM to different problem areas. The financial support was allocated depending on priorities set according to environmental risks (Figure 12). Prefered were R&D grant applications from industrial companies (No applicant, no R&D), which employ production technology. This approach led in many cases to practical use of the results (more than 80%).

In many cases it turned out, that radical redesign of existing production technologies could drastically reduce the amount of pollution in comparison to its original amount. Also economic advantages could be achieved or seem to be possible depending on size and capacity of the respective production plant.

The detailed analysis of these 295 projects was very encouraging with respect to the new approach of "production integrated environmental protection" which was published by the Federal Ministry for Research and Technology in January 1994.

The principles for the practical realisation of this new approach are indicated in the next transparency (Figure 13).

- The **product** as output of the production process must be benign to the environment. It should have a long lifetime, easy to repair, easy to recycle and should not pose any disposal problems.
- The **production process** has to be modified or totally changed. Different technological approaches have to be tried to find real breakthroughs or gradual improvements of the production process. The decisive criterion is minimizing the emissions and also wastes. At the same time material flows should be interlinked and minimized and material cycles should be closed.
- Each and every production process needs **input materials**. Searching for the right input materials is very important because many of the environmental characteristics of the product and production process are determined by the input materials.

Starting point and a very important selection criterion for R&D projects is the amount of emissions or wastes which should be avoided (Figure 14).

O.K., I admit all this sounds terribly abstract. Therefore, I would like to mention two very recent examples of R&D projects in the framework of Production-Integrated Environmental Protection.

One is the so called "Green Television-Set" (Figure 15). In this project a TV-Set should be designed according to the explained requirements of an environmentally benign product, it should have a long lifetime, it should be easy to repair, easy to recycle and avoid the special problems of electronic scrap.

The other is "dry grinding", "dry milling" or "dry lathing" to avoid the lubricating and cooling liquids normally used to achieve high working speeds and a long lifetime of the respective tools (Figure 16). The used lubricant/coolant contaminated by little metal pieces is a problematic waste. In Germany the annual amount of this special waste is 750,000 tons posing a hazard to the environment. Its disposal costs are more than 1 billion DM per year. Avoiding these lubricating and cooling liquids would reduce the burden on the environment and can save costs for the respective industries.

These examples are also prototypes for the way we organize and support projects.

Normally we initiate project clusters or collaborative projects. Each part of this cluster is a certain work package carried out by one partner in a well organized divison of labour. All the project partners exchange their work plans and their results according to an agreed schedule monitored by a coordinator. Everything is settled in a cooperation contract.

The project partners as a rule are several industrial companies, university institutes and research institutions like the National Labs or the Fraunhofer-Gesellschaft. The idea of this kind of interlinked projects, the Americans call it R&D-consortia, is to facilitate as much technology transfer between these partners as possible.

Why governmental financial support at all? Why not simply set rules by the government and let industry struggle to meet the requirements?

There is no conclusive answer. Our answer for the time being is: The environmental situation in Germany is very dynamic taking into account public demand, the environmental political discussion, and the resulting environmental legislation. Therefore within a few years the requirements to be met by industry have changed dramatically and became much more demanding.

Therefore, R&D supported by the government
- helps to find new solutions to meet new requirements,
- assists mainly small and medium sized companies,
- shows also new possibilities to increase environmental standards.

In the case of the production-integrated environmental protection, the challenge is much more comprehensive because new technical solutions can mean the change of major parts of the production infrastructure, of the material flows, and so on. That means, the system to be considered comprises more or less the whole production infrastructure. Therefore the incentive to consider these kinds of modifications at all must be strong enough, at least in the beginning, that this process of re-thinking can get off the ground.

Now lets consider the financial means provided by the Federal Ministry for Research and Technology (Figure 17). The over-all 1994 budget amounts to about 9.5 billion DM, of which 4 billion DM are spent to finance the big research centers (National Labs) on an annual basis, 3.8 billion DM go into R&D grants. Environmental research (Figure 18) is supported by about 725 million DM out of which environmental technology receives 266 million DM, ecology research 250 million DM and climate research 210 million DM. The 266 million DM for environmental technology consists of 165 million DM for R&D grants and 101 million DM for R&D in National Labs.

Only 30 - 35 million DM are earmarked for R&D in Clean Technologies. This amount is extremly small in comparison to the total turnover of the corresponding industries. That means we have to concentrate our activities on key problems, key or cross cutting technologies which possibly or hopefully are of importance for many industrial processes and many industrial branches.

Later during this workshop Dr. Günter Panzner will present to you quite a number of practical examples of our work in clean technology.

REFERENCES:

"Environmental Research and Technology"
Programme 1989 - 1994
published by The Federal Minister for Research and Technology
Bonn, 1989 (English Translation)

"Environmental Protection in Germany"
National Report of the Federal Republic of Germany for the United Nations Conference on Environment and Development in June 1992 in Brazil
published 1992 by the Federal Ministry for Environment

"Environmental Protection Technology"
selected projects supported by the Ministry for Research and Technology in the period 1975 - 1992
published 1993 by the Project Management Division "Environmental Protection Technology", DLR, Bonn (English Translation)

"Production Integrated Environmental Protection"
A promotion concept of the Federal Ministry for Research and Technology
published 1994 by the Project Management Divison "Environmental Protection Technology", DLR, Bonn (English Translation)

dying forests and extremly polluted rivers in the 70ties led to

**environmental policy (new laws, new regulations)**

and

**governmental financial support for R & D in environmental technologies**

main R & D areas were:

| | |
|---|---|
| water technologies, e. g.<br>- clean drinking water<br>- treatment of wastewater | since 70ties |
| technologies for cleaner air, e. g.<br>- reduction of $SO_2$, NOx<br>- reduction of dust | since 70ties |
| - phasing out of CFC | since 80ies |
| abandoned hazardous sites / landfillings | since 80ies |
| waste management<br>- thermal treatment, conditioning<br>- recycling<br>- prevention | since 80ies |
| low emission processes and products | since 1975 |
| production-integrated environmental protection / clean technologies | since 1994 |

Figure: 1

Sulphur dioxide emissions in the Federal Republic of Germany

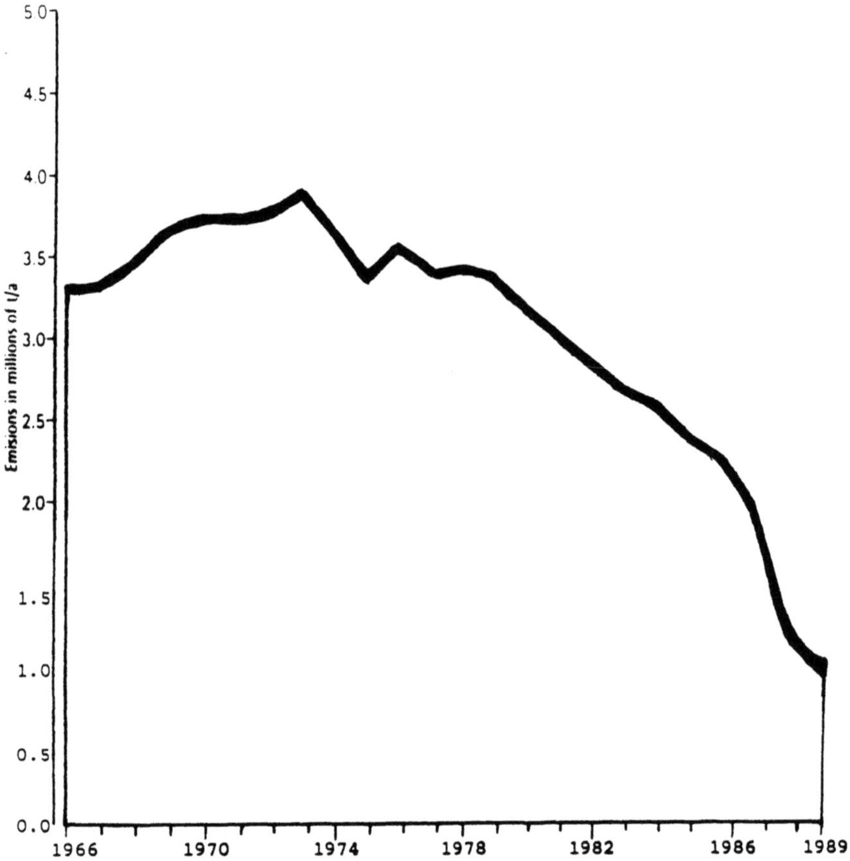


Following the 1973 oil crisis, emissions of sulphur dioxide in the Federal Republic of Germany dropped continously and, over the past few years dramatically. This success may be attributed particularly to the installation of flue gas purification facilities in power stations.

Figure: 2

**Dust emissions in the Federal Republic of Germany**

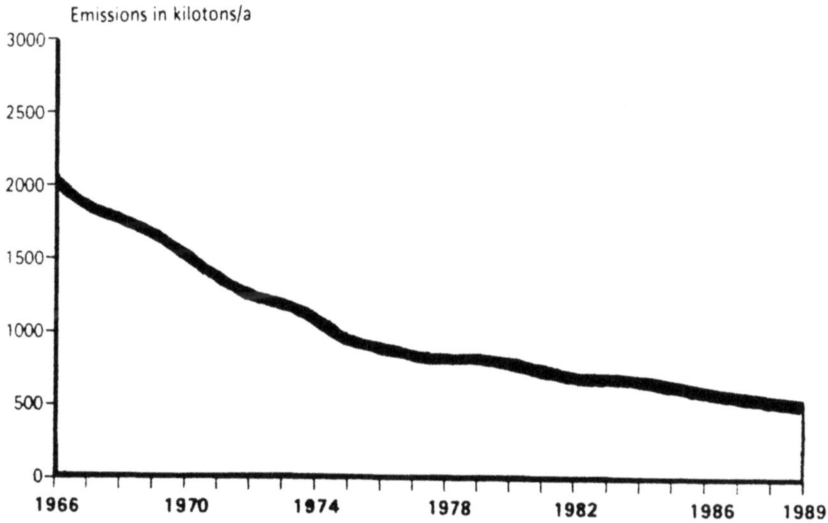


Thanks to ever more stringent environmental legislation, particulate emissions in the Federal Republic of Germany have dropped by 80% since the 1970s.

Figure: 3

Visual presentation of the annual volume of waste occurring in the Federal Republic of Germany, including sewage sludge and dredged material (Figures for 1987; based on the assumption that the average density of the total volume of waste is approximately 1 $t/m^3$)

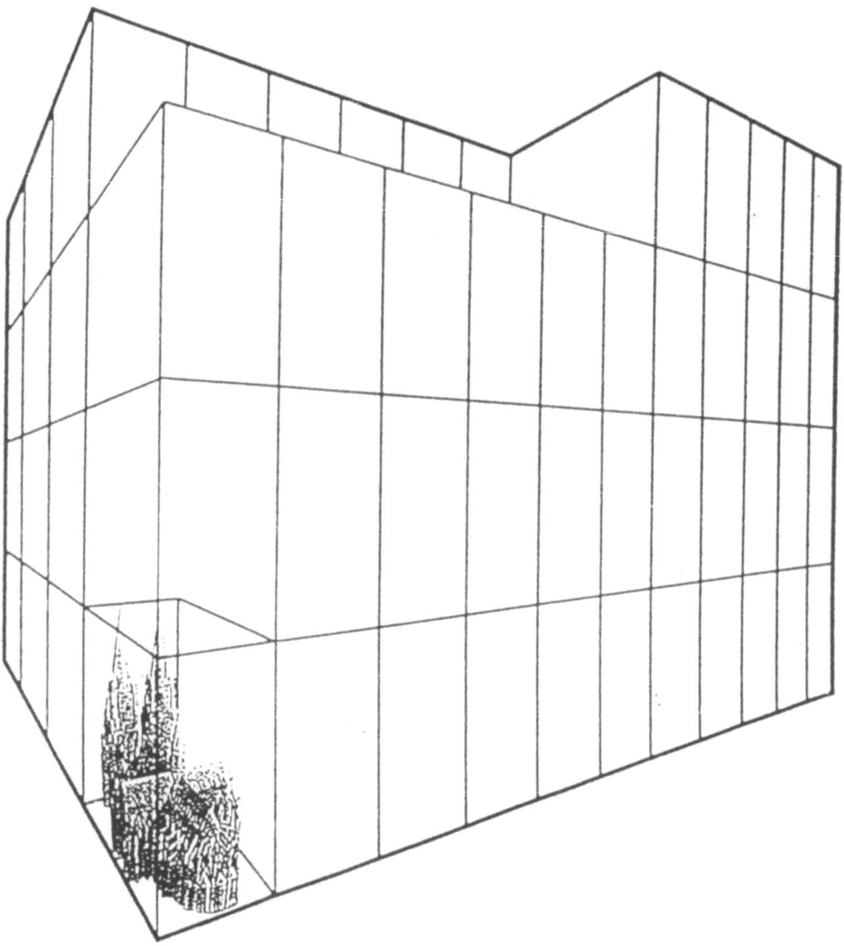

Figure: 4

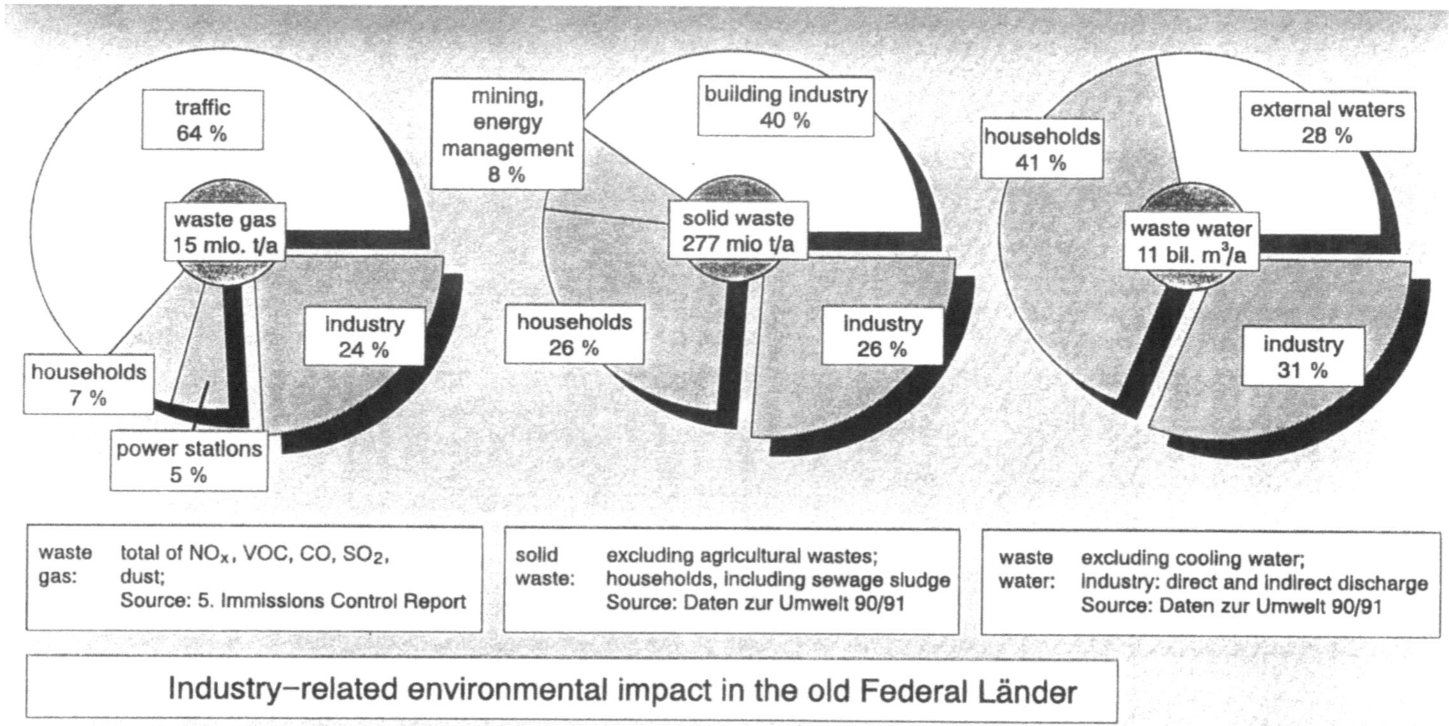
traffic
64 %
waste gas
15 mio. t/a
households
7 %
power stations
5 %
industry
24 %
mining,
energy
management
8 %
building industry
40 %
solid waste
277 mio t/a
households
26 %
industry
26 %
households
41 %
external waters
28 %
waste water
11 bil. m³/a
industry
31 %
waste gas: total of NOx, VOC, CO, SO2, dust; Source: 5. Immissions Control Report
solid waste: excluding agricultural wastes; households, including sewage sludge Source: Daten zur Umwelt 90/91
waste water: excluding cooling water; industry: direct and indirect discharge Source: Daten zur Umwelt 90/91
Industry–related environmental impact in the old Federal Länder

Figure: 5

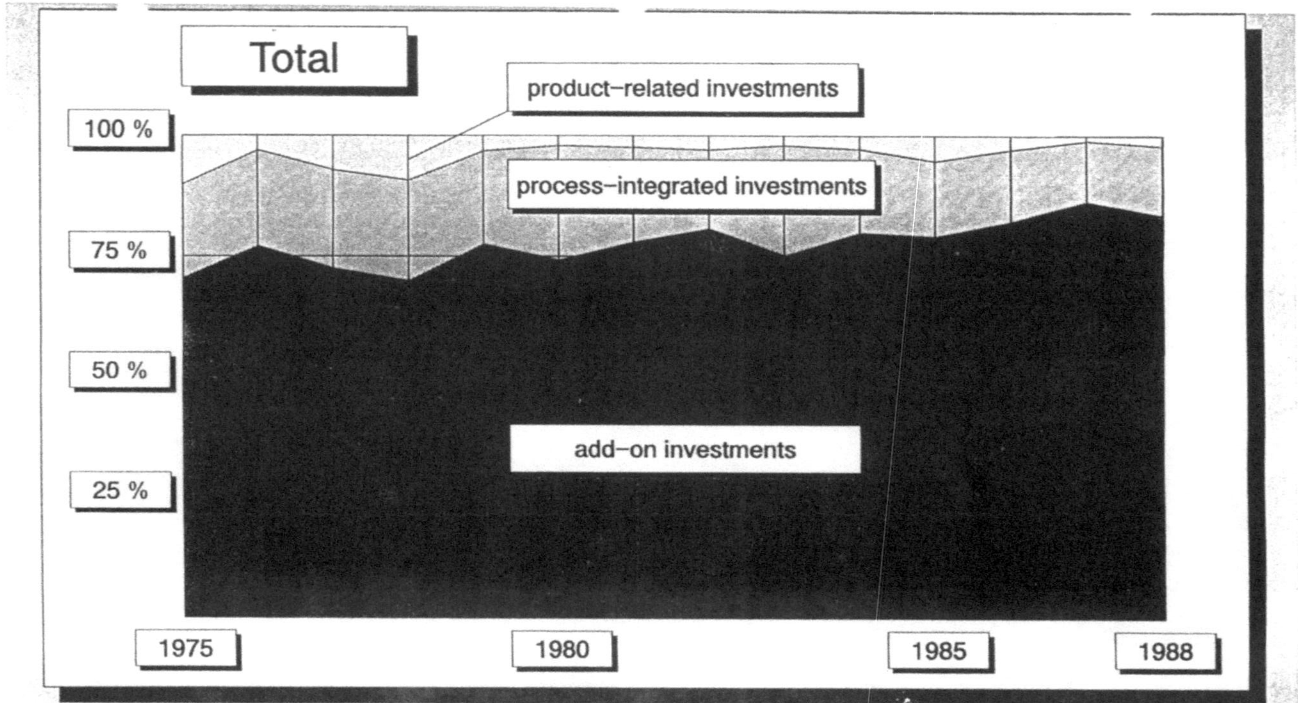


Distribution of environmental protection investments by industry

Figure: 6

**Pro's and Con's of end-of-pipe technologies**

**Pro's:**

they do their job
- very efficiently
- very specifically
- very reliable

the huge rest of the production infrastructure is unchanged
the marginal costs are confined

**Con's:**

shift from air and water pollution to more waste (potential burden to soil and groundwater)

when out of operation (or switched off) restauration of former polluting conditions

control and monitoring of the pollutants

additional costs for investment and operation

excessively increasing costs in case that higher standards will be required

Figure: 7

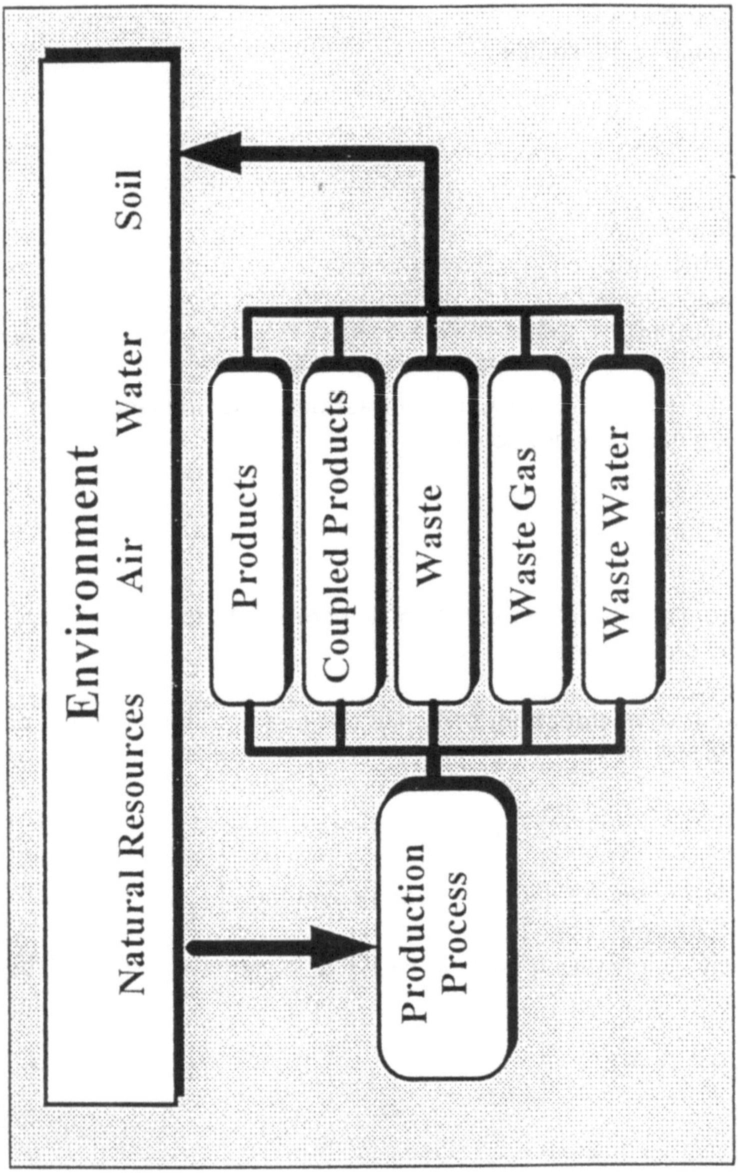
Environment
Natural Resources
Air
Water
Soil
Production Process
Products
Coupled Products
Waste
Waste Gas
Waste Water

Figure: 8

**product & production integrated environmental protection**

**objectives:**
(set by public demand, regulations, high costs)

- avoid or minimize emissions
- avoid or minimize wastes

with the product and during its production

**answers:**

- use of new technologies
- interlinking material flows
- minimizing of material flows
- closing of material flows (recycling)

**consequences:**

- change of production pattern
- new production equipment
  (gradually improved or drastically changed)
- new network for the flow of materials and goods
- little input of row materials
- little output of emissions and wastes

Figure: 9

Since the end of August 1994 a new law is in force

**"Waste Management and Product Recycle Act"**

will help to establish a Recycle Economy

Figure: 10

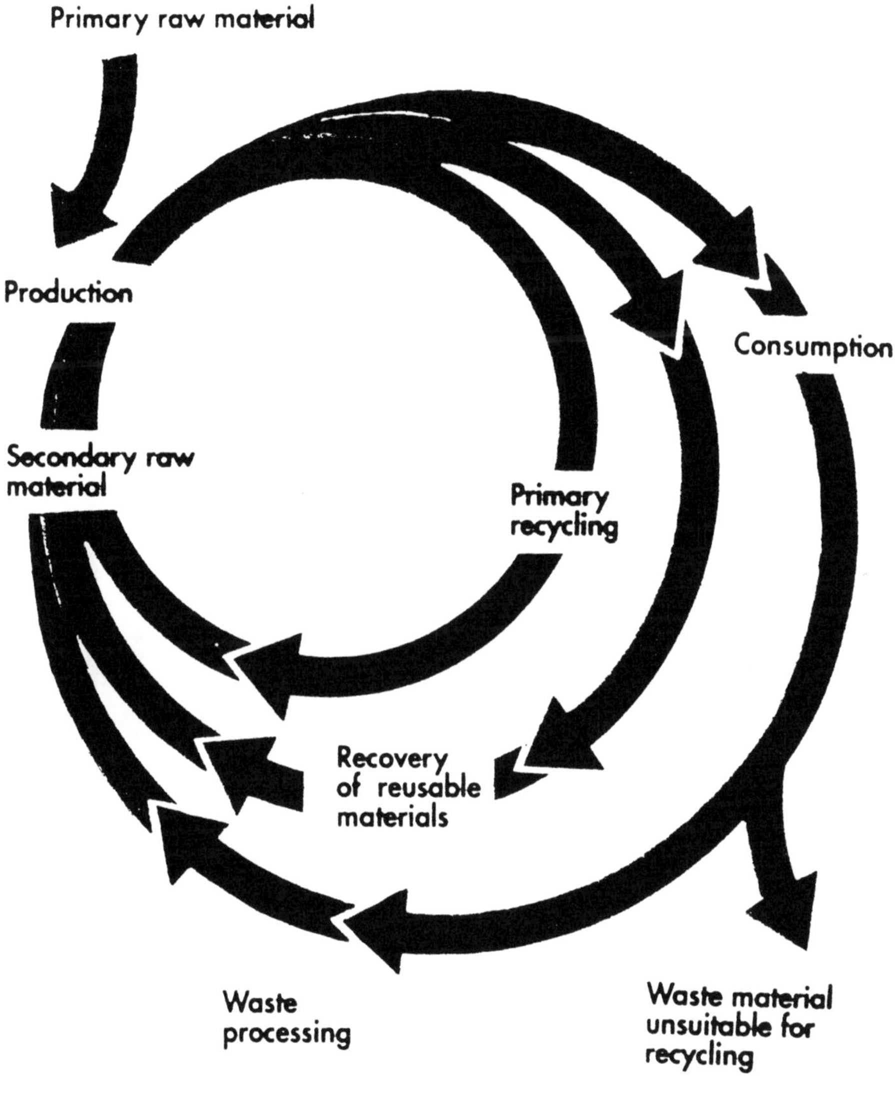
Primary raw material
Production
Consumption
Secondary raw material
Primary recycling
Recovery of reusable materials
Waste processing
Waste material unsuitable for recycling

Figure: 11

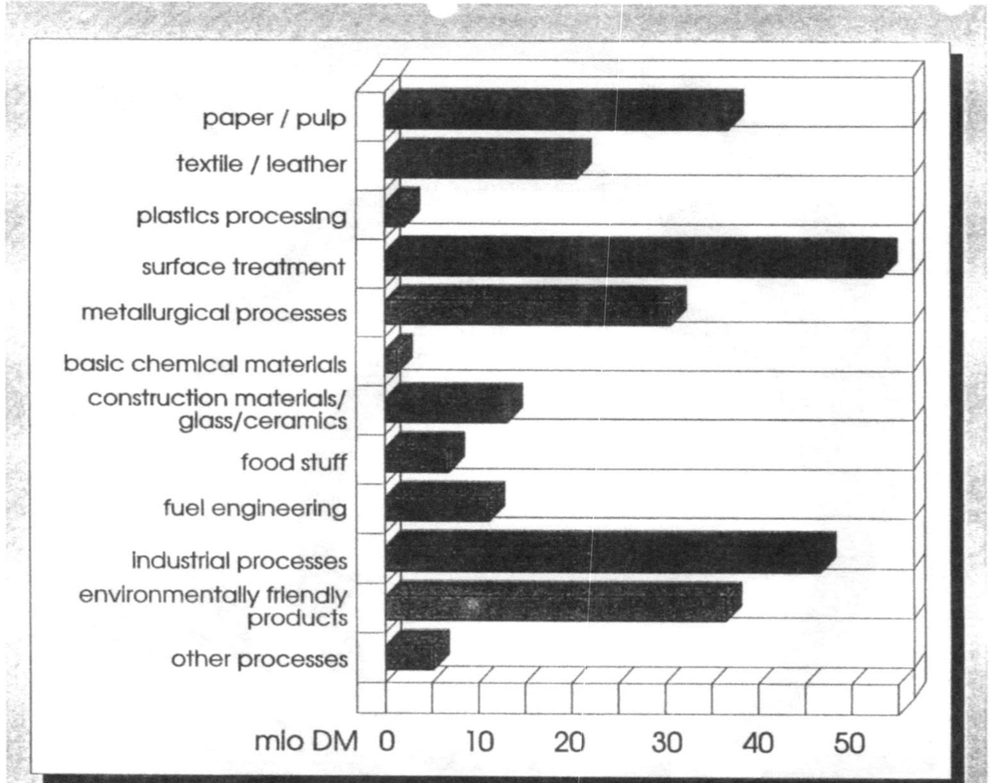


Figure: 12:
Allocation of financial support
295 R&D projects, 265 Mill. DM

principles for the practical realization of
**product and production integrated environmental protection**

° product as output must be benign to the environment (long livetime, easy to repair, easy to recycle, no waste problem)

° production process
seeking technological breakthrough's or gradual improvements by trying different technological approaches

minimizing emission/pollution by interlinking (and minimizing) material flows

closing of material cycles

° seeking the right input materials
environmentally benign
minimizing material flows

Figure: 13

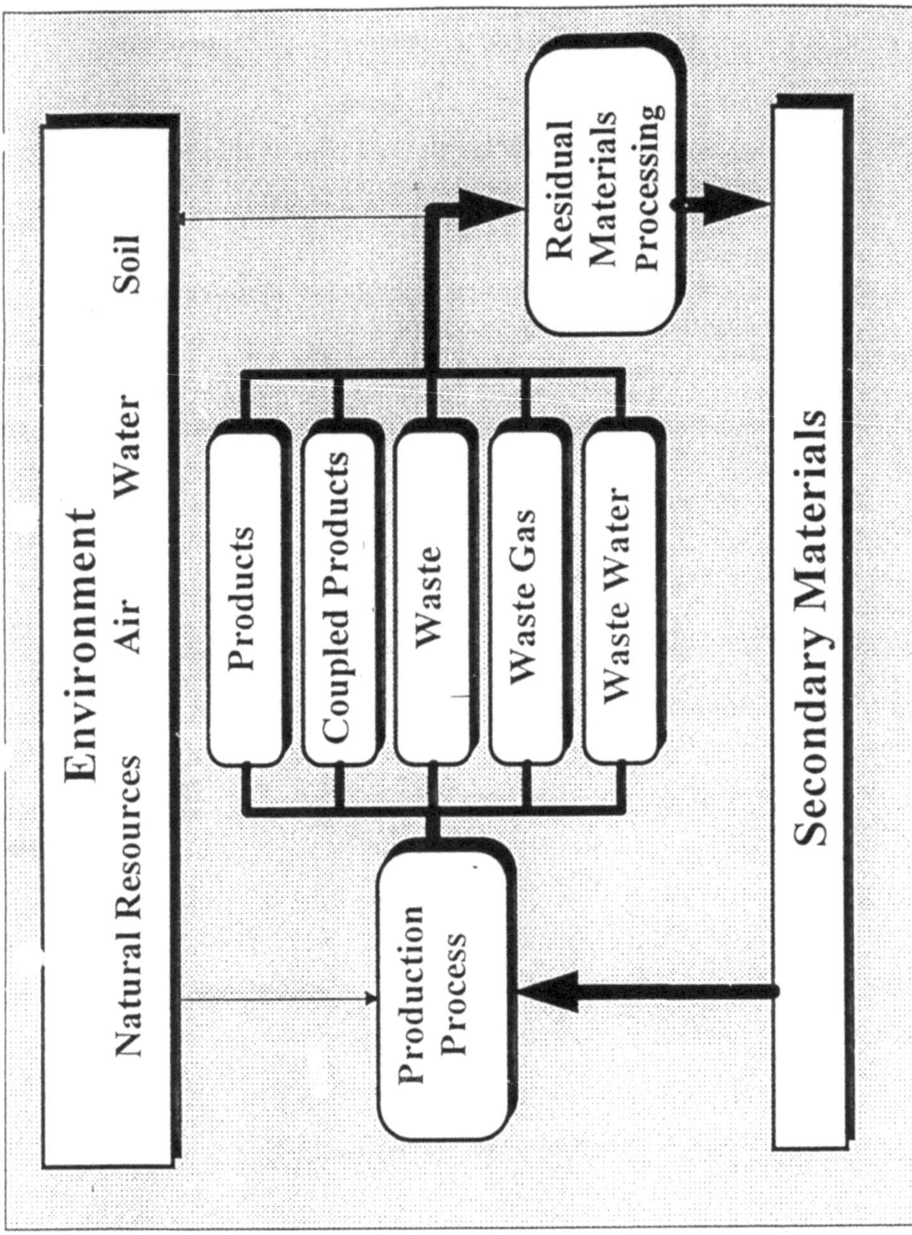
Environment
Natural Resources
Air
Water
Soil
Production Process
Products
Coupled Products
Waste
Waste Gas
Waste Water
Residual Materials Processing
Secondary Materials

Figure: 14

collaborative project (project cluster)

topic: **contribution to the development of a recycling economy: "green" TV-set**

| partners: | subtopic: | supported by % |
|---|---|---|
| Institute for Appl. Ecology | ecological monitoring | 100 |
| Deutsche Thomson-Brandt | circuits/ electrical connections | 40 |
| Grundig | coordination, plastic materials | 40 |
| Loewe Opta | metallic materials | 40 |
| Nokia Audio Electronics | loudspeaker | 40 |

total support by Federal Ministry of Research and Technology: about 1.1 Million DM

Figure: 15

collaborative project (project clusters)

topic: **potential of dry grinding**

| partners: | subtopic: | supported by % |
|---|---|---|
| University Hannover | basics and coordination | 100 |
| Lenze GmbH | experiments in grinding special profils | 22 |
| Diamant Werkzeuge | development of CBN-grinding discs | 22 |
| Krebs & Riedl | development of sintert discs | 22 |
| Dieselmotoren-werke | experiments in external round grinding | 22 |

total support by Federal Ministry of Research and Technology: about 0.5 Million DM

Figure: 16

| | |
|---|---|
| TOTAL | DM 9.500 MILL |
| INSTITUTIONS | DM 4.000 MILL |
| PROJECT FUNDING | DM 3.800 MILL |
| ENVIRONMENTAL R & D | DM 725 MILL |
| CLEAN TECHNOLOGY R & D | DM 30 - 35 MILL |

1994 BUDGET OF THE FEDERAL MINISTRY FOR RESEARCH AND TECHNOLOGY

Figure: 17

## Environmental Research

### Fed. Ministry for Research and Technology Expenditures 1994: 725 Mill. DM

| Ecological Research | Environmental Technology | Climate, Atmosphere |
|---|---|---|
| Supported Projects<br>85 Mill. DM | Supported Projects<br>165 Mill. DM<br>(30 - 35 Mill. DM for Clean Technologies) | Supported Projects<br>106 Mill. DM |
| Supported Institutions<br>GSF, UFZ, KFA, GKSS, IGB u.a.<br>165 Mill. DM | Supported Institutions<br>KfK, GKSS u.a.<br>101 Mill. DM | Supported Institutions<br>KFA, KfK, AWI, DLR GKSS, PIK, IfT, IAP<br>104 Mill. DM |
| 250 Mill. DM | 266 Mill. DM | 210 Mill. DM |

Figure: 18

## How to Prevent Nitrous Gas Emission in Nitric Acid Treatment of Metals

Janos de Jonge
Department of Chemical Technology
University of Veszprém
Egyetem utca 10 H-8201 Veszprém
Hungary

### *Introduction*

Nitrogen oxide pollution is a nuisance caused mainly by flue gases of power plants using fossile fuels and by the exhaust gases of internal combustion engines. The chemical industry, particularly the production of nitric acid is blamed for the pollution of the environment by its well known brown tail gas.

As Joshi et al. state, the cause of the latter pollution lies in the reaction of nitrogen dioxide and water, taking place in the absorption towers, in liquid phase:

$$3NO_2 + H_2O = 2\ HNO_3 + NO \qquad (1)$$

i.e one third of the nitrogen dioxide transforms to nitric oxide which does not react with water. In the presence of oxygen, nitric oxide oxidizes

$$2\ NO + O_2 = 2\ NO_2 \qquad (2)$$

and the formed nitrogen dioxide reacts again with water (see Eq. 1)

If a -theoretical- absorption tower is devided into $n$ ideal sections and in each section the transformations given in Equations (1) and (2) are complete, then the off gas leaving the $n$-th section contains $(1/3)^n$ mole of $NO_2$, if 1 mole of $NO_2$ entered into the absorption column.

It is evident, that $(1/3)^n \rightarrow 0$ if $n \rightarrow \infty$, therefore to the complete absorption of nitrous gases an absorption tower is needed, having endless ideal sections. This means that this tower must have endless hight or diameter, but this requirement is absurd. Therefore, the off gases leaving a nitric acid producing absorption tower contain always polluting nitrous gases.

NATO ASI Series, Partnership Sub-Series, 2. Environment - Vol. 2
Cleaner Technologies and Cleaner Products
for Sustainable Development
Edited by H. M. Freeman, Z. Puskas, and R. Olbina

*The Problem*

Several chemical reactions exist, having more or less practical importance, which produce nitrous oxides in *concentrated form, e.g.* if metals, characterised by positive electrode potential compared to hydrogen electrode, are dissolved in nitric acid.:

$$3\ Ag + 4\ HNO_3 = 3\ AgNO_3 + NO + 2\ H_2O \qquad (3)$$
$$3\ Cu + 8\ HNO_3 = 3\ Cu(NO_3)_2 + 2NO + 4\ H_2O \qquad (4)$$
$$3\ Hg + 8\ HNO_3 = 3\ Hg(NO_3)_2 + 2\ NO + 4\ H_2O \qquad (5)$$

In the production of nitrate salts of the above metals, the evolved nitrous oxides are either emitted into the atmosphere or, for reducing the air pollution, they are mixed with air and led through absorption columns rinsed with water or dilute nitric acid. This method seems to be the easiest to realise and maybe economic, although, the air pollution cannot be avoided and the produced nitric acid is generally weak (about 10 or 15 w/w % $HNO_3$) therefore mostly useless and must be neutralised before its discharge.

Numerous other processes were proposed for the absorption, adsorption, reduction etc. of nitrous gases, as e.g. absorption in alkaline solution, in urea containg nitric acid solution, adsorption on silica gel, on zeolites etc., thermal destruction in flame, catalytic reduction with methane, hydrogen or ammonia gases. The common drawback of these processes is either their limited efficiency or their expensive, mostly uneconomic character.

*The Idea*

If in a chemical process *nitrous gases evolve in concentrated form*, which are diluted only with water vapour, and the nitric oxide is oxidized with *pure oxygen* (instead of air) and *the nitrogen dioxide is absorbed in water,* then these reactions will be faster than in the presence of air as oxidizing medium, due to the higher partial pressure of reacting gaseous components. It is evident that also in this case the nitrous gases will not completely be absorbed in the absorption column because Equation (1) and (2) are valid, but in this case the off gas leaving the absorber will contain only small amounts of nitrous oxides, excess oxygen and water vapour.

Instead of emitting this gas into the atmosphere, pure oxigen is added to this stream to retrieve the consumed amount by the oxidation of nitric oxide, and this oxygen rich gas is *recycled* for the further oxidation of the inlet nitrous gases, as it is shown in *Figure 1.*

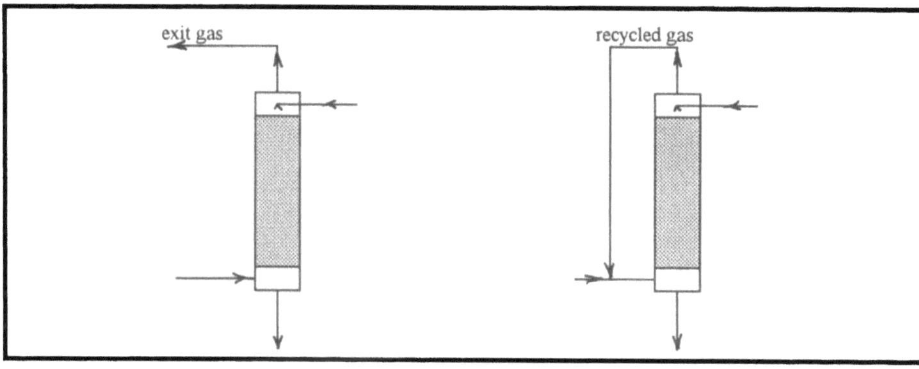


*Figure 1.* The traditional NOx gas absorption and the exit gas recycling.

With the recycling the $NO_x$ containg off-gases, their *complete absorption* can be realised in spite of the finite dimensions of the absorber, because the *residence time* of nitrogen oxides can be elongeted *ad libidum.* Albeit the infinite elongation of residence time can practically never be fulfilled, it can approached to the desired degree and with this to put into practice the *pollution free nitrous oxides absorption:*

$$2\ NO + 1.5\ O_2 + H_2O = 2\ HNO_3 \qquad (6)$$
$$2\ NO_2 + 0.5\ O_2 + H2O = 2\ HNO_3 \qquad (7)$$

Another advantage of the recycling is that with an absorber having well designed dimensions, more concentrated (50 to 60 w/w %) nitric acid can be produced, this acid can be applied again in the dissolution of the given metal. This means, e.g. in the production of silver nitrate, that about 25 % of nitric acid can be saved and the over-all reaction of

$$2\ Ag + 2\ HNO_3 + 0.5\ O_2 = 2\ AgNO_3 + H_2O \qquad (8)$$

can be realised.

*The Realisation*

Based on the idea discussed earlier, a pilot plant was built for metal dissolution. The flow diagram of this plant is given in *Figure 2*

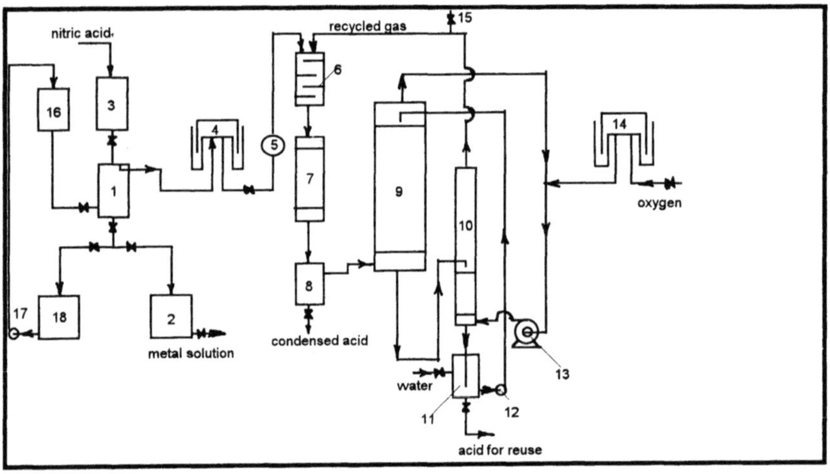


*Figure 2. Flow diagram of metal dissolver plant*
1 - metal dissolver (reactor); 2 - reservoir for metal solution; 3 - nitric acid tank; 4 - NOx gasometer; 5 - gas flowmeter; 6 - gas mixer; 7 - condenser - cooler; 8 - condensed acid tank; 9 - absorber; 10 - stripper column; 11 - acid reservoir; 12 - acid pump; 13 - gas blower; 14 - gasometer; 15 - purge valve; 16 -acidic water tank; 17 - pump; 18 - acidic water reservoir.

The dissolution of the metal is discontinuous. After the charging of metal scrap into the dissolver tank (1), to remove the air, water is led from the reservoir (16) into this reactor. The water is led off to the tank (18) and pumped (17) again into the reservoir (16). During this process NOx gas is sucked from the gasometer (4) into the reactor (1).

Measured amount of nitric acid is led from the nitric acid tank (3) into the reactor, where the metal dissolution begins and nitrous gases fill up the NOx gasometer (4).

From this gasometer regulated stream of NOx is led through the flowmeter (5) into the gas mixer (6) where the NOx is mixed with the oxygen rich recycled gas stream. The reaction heat of the exotherm NO-oxidation reaction is removed in the cooler-condenser (7) and the $NO_2$ containing gas enters into the absorber (9). The NOx lean gas mixture leaving the absorber is recycled by the blower (13) through the stripper column (10), where the gas stream removes the occluded nitrous gases from the nitric acid, this gas stream is then returned to the gas mixer (6). The nitric acid solution leaving the absorber (9) flows through the stripper column, from there to the acid reservoir (11) and an acid pump (12) recycles the acid to the top of the absorber.
As it can be seen, the NOx containing gas stream is recycled in the gas mixer - absorber - stripper column - gas mixer in a closed loop and, due to the volume reduction caused by the NO

oxidation (see Eq. 2), the closed gas loop "sucks" automatically the needed amount of oxygen from the $O_2$ gasometer (14).

*Table 1.* shows the theoretically needed volume of oxygen, if 1 kg of metal is dissolved in nitric acid.

*Table 1.* Oxygen consumption in metal dissolution

| Metal | $Stdm^3$ $O_2$/kg metal |
|---|---|
| Silver | 52 |
| Mercury | 56 |
| Copper | 176 |

The practice proved that this technology is sound and runs trouble free without the emission of nitrous gases.

Problems arise only if inert gases accumulate in the closed gas loop. This can be the case if air was not completely removed from the metal dissolver reactor, or if the metals contain considerable amount of alloying metals, e.g. zinc. In the latter case also hydrogen gas evolves during the metal dissolution. In the discussed case hydrogen is "inert" gas from the viewpoint of reactions given in Equations (1) and (2). The growing concentration of inert gases lessens the partial pressures of reacting gas components (NOx and $O_2$) and the needed reactions slow down or even stop.

This unfavourable situation can be terminated by the activation of the purge valve to replace a given volume of recycled gas with pure oxygen. Naturally, this process involves temporarily nitrous gas emission. This makes this "Pure Technology" to a "Low Waste Technology".

Reference

Joshi J.B. *et al.* (1985) Absorption of NOx Gases. Chem. Eng. Commun. 33: 1-92

# Emission Reduction of Chlorinated Hydrocarbons by a Mineral-Based Metal-Oxide Catalyst

Kántor, S.* - Adonyi, Z.:
Technical University of Budapest,
Department of Chemical Technology
H-1521 Budapest
Műegyetem rkp. 1-3.
Hungary

## Introduction

Although huge effort has been spent in the last decade on replacing chlorinated hydrocarbons with non-halogenated chemicals, they still play an important role in industrial production as well as in everyday life. Therefore, the application of low-cost process air cleaning technologies for existing production lines is necessary to meet recent air-emission requirements.

There are several ways of controlling chlorinated hydrocarbon emission including catalytic incineration [EPA,1991]. Unfortunately, catalysts which are used in incinerators are expensive, because they are mainly based on noble metals. Another two problems are that they are very selective and the concentration of contaminant in processed air is limited [Agarwal et al.,1992].

The key problem to be solved in this project is to find a proper catalyst, which can function in the presence of $Cl_2$ and HCl, which keeps its activity for a sufficiently long time to be used in an industrial plant and which has no tendency for quick deactivation or coking under higher concentration than allowed for today's technologies.

After extensive literature search and some experimental work, a mineral-based metal-oxide catalyst was found which fulfils the above requirements. This catalyst is produced in Hungary from the sludge of manganese ore processing. The average composition of the main components of the catalyst can be seen in Table 1. Of course, due to the source of the raw material, all of the elements of the Earth surface can be found in the catalyst.

NATO ASI Series, Partnership Sub-Series, 2. Environment – Vol. 2
Cleaner Technologies and Cleaner Products
for Sustainable Development
Edited by H. M. Freeman, Z. Puskas, and R. Olbina

Table 1.
Average oxide composition of the catalyst, wt.%

| | | | |
|---|---|---|---|
| $SiO_2$ | 32,0 | $K_2O$ | 3,0 |
| $Fe_2O_3$ | 26,0 | MgO | 4,0 |
| $MnO_2$ | 18,0 | $Na_2O$ | 0,25 |
| MnO | 3,0 | BaO | 0,075 |
| CaO | 7,0 | $P_2O_5$ | 0,45 |

## Experimental

The experimental apparatus is shown in Figure 1. The main component of this set-up was a quartz tube in electric furnace with temperature control. Temperature could be kept on the desired value within ±5°C. Stable and controlled gas-flow was achieved by the use of the carrier-gas system of a CHROM-4 gas-chromatograph. A part of the gas-flow bubbled through a thermostated, dark gas-washing flask, where it became saturated with model compound. Thus, concentration of model compound in inlet gas was controlled by saturation temperature and distribution ratio of air flows.

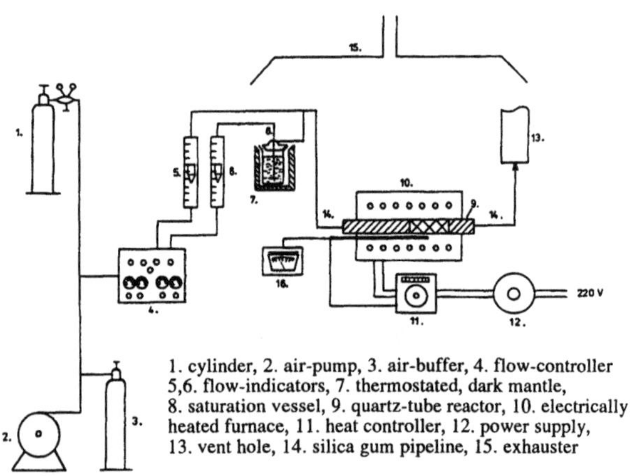


Figure1. Schematic diagram of the experimental apparatus

The first 2O cm section of the tubular quartz reactor, which was used as a preheater, was filled with glass beads cleaned by diluted HCl-solution and subsequently washed by distilled water. The length of the catalyst bed was 12,5 cm, followed by another 7,5 cm length of glass beads.
Gas samples were analysed by gas cromatography from gas samples which were taken in the same time from both inlet and outlet gas-flow.
The whole experimental set-up was built into a well ventilated glass hood to prevent the emission of the chlorinated hydrocarbons into the laboratory atmosphere.
In the investigations the most frequently used model compound was trichloroethylene mainly, because this takes a large portion of chlorinated hydrocarbon wastes (in Hungary it had a share of 36% in 1992). However, to study the dependence of the catalytic effect on different chemical structures, the following model compounds with different Cl:H ratios were studied. The decomposition behaviour of ethanol, which represented the non-chlorinated hydrocarbons and tetrachloroethylene with no hidrogen, were expected to be significantly different from molecules with both chlorine and hidrogen atom in their molecule.

Table 2.
Model compounds used for experiments

| Compound | Cl:H atomic ratio |
|---|---|
| Trichloroethylene | 3:1 |
| Tetrachloroethylene | 4:0 |
| l,l,2-Trichloroethane | 3:3 |
| l,2-Dichloroethane | 2:4 |
| Ethanol | 0:4 |

## Results and discussion

The aim of the first experiments was to compare the decomposition degree of the trichloroethylene in the presence of catalyst and glass-bead with similar hydrodynamic parameters. Figure 2. shows the basic existence of the catalytic effect of the selected catalyst on the decomposition of trichloroethylene. It can be seen that at 5OO$^{O}$C the conversion of trichloroethylene was not observed in the presence of glass-bead, which was considered to inactive to the decomposition, while a 70% conversion could be measured in the presence of fresh catalyst.

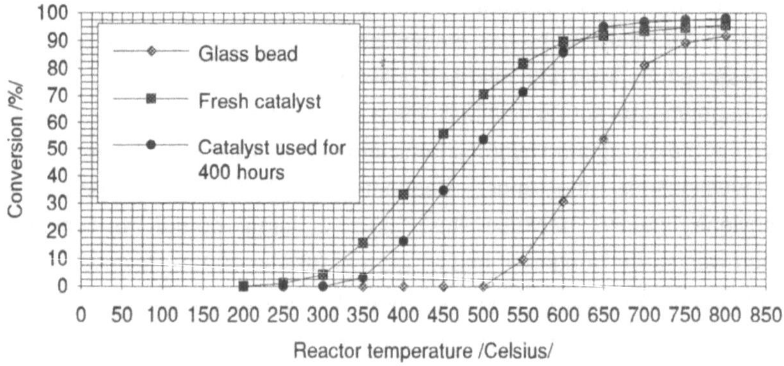


Figure 2. Conversion of trichloroethylene in air at 0,38 s residence time and 1000 ppm inlet concentration

The selected catalyst can be used industrially only if its activity is maintained for a sufficiently long time. Results of duration experiments are shown in Figure 3. Here, the pretreatment means contact with an air stream of 60 $dm^3$/hour at 500°C for 72 hours.

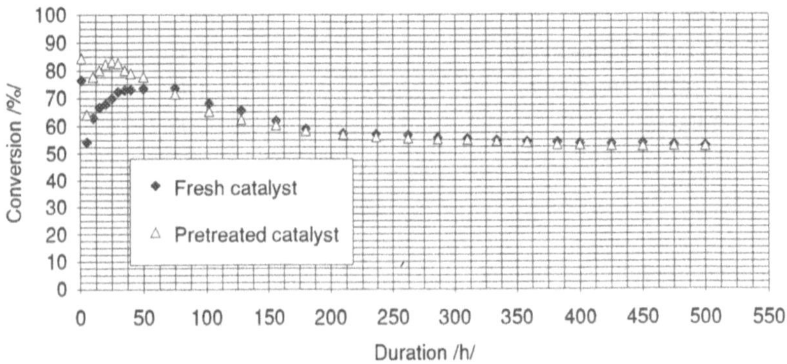


Figure 3. Conversion of trichloroethylene in air versus time on stream; at 500°C, 1000 ppm inlet concentration and 0,38 s residence time

As it can be seen from Figure 3. a slightly higher conversion was experienced on the pretreated catalyst than on the fresh one but the different performance of the two catalysts gradually converged after 5O-6O hours on stream. What is more important is that the activity of catalyst stabilised after 35O hours on stream.

Based on several analytical experiments with X-ray diffraction, secondary ion mass spectroscopy (SIMS), scanning electron microscopy and surface area measurements, the observed stabilisation of conversion is connected to the crystalline form changes of the surface of catalyst.

On the Figure 4/a. the electron scanning microscopic picture of the surface of the fresh catalyst can be seen. The mixed mineral structure can be realised but no sign of any symmetrical forms.

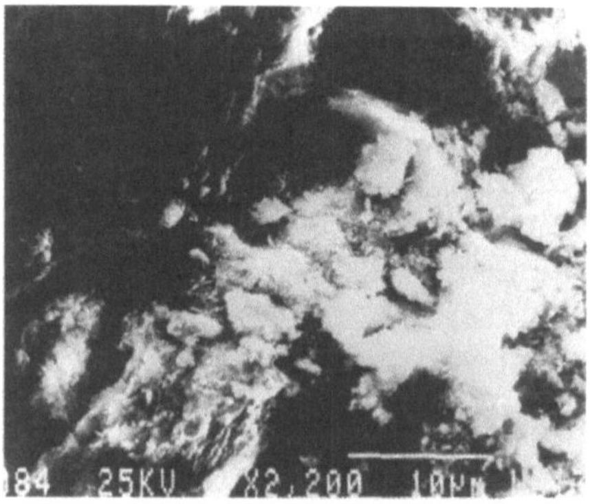

Figure 4/a. Fresh catalyst

Figure 4/b. Catalyst used for 44 hours

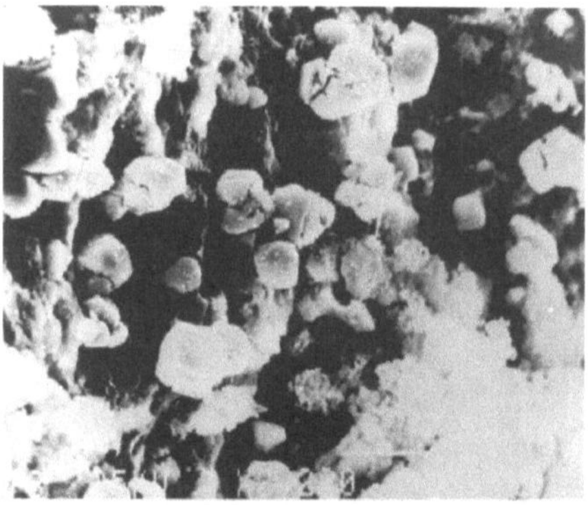

Figure 4/c. Catalyst used for 400 hours

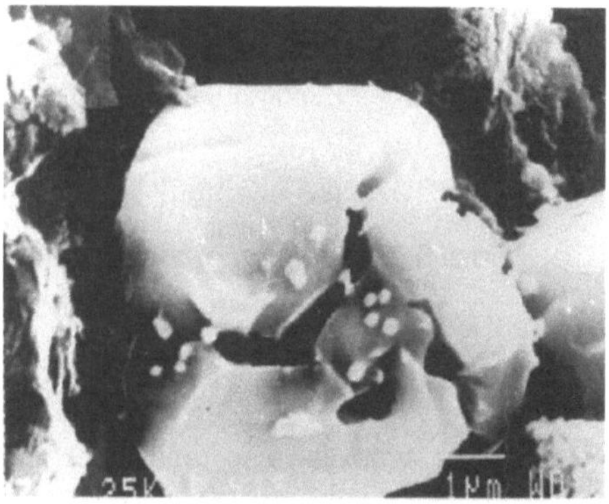

Figure 4/d. Catalyst used for 400 hours

Figure 4/a-d. All pictures were taken by JEOL 8404 electron scanning microscope with 25 KV and 2.200 (Figure 4/a-c.) or 15.000 (Figure 4/d.) times imagination.

On the Figure 4/b., which was taken from catalyst used for 44 hours, new cristals can be observed. Their number and size is increased with the lengt of use as it can be seen in Figure 4/c. On the Figure 4/d. one of the crashed surface cristals can be seen with more than six times bigger imagination.

SIMS mesurments indicated that the elementary composition of catalysts is similar in the case of fresh and used samples. Therefore, the formation of surface chloride can be precluded.

Next, experiments were carried out to study the effect of concentration of the model compound on catalytic conversion /Figure 5/. Here we encounter the encouraging fact that after an initial decrease in conversion in the concentration range of O-2OOO ppm, the increase of the concentration of trichloroethylene do not impair the conversion. Moreover, with catalysts which were previously on stream, an increase of conversion can be experienced. At the same time, no rise of the temperature of the catalyst-bed could be measured. This leads to the conclusion that HCl and $Cl_2$, which are the products of the catalytic decomposition, enhance the activity of the catalyst.

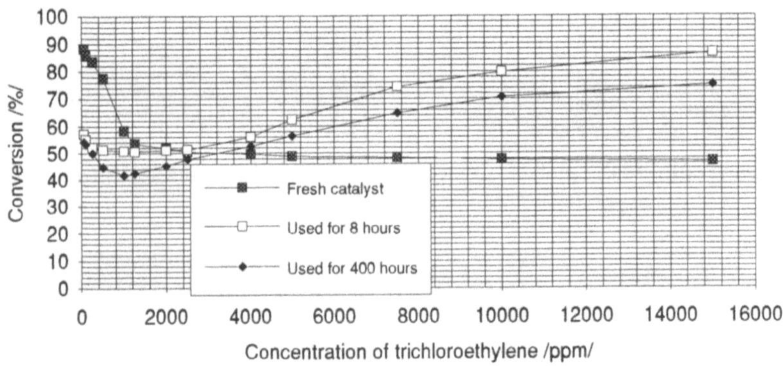


Figure 5. Conversion of trichloroethylene as the function of inlet concentration at 500$^{o}$C and 0,38 s residence time

Residence time is also an important parameter besides temperature in such technologies, so the effect of residence time on conversion was also measured. With a residence time of l,9 s, 95%, conversion could be achieved at 5OO$^{o}$C for trichloroethylene in the presence of catalyst stabilised after 35O hours on stream.

Finally, the decomposition of previously mentioned model compounds with different Cl:H atomic ratios was studied. These results are shown in Figure 6.

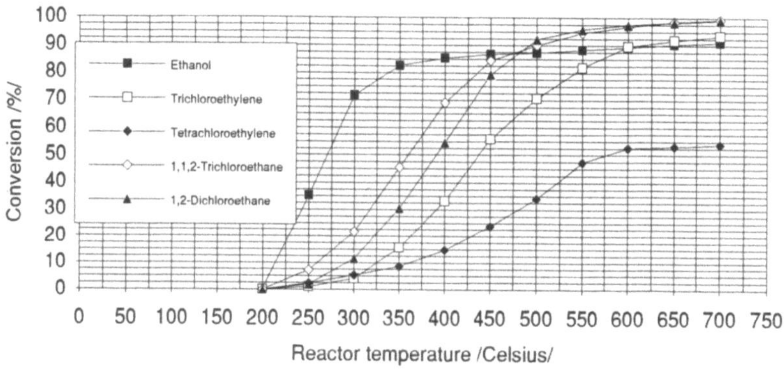


Figure 6. Conversion of different model compounds at 0,38 s residence time and 1000 ppm inlet concentration in the presence of fresh catalyst

As it could be expected, tetrachloroethylene has the lowest decomposition degree, but for the other three chlorinated hydrocarbons, nearly 90% or higher conversions could be achieved at 600°C.

## Conclusion

From the results of the experiments, it can be stated that a process based on investigated catalyst is suitable to clean air streams polluted by chlorinated hydrocarbons.

Before implementation, a more detailed study of the possible by-products and a deeper understanding of the decomposition mechanism would be desirable.

## References:


Agarwal, S.K., Spivey, J.J., Butt, J.B. (1992): Catalyst deactivation during deep oxidation of chlorohydrocarbons, Applied Catalysis A: 82:259.

EPA, Office of Research and Development (1991): Control Technologies for Hazardous Air Pollutants. EPA/625/6-91/014.

# Waste Reduction Strategies in the Chemical Industry

Peter Mizsey and Zsolt Fonyo

Department of Chemical Engineering,

Technical University of Budapest

H-1521 Budapest, Hungary

## INTRODUCTION

The industrial production is unfortunately associated with an inevitable and inconvenient waste generation. Historically, the waste treatment and process design were addressed independently due to the relatively low end-of-pipe treatment costs and environmental awareness. In the 1980s the emphases shifted from end-of-pipe treatment to the reduction of wastes at source as a results of the increased environmental awareness and regulations. Since the goal of zero waste seems to be unrealistic for most activities (Crittenden and Kolaczkowski, 1993), it is necessary to strengthen further research in the areas of reduction of wastes at source, to consider recycling on-site and off-site, and, to develop the methods of the end-of-pipe treatment, too.

The waste management hierarchy defined by EPA (US Environmental Protection Agency) ranks pollution prevention at the top (Rittmeyer, 1991). The hierarchy is: • *source reduction,* • *recycling,* • *waste separation and concentration,* • *energy and material recovery,* • *waste treatment,* • *disposal*, respectively. The source reduction has also an economic advantage, since it reduces or eliminates the burden upon the later steps of the technology, such as waste treatment.

Switching the emphasis from waste treatment to waste minimisation requires technological changes in the chemical process industries. Technological changes can be categorised into two areas: retrofitting on existing industrial activities and development of new cleaner processes. It is a heuristic rule that incorporating waste minimisation during process design is less complicated than modifying operations at an existing plant.

In the case of a chemical factory, which includes more chemical processes or plants, for efficient global waste minimisation it is not enough to study and develop individually the several processes because this might result in a local optimum. The improvements at the different processes are to be investigated and co-ordinated on a higher, second level, on the factory level and it is also necessary to study together the processes to minimise the waste emission of the entire factory.

NATO ASI Series, Partnership Sub-Series, 2. Environment – Vol. 2
Cleaner Technologies and Cleaner Products
for Sustainable Development
Edited by H. M. Freeman, Z. Puskas, and R. Olbina

**Waste minimisation on the process / plant level**

The developments of process improvements to reduce emission has historically been accomplished on an ad hoc basis by creative engineers. This generation of waste minimisation focuses on good housekeeping practices, inventory control and minor changes in operation practices.

Even more rapid progress is possible if systematic techniques are used to identify process improvement opportunities. The basis of these systematic strategies is the hierarchical approach for chemical process synthesis.

A major hierarchical approach for chemical process synthesis has been proposed by Smith and Linnhoff (1988). They have used the *onion diagram* to emphasise the sequential or hierarchical nature of process design, starting with the reactor and following with subsequent layers of separation system, a heat exchanger network, and a utility system.

Smith and Petela (1992) have developed their systematic procedure based on the onion diagram for waste minimisation in the process industries. They have identified two classes of wastes from chemical processes and located their origins to the layers of the onion diagram. The two inner layers of the onion produce *process waste* that is defined as waste by-products and purge streams. The outer two layers of the onion produce *utility waste*. Utility wastes are flue gas and ash from fuel combustion which are responsible for the global warming, waste from boiler feed water treatment, cooling water, etc. The utility waste is associated with the energy consumption and it is also a target to be minimised for a given process (Smith and O. Delaby, 1991). The systematic procedure proposed by Smith and Petela (1992) follows the way of process synthesis according to the onion diagram considering waste minimisation.

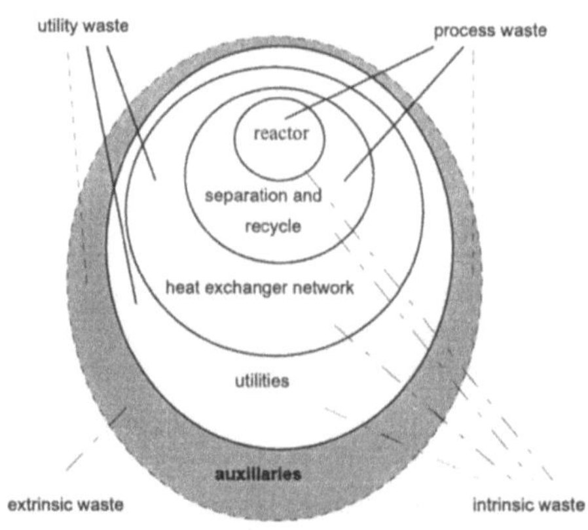


Figure 1. Extended onion diagram

Table 1

*Possible sources of process and utility waste and solution alternatives according to the extended onion diagram*

| LAYER | PROCESS WASTE | |
|---|---|---|
| | *possible source* | *solution alternative* |
| *Reactor* | difficult to separate and recycle unreacted feed material | increase conversion |
| | primary reaction produces waste<br><br>secondary reaction produces waste<br><br>impurities in the feed material become waste or react to produce waste<br><br>degraded catalyst | Consider:<br>• different reaction path<br>• reactor type<br>• reactor temp., pressure<br>• excess reactant<br>• product removal during reaction<br>• different feed material (consider separation)<br>• different catalyst |
| | poor control and operation, non-optimal conditions | improve unit operation and control |
| *Separation and recycle* | feed impurities | purify feed (consider reactor operation) |
| | extraneous material for separation | eliminate extraneous material by process modifications |
| | unrecycled waste streams | carry out additional separation of waste streams |
| | waste produced in reversible reaction | separate and recycle waste |
| *Auxiliaries* | start up and shutdown losses | minimise the number of shutdown by designing for high availability, plant maintenance, and flexible operation |
| | storing | • improve storing facilities,<br>• improve inventory management,<br>• better housekeeping |
| | equipment cleaning | • use reliable equipment<br>• design for a small in-process inventory |
| | fugitive emissions | • use reliable equipment<br>• improve maintenance |
| | sampling | closed-loop sampling |

| LAYER | UTILITY WASTE | |
|---|---|---|
| | *possible source* | *solution alternatives* |
| *Heat exchanger network, utilities*<br><br>*Auxiliaries* | direct emissions from the plant and its utility system (on-site emissions from utility system and off-site emissions from power station) | • increase the energy efficiency of the process (pinch technology, Linnhoff *et al.*, 1982)<br>• switch fuel<br>• desulphurise fuel and flue gas<br>• select low $NO_x$ burners<br>• recycle flue gas<br>• chemical reduction |
| | waste water | use conceptually based approach for waste water minimisation |

Berglund and Lawson (1991) have classified waste streams as *intrinsic* wastes that are inherent in the fundamental process configuration and *extrinsic* wastes that are associated with the auxiliary aspects of the operation. Extrinsic wastes may occur as a result of unit upsets, selection of auxiliary equipment, fugitive leaks, storing problems, process shutdown, sample collection and handling, etc. The onion diagram does not include the auxiliaries because they have not been involved in the hierarchical process design. However, when waste minimisation is considered we have to extend the onion diagram with a fifth fictive layer (Mizsey, 1993) which represents the auxiliaries (Figure 1).

In Table 1 possible sources of process and utility wastes and solution alternatives are listed according to the extended onion diagram based on the work of Smith and Petela (1992) and Mizsey (1993). It is to be noted that the decisions made for waste minimisation on the subsequent layers are not independent of each other and the interactions and interconnections are to be considered according to the basic theory of hierarchical process design.

*Extended systematic waste reduction procedure for retrofitting*

Douglas (1985) has also proposed a hierarchical procedure for chemical process design where heuristic rules are used to guide search directions to overcome the need to examine all possible structures in order to find a small number of near-optimal arrangements. Recently, Douglas (1992) has slightly modified his hierarchical procedure with the objective of reducing emissions. The procedure works in the case of new "cleaner" process design.

This hierarchical procedure is developed and extended for the case of retrofitting (Fonyo *et al.*, 1994). The extended systematic procedure considers a chemical plant and consists of the following steps.

*1. Defining and tracking wastes*

Establish a table showing what wastes are generated. After defining the wastes produced, find where and why they are produced (Table 2a). Draw a simple waste oriented flowsheet of the process highlighting the origin and route followed by each waste stream.

*2. Data collection*

For each of these waste streams, define quantities, degree of hazard, and the economics of managing waste. For the cost calculation, we should consider:

- the loss of money arising from the inefficient use of raw materials (and extra usage of utilities)
- general management costs (storage, shipping, extra labour etc.)
- waste treatment / disposal cost

*3. Producing waste minimisation alternatives*

Create modified process alternatives. Table 2b shows the extended hierarchical procedure of Douglas (1992) for retrofitting problems (additions are indicated by italics).

Table 2

*Possible sources and solution alternatives of wastes*

| INTIRINSIC (PROCESS) WASTES | EXTRINSIC WASTES |
|---|---|
| **1- Input-output structure of the process**<br>a) reaction chemistry<br>b) air oxidation to NOx<br>c) spent catalyst.<br>**2 -Recycle structure of the flowsheet**<br>a) adding reactor diluents to shift the product distribution or shift the equilibrium conversion<br>b) adding heat carriers<br>c) adding reactor solvents<br>**3-Specification of the separation system**<br>a) General structure<br>b) Vapour recovery system<br>i) absorber solvents<br>ii) regeneration of adsorption beds<br>iii) removing spent adsorbents<br>iv) the use of reactive absorber to remove toxic materials.<br>c) Liquid recovery system<br>i) stripping agents<br>ii) extraction solvents<br>iii) crystallizer (recycle and purge streams (almost pure water))<br>iv) crystallizer purge streams (not almost pure water)<br>v) reactive crystallization by-prod.<br>vi) spend adsorbents<br>*vii) waste gas washing*<br>d) Solid recovery system<br>i) problems caused by cake washing | *GENERAL*<br>- Equipment and tank cleaning, washing<br>- spills and leaks<br>- fugitives<br>- maintenance wastes and materials<br>*OCCASIONAL*<br>- off-specification reactants and products<br>- sampling and analytical losses<br>- start-up and shut-down losses<br>- unit upsets<br>*ENERGY*<br>- utility waste |

*a.) Possible sources of waste*

| INTIRINSIC (PROCESS) WASTES | EXTRINSIC WASTES |
|---|---|
| **1- Input-output structure of the process**<br>a) *search for optimum reaction conditions*<br>change the chemistry<br>*recycle, reuse, reclamation*<br>b) change to $O_2$ in recycle $CO_2$ oxidations<br>c) regenerate the catalyst<br>*change the catalyst*<br>**2 -Recycle structure of the flowsheet**<br>a) change the diluent<br>b) change the heat carrier<br>c) change the solvent<br>*change the chemistry*<br>*change the process*<br>**3-Specification of the separation system**<br>a) phase splits<br>b) vapour recovery system<br>i) change the solvent<br>ii) change the bed strippping agent<br>iii) change to absorption or condensation<br>c) Liquid recovery system<br>i) change the agent<br>*change the process*<br>*procedural changes*<br>ii) change the solvent<br>*change the process*<br>iii) reuse the purge water elsewhere in the process<br>iv) remove the contaminants and recycle the water<br>look for a different separation system<br>v) look for a different separation tech.<br>vi) regenerate the adsorbent<br>vii) *change the washing material, find a way to reuse it*<br>d) Solid recovery system<br>i) filter mother liquor streams<br>improve the existing process<br>look for a different process | *GENERAL*<br>- some simple recycling operations<br>- changes in cleaning operations, (improved, or dry cleanup techniques)<br>- better housekeeping procedures<br>- reduction and/or control of fugitive emissions (spill controls, leakless valves, closed-loop sampling)<br>*OCCASIONAL*<br>- new or improved maintenance procedures<br>- preventive maintenance<br>-improved material handling and inventory management<br>*ENERGY*<br>- energy optimization<br>*Some Other*<br>- administrative control<br>- some simple material and equipment changes<br>- segregate wastes to increase recoverability<br>- evalute the need for each step and eliminate steps that are unnecessary<br>- educate operator |
| consider alternative methods for recycling waste streams into the process or into by-product markets | |

*b.) Waste reduction alternatives*

*4. Technical and economical evaluations of alternatives*

The alternatives produced are evaluated from the technical and economical standpoints to see, whether or not they achieve the desired reduction in wastes, their effect on product quality, if they need important changes in the process layout/equipment, their profitabilities, implementation times, if raw material substitution is required, are the necessary alternative materials available, utility requirements etc.

*Case studies*

The extended systematic procedure for waste reduction is tested on 26 processes in the literature (DECHEMA, 1990). The waste minimisation solutions can be deduced from the possible alternatives (in Table 2b) for these retrofitting cases. Table 3 shows a summary of the investigations (Fonyo *et al.*, 1994). The source of the problem of 18 processes is the reaction chemistry i. e. by-products. 7 of these have eliminated/reduced the waste either by changing reaction chemistry, or by raw material substitution, or, optimising the reaction while the remaining 11 processes (61 % of 18) have used recycle, reuse, recovery techniques.

The evaluation shows that the techniques of source reduction and recycle, reuse, recovery have similar importance in process waste minimisation of retrofit situations. The waste problems of 12 processes (46%) have been solved by source reduction technique and in the other 14 cases (54%) recycle, reuse, and reclamation techniques have been used.

*Strategy for minimisation of utility waste*

The problem of minimisation of the utility waste can be classified into two groups according to the source of the utility waste: • *energy intensification problem,* • *waste water minimisation problem.*

If we can save energy, we can minimise the emission associated with the production of energy. A powerful tool for the minimisation of the energy consumption is the pinch technology developed by Linnhoff *et al.* (1982). There is no limit for the use of the pinch technology, individual plants or entire factories can be investigated by this powerful tool both in the case of retrofitting and in the case of design of new cleaner processes. The pinch technology has been successfully used several times to minimise utility waste via energy integration (e.g. Smith and Delaby 1991, Ferner, 1993). Korner (1988) reports about a successful energy integration at BASF, Ludwigshafen. They could save 790 MW energy that resulted in the following reduction in emissions from power stations on the site: carbon dioxide 218 t/h, sulphur dioxide 1.4 t/h, nitrogen oxides 0.7 t/h, waste water 70 t/h.

A favourite energy integration method is the heat pump. Laue (1994) has investigated the impact of heat pumps on global warming and ozone depletion. The heat pump has the potential to reduce primary energy consumption and the corresponding utility waste emission by 20 - 85% depending on the circumstances of the energy

Table 3

*The number of processes classified according to the sources of waste and the waste minimisation techniques applied*

| WASTE MINIMIZATION TECH. \ SOURCES | | 1 Input output structure of the process (a) reaction chemistry | (c) spent catalyst | 2 recycle structure of the flowsheet (c) adding reactor solvents | 3 separation system (c) liq. rec. sys. i | ii | iv | vii | (d) solid rec. sys. | | |
|---|---|---|---|---|---|---|---|---|---|---|---|
| SOURCE REDUCTION | optimizing reaction | 2 | | | | | | | | 2 | 7.7 % |
| | improved process | | | | | | | | 1 | 1 | 3.8 % |
| | changing chemistry | 4 | 1 | 1 | 1 | 1 | | | | 8 | 30.8 % |
| | new separation process | 1 | | | | | | | | 1 | 3.8 % |
| | *sub total* | 7 | 1 | 1 | 1 | 1 | | | | 1 2 | 46.1 % |
| RECYCLE REUSE RECOVERY | gas | 3 | | | | | | | | 3 | 11.6 % |
| | liquid | 6 | | | 1 | | 1 | 1 | | 9 | 34.7 % |
| | solid | 1 | | | | | | | | 1 | 3.8 % |
| | energy | 1 | | | | | | | | 1 | 3.8 % |
| | *sub total* | 11 | | | 1 | | 1 | 1 | | 14 | 53.9 % |
| | **TOTAL** | **18** | **1** | **1** | **2** | **1** | **1** | **1** | | **26** | **100%** |

production. The proper selection of the working fluid has also an influence on the ozone depletion.

Annakou and Mizsey (1994) have rigorously investigated a heat pump assisted distillation and compared that with the common distillation. According to their calculation the heat pump minimises about 60% of the flue gas emissions, in addition to lowering energy consumption and reducing the total cost.

For the *waste water* minimisation in the process industry Wang and Smith (1994) have recently published an approach. Targets are first set for freshwater, regeneration and waste water flowrates using limiting water profiles. These profiles allow constraints due to the minimum mass transfer driving forces, equipment fouling, corrosion limitations, etc. to be included in the problem formulation. Both single and multiple contaminants can be addressed. This conceptually based approach allows the designer also to identify alternative structures for the same problem where this is possible.

**Waste minimisation on the factory level**

The systematic techniques developed and used for process improvements are very powerful tools for waste minimisation on the process or plant level. Current technologies are being used to develop processes for effluent reduction. However, these operations will also reach their limits and a third generation of waste minimisation is inevitable. New design methods for process synthesis will be developed to further minimise the process waste emission by maximising mass efficiency. These methods are to be able to include not only one chemical process or plant but they can investigate more processes or plants trying to minimise/eliminate the global emissions of the plants included in the entire factory.

It is easier to extend the investigation in the case of minimising the utility emissions. The pinch technology and the conceptually based approach for waste water minimisation are able to handle more plants, practically the entire factory at the same time.

*The problem of process waste seems to be more complicated* although the design problem is somewhat analogous to the design of energy efficient processes. El-Halwagi and Manousiouthakis (1989) have developed the concept of mass exchange networks (MEN's) and proposed a systematic procedure for the synthesis of MEN's.

A mass exchange network consists of separators and mass transfer units that achieves, in a cost effective manner, minimal emission of process waste. The synthesis problem can be formulated as: given a set of pollutant rich streams and a set of pollutant lean streams, synthesise a MEN, that is a set of separators and transfer units, that can transfer the pollutant species from the rich streams to the lean streams at minimum venture cost. A key feature of this approach is that it combines thermodynamic and driving force constraints into the optimisation. Cohen and Allen (1992) have implemented this approach in a refinery for the recovery and reuse of phenol from waste streams. By using linear programming and mixed-integer non-linear programming formulation the optimal mass exchange network is obtained that recovers 96.7% of the phenol present in the rich streams.

The aim of third generation modification is to design such processing that only products leave the global system. Within the frame of a factory we integrate plants if certain component(s) of the waste streams leaving one of the plants can be utilised in another plant as raw material. This on-site recycle in the factory is the so called closed-cycle processing that is an effective method for waste elimination incentives and it is practically an effort to copy the processes taking place in the nature, which are all closed-cycle processes.

## A systematic procedure for process waste minimisation of a factory

The waste minimisation incentives at the chemical plants should be co-ordinated on a higher level, on the factory level. A systematic procedure is proposed for this two level problem (Mizsey, 1994). The systematic procedure (Figure 3) investigates the plants of the factory. During the operation of the plants good housekeeping practices, inventory control, and minor changes in operation practices should be considered. Afterwards, the systematic procedure identifies process waste and the raw material requirements of every plant in the factory.

The investigations always include an economical analysis, too. If there are no special environmental restrictions, the economic trade-offs should be also determined between the different alternatives, closed-cycle processing, plant renewals or end-of-pipe treatment. Dyer and Taylor (1994) have published a balanced approach of the waste management with several helps for the cost estimations. For this reason, first, the costs of end-of-pipe treatment should be estimated and in the course of the following investigations and decisions they are also considered.

It is necessary to investigate before making some further major improvements on the plant level for waste minimisation whether the waste streams and the raw material requirements can be matched. If the sources and sinks cannot be matched it is to be investigated whether some modifications could still help to realise a match of a waste stream and a raw material requirement, e.g. Katin (1991).

*The main goal is to utilise the waste of a plant in another plant that is to design closed-cycle processing.* MEN synthesis can be completed if necessary. If MEN is not acceptable, the investigation is to be continued on the plant level.

If the waste minimisation or elimination cannot be completed by closed-cycle processing, the waste minimisation activities are to be continued on the plant level again considering major process improvements. These activities can be carried out according to one of the systematic procedures developed for waste minimisation on the plant/process

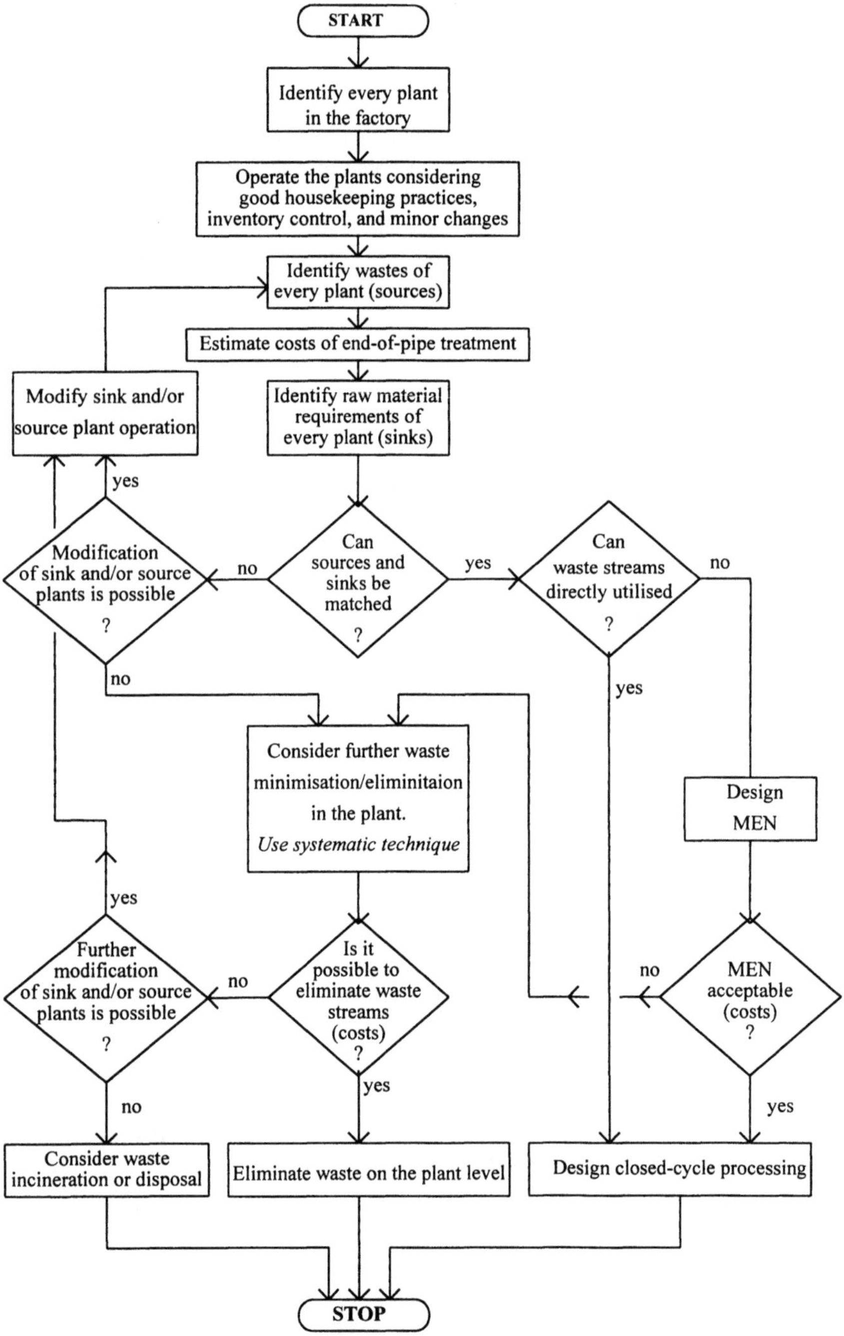


Figure 3. Systematic procedure for process waste minimisation of a factory

level. If the waste cannot be eliminated it can be investigated again whether further modification of the sink and/or source plants could help or not. This investigation is necessary because the major process improvements carried out on the plant level can change the waste and the raw material requirements of the plants. During these activities the costs of the different alternatives are also estimated and considered if there is no special environmental restriction.

Process waste that cannot be eliminated should be incinerated, treated, or disposed.

## Summary

If cleaner technologies are to be designed for sustainable environment there are several strategies for the process and utility waste reduction. In case of the utility waste the solution seems to be easier compared to the problem of the process waste: both the pinch technology for the minimisation of the energy consumption and the conceptually based approach for waste water minimisation are able to handle and solve the problem of individual plants or the entire factory. In case of the process waste the situation and the alternatives for the waste reduction are more complicated. The systematic techniques developed and used for process improvements reach their limits at plant level. For further development, however, the decisions made on the plant level should be co-ordinated on the factory level considering the possible interactions among the different plants. With the use of the proposed systematic procedure for process waste minimisation of a factory closed cycle processing and the cost effective minimal global emission can be realised.

## References


Annakou, O. and P. Mizsey (1994) Rigorous investigation of heat pump assisted distillation, accepted for publication in *Heat Recovery Systems & CHP*

Berglund, R. L., and C. T. Lawson (1991) Preventing pollution in the CPI, *Chemical Engng*, pp. 120-127

Cohen, Y. and D. Allen (1992) An integrated approach to process waste minimisation research, *Jl of Hazardous Materials*, **29**, pp 237-253

Crittenden, B. and S. Kolaczkowski (1993) Waste reduction and pollution prevention, *School of Chemical Engineering*, University of Bath

DECHEMA (1990) *Produktionsintegrierter Umweltschutz in der chemischen Industrie*, Frankfurt am Main DECHEMA, *Produktionsintegrierter Umweltschutz in der chemischen Industrie*, Frankfurt am Main

Douglas, J. M. (1985) A hierarchical design procedure for process synthesis, *AIChE Jl*, **31**, pp. 353-362

Douglas, J. M (1992) Process synthesis for waste minimization, *Ind. Eng. Chem. Res.*, **31**(1), pp. 238-243

Dyer, J. A. and W. C. Taylor (1994) Waste management: a balanced approach, *AIChE Spring Meeting*, paper S71b, Atlanta

El- Halwagi, M. M. and V. Manousiouthakis (1989) Synthesis of mass exchange networks, *AIChE Jl*, **35** pp. 1233-1244

Ferner H. M. (1993) Energieeinsparung in Industriebetrieben durch optimierte Wärmeintegration - Ein Beitrag zur Reduktion der $CO_2$ Emissionen, *Integrierter vorsorgender Umweltschutz - VT-Newsletter 8*, pp. 103-111

Fonyo, Z., S. Kürüm and D. W. T. Rippin (1994) Process developments for waste minimisation: the retrofitting problem, *Computers chem. Engng*, **18**(Suppl), pp S591-595

Katin, R. A. (1991) Minimize waste at operating plants, *Chem. Eng. Progress*, 87(7), pp. 39-41

Korner, H. (1988) *Chem. Ing Tech*, **60**, p. 511

Laue, H. J. (1994) The impact of heat pumps on global warming and ozone depletion, *Heat Recovery Systems & CHP*, **14**(3), pp. 323-335

Linnhoff, B. et al. (1982) User Guide on Process Integration for the Efficient Use of Energy, IChemE, Rugby, United Kingdom

Mizsey, P. (1993) The actual problems of the environmental protection in the chemical industry (in Hungarian), *Magyar Kémikusok Lapja*, **48**(10-11), pp. 411-420

Mizsey, P. (1994) Waste reduction in the chemical industry: a two level problem, *Jl of Hazardous Materials*, **37**, pp 1-13

Rittmeyer, R. W. (1991) Prepare an effective pollution-prevention program, *Chem. Eng. Progress*, **87**(5), pp. 56-61

Smith, R. and O. Delaby (1991) Targeting flue gas emissions, *Trans IChemE*, **69,** Part A, November pp. 492-505

Smith R. and B. Linnhoff (1988) The design of separators in the context of overall processes, *Chem Eng Res Des*, **66**, pp. 195-228

Smith, R. and E. Petela (1992) Waste minimisation in the process industries, *The Chemical Engineer*, No. 517

Wang, Y. P. R. Smith (1994) Wastewater minimisation, *Chem. Engng Sci.*, **49**(7), pp. 981-1006

# To the reduction of Environmental Pollution during reproductions used by Deads and Wastes in Hungary

János Mujzer
Veterinarian of the "ATEV"
Protein Processing Share-Holding Co.,
1097 Budapest
Illatos str. 23.

I. Introduction The ATEV /Animal Protein Producing Company/ established as a state one in 1949 with an obligation to solve any epidemic difficulties which can be arrisen in Hungary, and annihilate the wastes of slaughtering hauses. In 1993. the position of the company legally changed and called ATEV Protein Processing Share-Holding Company keeping the original mission. The basic activity of the company is the annihilation of the animal carrions and wastes. This annihilation is effected by the industrial processing - based on the international regulations - and producing different kind of commodities like meatmeal, technical animal fat, blood meal ect., which can be used as a row material for fodder cosmetics, or unsuitable parts, which must be entombed by our factories. We remark that the Hungarian regulations are absolutely conformed to the rules of EEC 27611/1990, No.: 90/667/EEC prescriptions. While 9 factories were in function upto 1992, we make the service - keeping the regulations - on the same level, with 7 factories for the time being.

Foot note: Privat: 1133 Budapest, Pannónia u. 68. HUNGARY

NATO ASI Series, Partnership Sub-Series, 2. Environment – Vol. 2
Cleaner Technologies and Cleaner Products
for Sustainable Development
Edited by H. M. Freeman, Z. Puskas, and R. Olbina

*II. The technology* Meatmeal and animal fat production The crude materials are delivered by running and droppingless tanks from the spot to the processing factories. This mass from the tanks or containers are depleted by mechanical way and fully sterilisied and hydrogenated on two bars and 132 C degre s in 20 minutes period. The heating effects, the pressures and the period of the handling are controlled by authomatically. The heated materials are passed to a cooking-drying unit /6,3 cubic meters/ through a pipeline to reduce the water content from 65 pct to max. 14 pct and the fat content 15 pct below. The milling, classification, packing and storing the meal or animal fat are effected by our factories. The quality and the bacteriological controll is effected in the laboratory of our factories and with the official authority. *Technology for the stink reduction* - The technology of our company are placed in fully confined places and the stink goes through an intermediary tank to the biofilter where we use vine-tendrils or peat as purificants. - The liquid from the stink vapours and gases arrisen during the production is partly precipitated partly transmitted by vacuum pumps in pipelines into the biofilter mentioned before. - Between the intermediary tank and the biofilter the transmitted gases are having further cleaning by mechanical, physical way and saturated by steam. - In the biofilter, different kind bacteriums finished the last purification. *Water and cleaner system* The factories are using their own wells-system and handling the water consuming as per the actual rules in Hungary. Every factory has their own cleaners of the waste-water which are laid on the natural waters like rivers or lakes. *The vapours condensation* Where as the row material has water in 65 pct nearly, it is necessary to use electric coolers for condensation of the vapours. It's well known that the handling of this pollution causes more difficulties.

*III. Annihilation* - The last station of the full neutralisation is the annihilation basis where the carrions and wastes will be entombed. - According to the Hungarian public, veterinary and enviromental regulation furthermore the Directives of the EEC Council dd. 27/11/1990 every row material arising from carrions or animal wastes in case any dangerious situation supposed in course of the production or during the consumption lateron must be annihilated /entombed/. - Our factories are having the disposal of these special territories.

*IV. Biogas program* In 1994 the ATEV started with a new experimental programm in cooperation of some Hungarian Universities how the non- utilizable material can be changed to reproductive one. This new idea is called "biogas programm" as during the procedure /Ch4, H2/ is arrisen in big volume suitalbe for production of the electricity in our factories and the by-products /biomass/ should be windable back to the agricultural area. The target of our several tests was to check the measure of the biogas production, what kind of crude animal materials can be mixed optimaly with vegetable carbons. The most favourable recpture are separately considered and our laboratories study the facilities to increase the optimal quantity of the biogas production. The biomass are very hardly controlled as per the Hungarian rules to avoid any infection. The evaluation of this researching programm is continuously controlled by ministries, official organisations, which are interested and involved in the solution. This practice is conformed to the West-European economical rules.

*V. The utilizable products on based of the row materials* The yearly processed row materials are about 200 thousand metric tons. The dried content is roughly 30 pct the rest is 70 pct water from this quantity. In the course of the production the water content pushed down to 10 pct with the meatmeal and below 15 pct with the animal fat. This figures are based on data between 1980-90.

# Environmental Education and Some Examples of Teaching for Sustainable Development in Chemical Engineering at the Technical University Budapest

***Gábor Széchy - Imre Szebényi***
*Department of Chemical Technology*
*Technical University of Budapest*

Before a few thoughts to teaching some elements of sustainable development for chemical engineering students are discussed, I think it necessary to give a short review of the environmental education at the Technical University of Budapest.

Environmental education has substantial tradition at the Technical University of Budapest, although earlier, the word "environmental" was not used for the characterization of this type of educational activity. Water quality management, air pollution controll, acoustics and many other fields have played a very important role in engineering education already a long time ago. A lot of discussions were and are still led, whether environmental knowledge should be included and integrated into the body of material of different engineering subjects or should it be taught in separate subjects. Our experience is, that the first solution is usually more effective in teaching actual and particular knowledge, but there should be a few separate subjects explaining the interconnected nature and complexity of environmental problems and enhancing environmental consciousness. A difficulty frequently encountered here is that engineering student are usually in lack of a fair knowledge in ecology, biology and sometimes even in chemistry.

Besides this, it is also questionable, whether in five years it would be possible to acquire the necessary knowledge required for an engineer and simultaneously to get a sound knowledge of the basic material of environmental science.

NATO ASI Series, Partnership Sub-Series, 2. Environment – Vol. 2
Cleaner Technologies and Cleaner Products
for Sustainable Development
Edited by H. M. Freeman, Z. Puskas, and R. Olbina

As the result of the above mentioned opinion, no separate undergraduate curriculum is offered at the Technical University of Budapest in environmental engineering. There are, however, a few specialization branches, which may be chosen by the students of the 3-rd and 4-th year. Such specialization opportunities are offered by the School of Mechanical Engineering in environmental equipments, by the School of Civil Engineering in water supply and canalization, and by the School of Chemical Engineering in cooperation with the Eötvös Lóránd University in Budapest in environmental engineering, as a branch of bioengineering.

In addition to the specialization branches, a new form of curriculum is now being offered, which is a transition towards postgraduate education and this is the so called supplementary education to acquire a secondary diploma in environmental engineering. This curriculum is offered by the Budapest University Association, which consists of the TUB, the Budapest University of Economics, University for Veterinary Sciences and the College of State Administration. The planned structure of this curruculum is shown in Table 1.

It is the postgraduate education, where separate currcula for environmental engineering can be offered in a very effective way.

Postgraduate education in engineering has a nationwide, more or less uniform format in Hungary, a quite strictly organized 4 semesters long, usually about 12 hours/week course, leading to a "specialist engineer" degree. Training is offered in many engineering fields, among them in environmental engineering.

The TUB has started the postgraduate "specialist engineer" course in environmental engineering in 1974, and since that time about 1000 persons have acquired their degree in this field. There was and is a continuing interest towards this course, in spite of the fact, that two years ago the tuition fee had to be raised very significatly. The curriculum of the first two semesters of this education is the same for all participants, while for the next two semesters the curriculum is divided into three branches: water quality management, air pollution controll and noise abatement. As an example, curriculum of the branch of water quality management is shown in Table 2. This table also shows the first two (common) semesters.

Another important "specialist engineer" course is offered at the TUB since 1992 in "environmental analytical chemistry". The list of subjects taught in this course is shown in Table 3.

After this general introduction, it is of use to talk about some more special aspects of chemical engineering education, which are connected to sustainable development.

It is very important, that at the evaluation and discussion of different chemical technologies not only production costs, product spectrum etc. are discussed, but also environmental aspects, like the occurence of hazardous materials in the technology, the quantity and quality of wastes, the potential upgrading of wastes etc. It must be made clear for the students, that environmental considerations become more and more decisive in the feasibility and practicability of a given technology.

A good example for this is brine electrolysis, where the mercury-cathode technology was already ruled out in Japan and is under pressure in other countries.

It should be emphasized to the students, that aspects of sustainability must be considered right at the beginning of the development of a new technology, at the selection of the basic reactions and feedstocks of the technology, at the development of the catalysts etc.

Unfortunately, very few large chemical technologies are developed in this part of Europe at the present time, but hopefully still in the lifetime of those, who are chemical engineering students now, this situation will change, and it is very important to teach this attitude of integrated environmental protection.

Presently, additive environmental protection plays a more important role in this part of Europe, than the integrated protection.

A very important and typical example of these additive technologies is flue gas desulphurization. None of the existing Hungarian power stations has flue-gas-desulphurization, but I think the situation is not much better in the other former Eastern Bloc countries of the region.

Thus, the importance of these technologies must be taught for the students, but it should be also stressed for them, that from the existing 20-25 technologies possibly those should be selected which do not produce further wastes, the deposition and handling of which is a problem again.

And when we are at power stations and energy production, it is also very important to stress the importance of energy-effectivity for the students, because this is a very basic element of sustainability and in this part of Europe this has a tremendous potential. It is probably well known that a one USD unit of the GDP is produced with the consumption of 3-4 times as much energy in Hungary, in Poland or in the Czech Republic than in the Western and Northern European countries.

Of course, there are many other technological examples which can be used, and should be used to develop a feeling for sustainability in chemical engineering students. Just to mention two more:

- the recovery of sulphur from petroleum fractions via hydrodesulphurization and the Claus-process

- and the use of rape seed oil methyl-esther as Diesel fuel.

It is also very interesting to explain the development trends of motor fuels as a consequence of environmental needs and requirements.

These were all examples, where environmental aspects and knowledge were incorporated into the material of a certain subject. As an example of a separate environmental subject, we would like to mention that a new separate subject is now being included into the chemical engineering curriculum of the TUB, with the title "Environmental Chemistry and Technology". This is a 2 semester subject, compulsory for all chemical engineering students.

Important is, that we should use every opportunity to stress the environmental and sustainability aspects for the students, because these will become more and more important in their lifetime - at least we hope so.

Table 1.

**Curriculum of the secondary diploma course in environmental engineering (Number of contact hours)**

| **Subject**<br>Teaching institution | **Semester** | | | | | |
|---|---|---|---|---|---|---|
| | 8 | 9 | 10 | 11 | 12 | 13 |
| I. ***Basic environmental subjects***<br>- Civil engineering aspects of environmental protection<br>TUB Faculty of Civil Engineering | 30 | 30 | | | | |
| - Environmental chemistry and analytics<br>TUB Faculty of Chem. Engineering | 30 | 30 | | | | |
| - Basic environmental biology and ecology<br>Univ. of Veterinary Sciences | 30 | 30 | | | | Diploma |
| - Environmental health science<br>Univ. of Veterinary Sciences | | | | 30 | 30 | |
| - Principles of environmental materials and energy management<br>TUB Faculty of Natural and Social Sciences | | 30 | 30 | | | |
| II. ***Special environmental engineering subjects***<br>- Mechanical engineering aspects of environmental technology<br>TUB Faculty of Mech. Engineering | | | 30 | 30 | | |
| - Chemical technologies<br>TUB Faculty of Chem. Engineering | | | 30 | 30 | | |
| - Transportation and Environment<br>TUB Faculty of Transportation | | | 30 | 30 | | work |
| - Environmental-friendly architecture<br>TUB Faculty of Architecture | | | | 30 | 30 | |
| - Electromagnetic Environmental protection<br>TUB Faculty of Electrical Engineering | | | | 30 | 30 | |
| III. ***Environmental management and law***<br>- Environmental economics. evaluation and impact asessment<br>Budapest Univ. of Economics | | | | 30 | 30 | |
| - Management of natural resources<br>Budapest Univ. of Economics | | | | 30 | | |
| - Environmental systems<br>Budapest Univ. of Economics | | | | | 30 | |
| - Environmental law<br>College of State Administration | | | | 30 | 30 | |
| **IV. *Elective Subjects*** | | | | 30 | 120 | 120 |
| **Total number of hours** | 90 | 120 | 120 | 300 | 300 | 120 |

Table 2.

## Curriculum of the post-graduate specialist-engineer course in environmental protection
### at the Technical University of Budapest

### Branch of water quality management

| Subjects | Hours per semester in the 1st semester (lecture+practical training) | 2nd | 3rd | 4th |
|---|---|---|---|---|
| Fundamentals of Environmental Protection | 48+0 | | | |
| Nature Conservation | 24+0 | | | |
| Siting and Construction of Industrial Plants | 12+12 | | | |
| Town Planning and Reconstruction | 24+0 | | | |
| Computing Methods | 12+12 | | | |
| Man and his Environment | | 36+0 | | |
| Biology-ecology | | 36+0 | | |
| Radioactive Pollutants and their Elimination | | 24+24 | | |
| Waste Treatment and Reutilization | | 24+0 | | |
| Water Chemistry and Technology | | | 12+12 | |
| Technical Microbiology | | | 12+12 | |
| Water Quality Management and Control | | | 24+0 | |
| Water Analysis | | | 24+36 | |
| Waste Water Treatment | | | 12+0 | 24+48 |
| Disposal of Industrial and Agricultural Waste Waters | | | | 24+0 |
| Optimization of Water Management Systems | | | | 12+0 |
| Legal Aspects of Water Protection | | | | 12+0 |
| Optional subject | | | | 24+0 |
| Total of hours per semester | 120+24 | 120+24 | 84+60 | 96+48 |

Table 3.

## Subjects of the post-graduate specialist-engineer course in environmental analytical chemistry
### at the Technical University of Budapest

| | Total No. of hours Lecture | Laboratory |
|---|---|---|
| I. Physical chemistry | 24 | 0 |
| Computers/chemometry | 24 | 24 |
| Sources of pollution: chemical nature of contaminants | 24 | 0 |
| Sample preparation | 24 | 24 |
| II. Regulations on pollution | 24 | 0 |
| Air pollution analysis, emission | 24 | 24 |
| Air pollution analysis, immission | 24 | 12 |
| Water/soil polln. anal., inorganics, speciation | 24 | 24 |
| III. Water/soil polln. anal., organics, particles | 24 | 24 |
| Waste materials assessment | 24 | 12 |
| Pesticide residue analysis | 24 | 12 |
| Analysis of radioactive pollution | 24 | 12 |
| IV. Analysis of contamination at working places | 24 | 0 |
| Biological test methods | 24 | 12 |
| Quality assurance of analytical measurements | 24 | 6 |
| Individual project | 0 | 36 |
| Intensive short course abroad (A or B) | 25 | 45 |
| A) Phys. chem., chemometry, agricultural/soil/waste fields | | |
| B) Trace metals/water/air pollution | | |

# Cement Industry - Is It Clean?

**Ferenc Tamás**
**Department of Silicate Chemistry and Technology**
**University of Veszprém**
**H-8201 Veszprém, P.O.B. 158**
**Hungary**

## *First Question: The Product - Is It Clean?*

Cement is manufactured of different mineral raw materials by quarrying, mixing, grinding, homogenization, followed by a high-temperature burning and finish grinding. Annual cement production of the world is over 1.2 billion metric tons! Cement manufacturing plants are gigantic, biggest lines can produce 6000-7000 tons of cement daily. Although manufacturing technology is easy, the uniform quality of such vast quantities can be maintained by high-tech control only.

Cement, however, is an intermediate product only. Almost all of the cement produced is used to make the final product: concrete, which contains only 10-20% of cement, the remainder is a filler, as gravel or crushed stone. Calculated from production data, the annual world production of concrete can be around 10 billion tons. This is the man-made solid material produced in highest quantities!

Cement-based materials, thus concrete, are durable: concrete products, made according to specifications have a service life of several hundred years (structures erected by the Romans, using "natural" cement are still in a good shape, and the oldest buildings, bridges etc. made with artificial cement are 150 years of age and are intact). Demolished concrete can be used as an aggregate for new concrete. Obviously, concrete is not everlasting: after geological times it will weather; weathering products however are limestone and clay, i.e. natural, nonhazardous materials.

## *The Answer: The Product Is Clean!*

NATO ASI Series, Partnership Sub-Series, 2. Environment – Vol. 2
Cleaner Technologies and Cleaner Products
for Sustainable Development
Edited by H. M. Freeman, Z. Puskas, and R. Olbina

## ***Second Question: Is Cement-Making Technology Clean?***

### *General Technological Description*

Several sorts of raw materials are used for the making if cement: various abundant natural rocks, as clay, sand, marl, limestone, but also industrial wastes, as fly ash, slags, etc.

Main steps of cement technology include: the winning and comminution of raw materials, preparation of the raw mix, pyro-processing (burning) , finish grinding (in the presence of additives, as gypsum), packaging of final product, despatch. Pyro-processing, being the most important step, requires a special paragraph.

Raw materials, ground, mixed in predetermined ratios and homogenized are burnt at very high temperature (1450$^{o}$C), among oxidizing conditions, in big rotary kilns. Due to the large dimensions, solids and flue gases spend long times in the high-temperature zone of the kiln. The entire milieu of cement making is strongly basic (because of the large excess of calcium oxide).The intermediate product of burning is called *clinker*.

There are two basic processes of cement making: the wet and the dry technology (other solutions, as the semi-dry process, or the one involving a shaft kiln instead of a rotary kiln, exist too, but are of lesser importance). In the wet process, the raw mix is introduced into the kiln as an aqueous slurry, while in the dry process as a dry powder. From the point of energy consumption, the wet process is obsolete: high amounts of water must be vaporized, requiring excess energy. The dry process enables also very effective heat exchange: hot gases preheat even precalcine the raw mix, thus the real high temperature device, the rotary kiln can be shorter. Flow sheets of the raw material preparation, and of the wet and dry process technologies are shown in Figs. 1., 2. and 3. The primary fuel can be powdered coal, or hydrocarbons (natural gas, fuel oil, etc.). Dust collection, usually by electrostatic precipitators is an integral part of all technologies.

Hot clinker is rapidly cooled, subsequently ground to a fine powder, in the presence of additives (gypsum, usually also wastes, as fly ash, blastfurnace slag, etc.). Despatch can be done in loose form or in sacks.

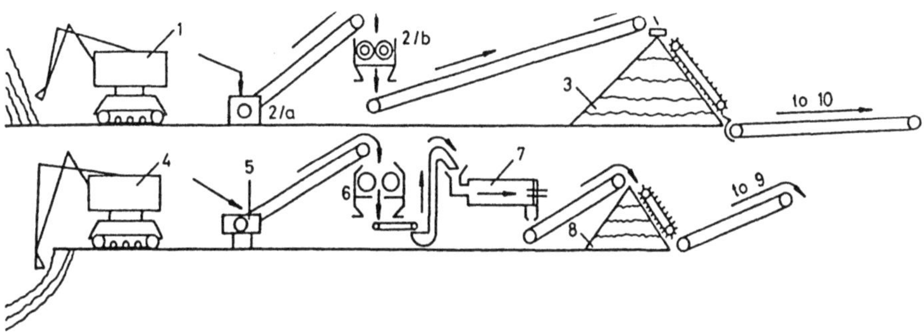


**Fig. 1. The raw material side of clinker manufacture**
*1* Excavator for blasted limestone, *2/a* Coarse size reduction of limestone, *2/b* Secondary size reduction of limestone, *3* Pre-homogenization and storage of limestone, *4* Excavator for clay, *5* Coarse size reduction of clay, *6* Secondary size reduction of clay, *7* Drying of clay, *8* Pre-homogenization and storage of clay, → Flow direction of solids

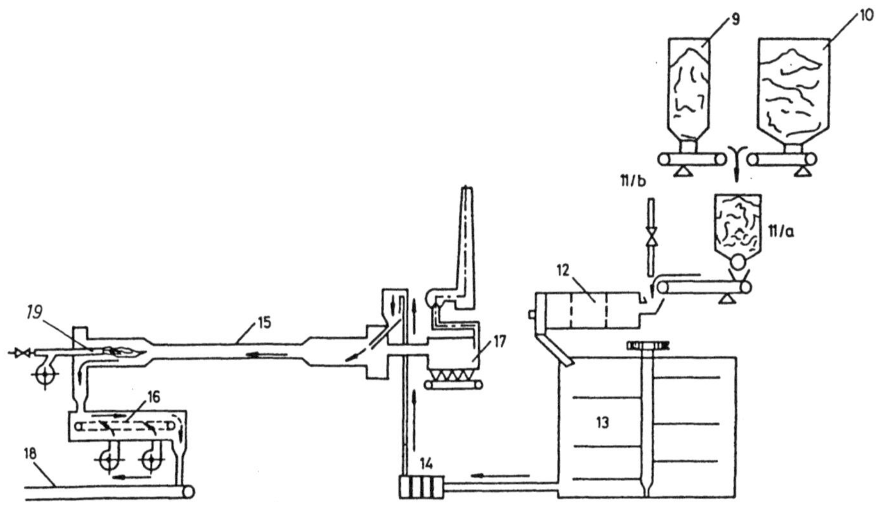


**Fig. 2. Wet-process technology of clinker manufacture**
*9* Clay storage, *10* Limestone storage, *11/a* Raw mix storage, *11/b* Water addition, *12* Wet ball mill, *13* Homogenizing basin, *14* Slurry pump, *15* Wet-process rotary kiln, *16* Grate cooler, *17* Dust collector, *18* Clinker transport, *19* Burner, → Flow direction of solids

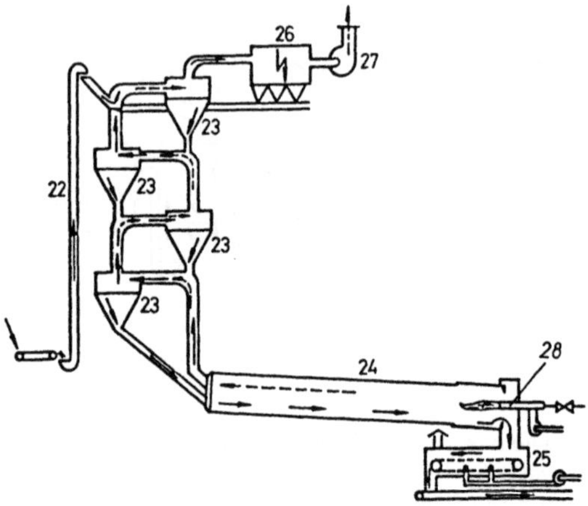


**Fig. 3. Dry-process technology of clinker manufacture**
*22* Elevator, *23* Heat exchange cyclones, *24* Dry-process rotary kiln, *25* Grate cooler, *26* Electrostatic precipitator, *27* Fan, *28* Burner
→ Flow direction of solids
--> Flow direction of gases

*Energy Considerations*

Cement is an energy intensive material due to high temperature burning and the high energy requirement of the grinding operations. But, as mentioned earlier, cement is not a final, but an intermediate product. Concrete, on the other hand contains only low amounts of cement and the preparation of aggregates (gravel, crushed stone) requires a low input of energy only. Thus concrete is much less energy intensive than other usual structural materials (steel, brick, plastics, etc.)

*Emissions*

Emissions of the cement industry are high, but can be decreased by efficient technical measures. Dust emissions are negligible if up-to-date electrical precipitators (ESP) are used. The $NO_x$ content of gases is high (due to the high temperature, thermal nitrogen oxide formation is probable), but can be decreased e.g. by adding a tertiary air duct or by a catalytic or non catalytic removal. Emission of carbon dioxide is high (as in all cases when fossil fuels are burnt, and the decomposition of limestone is also a major source of $CO_2$). Only insignificant amounts of carbon monoxide are emitted during cement manufacture, as oxidation is a prerequisite of the preparation of high-quality cement.

***The Answer: Cement-Making Technology Is Clean!***

***Third Question: Cement / Waste Interactions***

*Can Cement Be Used for Waste Solidification?*

The cement-based solidification is a simple and reliable method for the immobilization and thus disposal of most of the hazardous wastes (thus e.g. heavy metal containing or radioactive wastes). Cement is a well known material, its properties are standardized, it is readily available; cement and its products are inflammable, non-toxic. Cement based solidification requires low investment costs; no high-temperature treatment is required, no danger of gas- or secondary waste formation, thus mobile disposal stations can be operated too. Setting is fast in the total cross section of the body. The solidified product has a high pH values, thus most heavy metals or radionuclides are precipitated as solid hydroxides. The high pH ensures a better corrosion resistance of the (usually stedel) container and reduces the risk of micro-biological corosion. The solidified body has a high compressive and impact strength, and is resistant against radiolysis (unless the radiation intensity is very high).

Cement-based solidification has only slight disadvantages: the permeability of the product is rather high (but this can be decreaseed by finely grained additives, as silica fume). The main disadvantage is that the waste/matrix ratio is higher than in case of other solidification, thus high volumes of solidificates must be transported and deposited.

The answer: Yes, cement is an excellent material of waste solidification.

*Is Cement Industry Waste-free?*

Up-to-date cement manufacture is practically free of solid wastes (a proper maintenance and a rigid adherence to technological directives presupposed). The dust, precipitated in the dedusting equipment can be recycled. However, some technological changes (if wastes are used as fuels, to be outlined later) may lead to solid waste formation. This is only an exceptional case; the wastes formed this way are low and nonhazardous.

The answer: Yes, cement industry is practically waste-free.

*Can Cement Industry Utilize Wastes?*

Cement industry can utilize wastes in several ways: as raw or additive material, or as fuels. Examples for the use of wastes as raw materials: slags, fly ash, by-product lime, coming from several processes of the chemical industry (their use at the same time reduces carbon dioxide emission). Slags and fly ash can be used as additives too, and besides them also by-product gypsum, the waste of other chemical processes and also flue gas desulfurization. Thus wastes can not only a disposed of properly in the cement industry but can also be utilized.

Several industrial wastes can be used as fuels in the cement industry, both solid and liquid ones. Examples to the former and the latter, resp.: acid resin of hydrocarbon processing or waste tires, and used solvents from the pharmaceutical industry, or oil wastes. The disposal of hazardous organic wastes is perhaps the best demonstrated available technology, as the very high temperature, strong oxidizing atmosphere and the long residence time of gases destruct even the most stable organics. Conditions in the sintering zone of the rotary kiln are more severe as the circumstances prescribed even in the most drastic standards. Products of complete destruction: carbon forms carbon dioxide, hydrogen forms water, nitrogen compounds are destructed to elementary nitrogen, all normal ingredients of Earth's atmosphere. If the waste contains sulfur and/or chloride compounds, they are decomposed to sulfur dioxide and hydrogen chloride, resp.; although these are hazardous, but definitely many times less hazardous than the parent organics; and besides these acid compounds are readily absorbed by the acid milieu in the cement kiln.

This refers to products of *complete* destruction only; PID's, i.e. Products of Incomplete Destruction can be very hazardous, in many cases more toxic than the parent material. The formation or reconstruction of PID's however can, and must be avoided, by a rigid adherence of technological and safety precautions.

As a result, the exacting Environmental Protection Agency of the US considers the disposal of hazardous organic wastes by incineration in cement kilns as the Best Demonstrated Available Technology. As the calorific value of the wastes reduces primary fuel consumption, the method can be considered as waste utilization too.

The answer: yes, cement industry can dispose of and utilize wastes.

## *Problems Associated with the Use of Waste Derived Fuels (WDF)*

### *The Environmental Side*

The use of WDF's in the cement industry raises special problems, some of them being of environmental nature, while others would affect cement technology. One of the most important environmental question is:

*The fate of hazardous constituents in the WDF*

Organic constituents are practically fully destructed in cement kilns. Destruction and Removal Efficiency (DRE) values are very high indeed; but hazardous organic constituents are contained not only in the WDF, but, to some extent also in the raw material and the main fuel. If these substances contain the Principal Organic Hazardous Compound (POHC) selected to calculate the DRE, then this fact will cause an underestimation, and thus an artificially low DRE. The selection of the POHC for the calculation of DRE is essential to avoid a possible underestimation, esp. in case of kilns where coal is the main fuel: several organics, as toluene and benzene are formed of coal, and this may result in an unrealistic DRE, even in the complete absence of WDF. This is shown by the fact that 100% DRE is never detected, because there exists no "zero concentration" of any constituent of stack gas; the lowest possible value is "less than detection limit".

Even a high DRE for a selected organic constituent does not ensure the absence of PIDs in the stack gas. A safe method to avoid the formation or recombination of PIDs is to keep a significant air excess in the kiln, to avoid reducing circumstances. The best measure of this condition is the continuous monitoring of carbon monoxide in the stack gas. A too high amount of CO, however, is dangerous from the points of explosion hazard in the ESP and product quality too. A 100 ppm limit of CO in the exhaust gas is definitely on the safe side. A too high share of excess air, on the other hand, may bring other problems: besides less efficiency of combustion (by lower flame and exhaust gas temperatures) a new hazard: the formation of very toxic elementary chlorine, by the reaction $4HCl + O_2 = 2Cl_2 + 2H_2O$, may come into existence.

The anionic part of the WDF, mainly chlorides and sulfates form HCl and $SO_2$ (or even $SO_3$) after destruction in the hot zone. It has been quoted above that these acid gases will be readily neautralized in the basic milieu of the clinkering process. HCl reacts primarily with Ca, K and/or Na oxides in the kiln, to produce $CaCl_2$, KCl and NaCl. These salts are volatile and typically removed with the kiln dust, and only a slight part is emitted with exhaust gases. However, the input of Cl-containing WDF must be kept low, because otherwise serious technological problems may arise. WDFs, a contrary to fossil fuels, and especially coal, contain no, or only negligible amounts of sulfates, consequently a partial replacement of WDF for coal effectively decreases $SO_2$ emissions, thus reducing the environmental hazard of creating condensation nuclei for acid rain precipitation.

Metals contained in the WDF are not destroyed in the kiln: they will be either emitted, or absorbed in the clinker (and, to a lesser extent, in the kiln lining), or accumulated in the ESP dust. The U.S. EPA has set emission standards on 10, possibly hazardous metals: As, Be, Cd, hexavalent Cr, Pb, Hg, Ag, Sb, Ba and Tl. The first 4 of them are considered carcinogenetic, the others impose other hazards.

As lead, the most frequent heavy metal contained in WDF, is very toxic, and lead chloride is quite volatile, special emphasis was given to detect the fate of this element. Experience showed that the majority of Pb is retained in the clinker as an incorporated compound, but only in case of low-halogen WDF; the presence of chlorinated wastes increased the quantity of emitted Pb, and the main place of Pb deposition was the ESP dust. It is however worthwile to mention that the hazardous metal is not emitted to the air, but remains in one of the process solids even in that disadvantageous case.

The fate of other metals can follow two routes: those which are volatile or have volatile chlorides (besides Pb, these are Se, Sb, Ag, and Zn), behave lead-like; all others are retained. Emissions of those belonging to the Pb group should be carefully and regularly monitored, while those of the other groups represent no emission hazard, a proper operation of the kiln provided. There are however two very dangerous trace metals, which are insufficiently retained in cement manufacture process solids: these are Tl and Hg, with retention values of 90 and 61% only. Fortunately neither of these is a regular constituent of WDFs.

Several metals are precipitated in cement kiln dust (KD). This dust is partly or fully recycled. Practical tests have shown that metal concentrations in KD in case of conventional fuel or WDF are not significantly different. The only exception of this rule is Pb, and to a lesser extant, Se again, the concentrations of which are usually increased in case of WDF burning (by approximate factors of 3 and 2, resp)

Obviously not the absolute concentration of a metal in KD is of importance, but its ability to create a health risk. To evaluate this, the EPA of the United States has elaborated the TCLP (Toxicity Characteristic Leaching Procedure) method, consisting of the acid leachout of the sample of specified size and for a specified time, followed by filtering and analysing of the extract. Experience showed that (with two exceptions) metal concentrations in TCLP extracts are below the health-risk standards. The two exceptions are again Se and Pb, which, according to a recent research, sometimes slightly exceeded specified values, but never in case of WDF kiln operation, only if conventional fuels were used. Another interesting result is that there is no correlation between the Pb-content of the KD and of its TCLP result.

A part of hazardous metals is bonded in clinker. Usually there is no substantial difference in the metal concentrations of clinkers produced with and without WDF. Again Pb and Se, and also Zn are more, and V significantly less if WDF is used. None of the metals in TCLP extracts exceeds limits. The risk is still lower, as cement is used in concrete in 10-20 wt.% only, thus aggregates "dilute" cement, and the potential of metals to be leached out from a concrete structure is much lower by the fact too, that the TCLP process requires a grit, i.e. a material having a surface area several orders of magnitudes higher that the leachout surface of a concrete unit.

## The Cement Technological Side

In this chapter the main considerations for liquid fuels will be described, but similar procedures apply -- *mutatis mutandis* -- for solids as well.

### *Prequalification*

Prior to start a WDF programme, safe waste management requires the prequalification of candidate wastes. The waste generator should provide Material Safety Data Sheets of all materials which could occur in the WDF. After that the candidate sample is analyzed for all important quality characteristics (calorific value, viscosity, solids content, ash content, water content, halogens content, metals content); besides these, gas chromatographic tests are also required to determine volatile organics, and the absence of undesirable constituents, as some authorities do no permit the use WDFs with PCB, vinyl chloride, allyl chloride or pesticide contents. In case of blended WDFs, all blending ingredients must go through these long-lasting and careful prequalification steps.

The consistent quality of WDF can be ensured in most cases by a careful blending only. This can be done in the cement factory, although most cement producers welcome the help of special companies who take the responsibility of pre-qualifying, blending and safe transportation of the WDF, so the cement manufacturer makes only some simple acceptance tests. In other works the blending of the WDF is done within the factory.

### *Transport*

The safe transport of hazardous WDF is almost a bigger problem than its safe incineration in the kiln. Only licensed haulers, with a vehicle fleet having all necessary safety equipment, and carrying appropriate insurance should ship such loads.

Safety first! is the chief motive in solving the logistical tasks of transportation: routes should ensure minimum contact with the public.

### *Acceptance of WDF; safety precautions*

Although the supplier usually gives some data on fuel composition, these must be strictly controlled. Therefore the first task is to supervise all documents (bill of delivery, analytical reports, quantity, etc.) and make some rapid tests to control the accuracy of analyses (this can be done in approx. half an hour). If these are OK, the batch can be temporarily accepted, otherwise the inspector *must* reject the whole batch.

Before unloading of the WDF from the delivery vehicle to a temporary container all connections must be checked to prevent leakage. Small representative samples should be drawn during unloading and sealed for the environmental authorities and a possible later decisive analysis. The liquid must be filtered before transferring it to the final containers, as plastic bags, rubber gloves, paint skin etc. may occur in the fuel. The blended fuel storage tanks should be equipped with mixers or agitators to ensure homogeneity and prevent settling of solids or segregation.

All pipelines, containers, etc. should be equipped with emergency valves to stop liquid flow in case of leakage or overpressure; authorities sometimes prescribe (in case of very hazardous, flammable or explosive fuels) that all handling procedure should be done in an inert atmosphere. The liquid level must be continuously monitored and the flow stopped if a threshold level is reached.

Now the fuel -- if everything OK -- can be prepared for introduction to the kiln, provided that a system of safety valves immediately stops pumping if some irregularities take place in raw meal introduction, in the temperature of the sintering zone, in emissions (too high CO, HCl, $NO_X$ or particulates), power supply failure and some other cases. Workers must know security procedures, the use of safety equipment; in addition all personnel must be trained both in routine operations and emergency response, and this training updated at regular intervals. The keeping out of unauthorized persons is an obvious must.

*Introduction to kiln*

WDF can be introduced to the kiln jointly with the main fuel, in a suitable burner (in case of liquid or finely ground fuel) at points *19* or *28*, on Figs. 2. or 3., resp. The other method is to add WDF with the kiln feed at point *14* in Fig. 2., or, in case of dry-process technology, into one of the *23* cyclones, usually into the lowest one (here coarse particulates, or even large chunks, as unshredded tires can be introduced). Several patents exist however to introduce unground material directly to the kiln via complicated feeding systems. Some other solutions exist too, as e.g. a special "throwing device" for tires, from the burner end directly into the sintering zone, or a pyrolysis drum, in which the tires are heated by tertiary air, and gas products transported to the calciner.

The safest, from an environmental point of view, is the addition from the burner end, through a separate burner or throwing device. In that case hazardous organics pass through the high-temperature zone in any case, and get completely destructed. A good atomization (small droplets) is essential to achieve good and fast burning conditions.

*Effect of WDFs on kiln and cyclone operation*

The chloride-, sulfate- and alkali-contents of the WDF, in combination with the raw materials may bring some unusual disturbances in kiln operation. The most common of these are the formation of adherent rings in the kiln or cyclone which, in the worst case, may fully block material flow. The presence of potassium chloride in rings has been reported, if the chloride intake was high: even slight amounts of $Cl^-$ may mobilize the alkali coming from the raw material.

This is caused by the volatility of alkali (and, to a still higher extent, thallium) chlorides. At sintering zone temperatures, i.e. 1450°C the vapor pressure of these chlorides is well over atmospheric pressure, therefore they are fully volatilized; and, as their melting point is about 800°C, they would condense again, somewhere between the sintering and preheating zones, thus causing ring formation. Sulfates are less dangerous from that point, because of the lower volatility of alkali sulfates: under 1% $Cl^-$ and 2% $SO_3$ there is no danger of ring formation; the critical range is between 1 and 2% $Cl^-$ and 2 and 4% $SO_3$, while over these limits no stable kiln operation is possible (concentrations refer to mass-%).

The operation of the heat exchange system becomes also unstable at too high anionic loads. As a rule-of-thumb, it can be stated that that the combined chloride content, coming from the raw material *and* fuel must not exceed 0.02%, based on clinker mass. The sulfate content is not so decisive, but the molar ratio between $SO_3$ alkali oxides (this is called usually "degree of sulfatization", the amount of sulfates in form of alkali sulfates) should be less than 1. Over these values material cycles would develop, as alkalies and anions would not be absorbed fully by the clinker.

In case of higher loads, material cycles must be interrupted; i.e. the materials which would impose disturbances, must be removed by a suitable bypass. A gas bypass means the partial removal of salt-laden exhaust gases. Bypass gases are quenched by cold air and can be eliminated. Bypass operation is not easy: a compromise must be found between chloride removal and kiln economy. Up-to-date kilns, with tertiary air duct to the precalciner are more sensitive from this point, as here the gas volumes between sintering and preheating zones are less, consequently their chloride concentration higher. A part of hot raw meal can be bypassed too, in order to remove a part of kiln feed which has been enriched in volatile components by condensation from the hot gas.

A dust bypass means the partial removal of dust. Bypass systems for chloride removal differ from conventional ones in special requirements: dust removal must be of low velocity to bring a fractionation of dust in the extraction cross-section already; an adequate cooling which would allow dust precipitation at 450-500°C; and the recycling of hot bypass gases to a thermally preferred location of the preheater to avoid significant heat losses.

The commissioning of bypasses requires usually quite long times, until the meal and clinker coatings disappear. In this period the bypass system must temporarily discharge not only the chlorides introduced by the WDF, but also the additional chloride load arising from the dislodged rings.

Bypassed materials can be utilized in most cases. The heat contents of bypassed gases are suitable for heating. Part of the bypassed meal and dust can be added to cement, others, esp. bypassed meal, with a high CaO content can be marketed as a lime binder, or, if KCl content is high, as a fertilizer. In spite of these utilization possibilities usually there remains a portion of dust which must be landfilled.

*Effects on cement quality*

The use of WDF does not deteriorate clinker and cement quality; on the contrary, improvements are frequently reported. The cause of that is probably the stabilization (less fluctuation) of free lime contents, due to the high calorific value, and a consequent more intensive sintering. Under the microscope, the WDF-fired clinkers differ from normal ones by a smaller alite size, but with more definite outlines. This, in turn, brings a better grindability. In some cases, however, a darker color of the cement was reported in case of burning waste tires. This is probably caused by the steel belts.

*Pros and Cons of Waste Fuelling*

A safe disposal of hazardous wastes by using them as a secondary fuel in cement kilns can be interpreted, in terms of pros and cons:

Pros:

- For hazardous organics this solution is the Best Demonstrated Available Technology.
- The use of Waste Derived Fuels can bring considerable savings, by reducing costs of primary fuels and by collecting disposal fees from the waste generator.
- The quality of cement is usually improved by the use of WDFs.Cons:

Cons:

- The use of WDF for co-firing may require considerable capital investment.
- WDFs usually brings difficulties in normal kiln and preheater operation, unless special precautions and additional devices (e.g. bypass systems)
- The removal of bypassed solids may bring additional problems (landfilling, etc.)

- The permitting processes by environmental agencies require long time, even in case of exhaustive preliminary data collection.

- Local population always opposes the introduction of WDF materials. Open discussion with activist groups is essential!

***The Answer: Cement Manufacture is Practically Waste-Freee; Wastes coming from Other Industries Can Be Safely Disposed Of And At The Same Time Utilized as Raw- or Additive Materials, or as Fuels in Cement Manufacture***

## ACKNOWLEDGEMENT

The financial support of the Hungarian National Scientific Research Foundation (OTKA), Ref. No. 2556 is gratefully acknowledged.

## LIST OF ABBREVIATIONS

| | |
|---|---|
| BDAT | Best Demonstrated Available Technology |
| DRE | Destruction and Removal Efficiency |
| EPA | United States Environmental Protection Agency |
| ESP | Electrostatic Precipitator |
| KD | Kiln Dust |
| PID | Product of Incomplete Destruction |
| POHC | Principal Organic Hazardous Constituent |
| TCLP | Toxicity Characteristic Leaching Procedure |
| WDF | Waste Derived Fuel |

# Lithuanian Experience in Disseminating P2 Ideas and Techniques in Industry

Jurgis Staniskis and Audrius Sileika
Institute of Environmental Engineering
Kaunas University of Technology
Donelaicio st 20
LT - 3000, Kaunas
LITHUANIA

This paper is aimed to show how the pollution prevention ideas are making attempts to become prevalent in the way of manufacturer's thinking in Lithuania. The second objective is to discuss the role of academic institutions (emphasizing Kaunas University of Technology) in this research area. It is not an easy task, especially for the people and institutions disseminating this strategy among industrial companies and the society. Though the main ideas of PP are obvious for the major part of manufacturers, but putting this concept into practice is a rather complicated problem.

The occurring problems are mostly conditioned by the nowadays situation in Lithuanian industry. Planned economy and strong dissociation from the rest industrial world left a very deep sign in all spheres of live in the country. From that point of view Baltic States suffered much more (due to incorporation in the former Soviet Union) than other countries in Eastern and Central Europe

It should be incorrect to allege that cleaner production concept is a completely new subject in Lithuanian industry (as well as in former USSR respectively), but it was not a comprehensive and prevailing industrial strategy. As the whole industry was planned, only single examples of imaginary 'non-waste technologies' were created by directions from the governmental institutions. In most cases these examples had only demonstrative nature, and inefficient production or enormous energy consumption were hidden under the nice veil (with very few exceptions). All investigations and scientific research work were concentrated on the creation of specific technologies (chemical processes were in priority as usual)

NATO ASI Series, Partnership Sub-Series, 2. Environment - Vol. 2
Cleaner Technologies and Cleaner Products
for Sustainable Development
Edited by H. M. Freeman, Z. Puskas, and R. Olbina

and/or on the development of various analytical methods which were often left without practical implementation. Nobody paid attention to the problems related with industrial management systems as well as to the social and psychological questions, though the importance of these issues is obvious.

Under the such circumstances we have a lot of skilled and technically educated professionals in the industry and universities who are able to solve specific problems related to environmental technologies (unfortunately it means 'end of the pipe' in their thinking rather often). On the other hand number of specialists on industrial management and business economy as well as in social sciences gradually increases in Lithuania, but the level of environmental management in industry is still very poor. It is because of the lack in practical experience of people dealing with mentioned problems.

Another problem is that communications between the above mentioned 'technicians' and 'humanitarians' is very complicated because of different way of understanding the same problems. So, it is necessary to create an intermediate link between these opposite poles, and specialists on environmental systems may be this link. Such specialists ought to be trained both in decision making system and conventional technologies of environmental protection. It means that more systematic approach should prevail in their thinking.

The role of universities is very important in dissemination P2 ideas because of their possibilities to influence to the industrial specialists through the educational process. Institute of Environmental Engineering at Kaunas University of Technology is one of the institutions actively working in that area. The institute is organized in 1991 and consist of six sectors:

1. Water quality modeling and management.
2. Cleaner technologies and life cycle assessment.
3. Waste water treatment technology, process modeling and control.
4. Solid waste management.
5. Key elements in the environmental protection policy.
6. The use, management and sustainable development of wetlands.

The pedagogical work at the institute concentrates to the education in M.Sc. and Ph.D. levels and the following courses are delivered for undergraduate students:

- Environmental engineering - 1.
- Environmental engineering - 2.
- Modeling and forecasting of environmental pollution.
- Water quality modeling for open water bodies - 1.
- Water quality modeling for open water bodies - 2.
- Cleaner technology (theory and implementation).

In addition to mentioned activity the institute participate in network of environmental centers in Lithuania as the part of Kaunas University of Technology. The network is aimed to train M.Sc. and Ph.D. students in Environmental Sciences and the structure of the network is following:

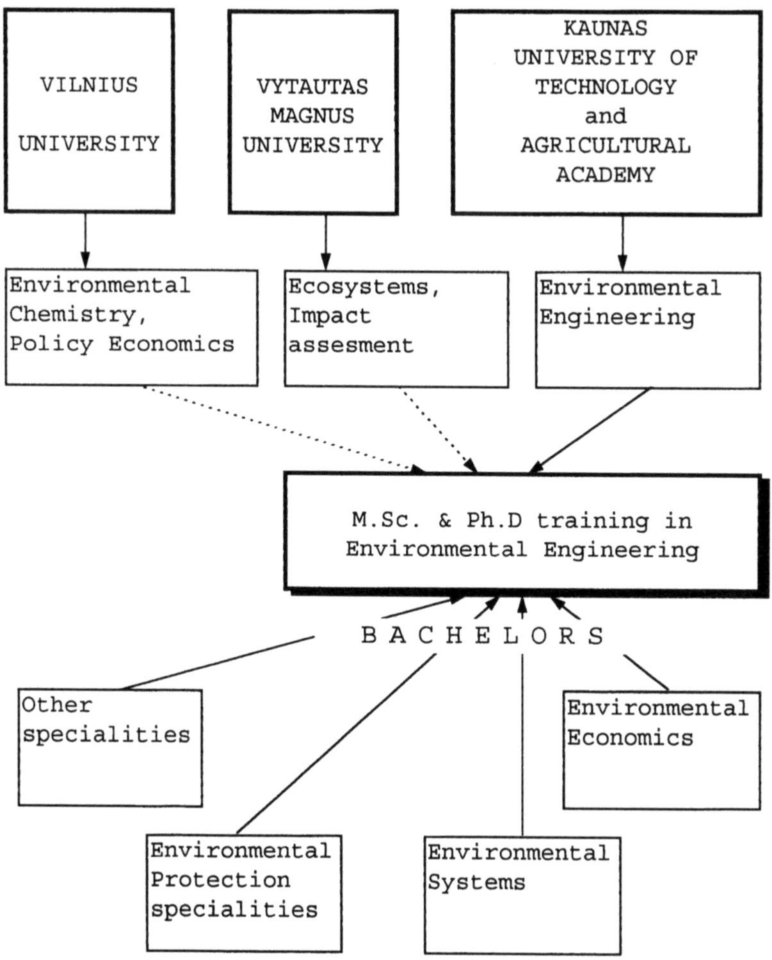

The working area of the institute will be training in environmental engineering. The problem is shortage of mentioned above bachelors of environmental systems specialty but preliminary curriculum for this course is already prepared.
The institute takes part in education process of the manufacturers as well. The joint project entitled Waste Minimization Opportunity Audits to Introduce Cleaner Technology in Lithuanian Industry has started in May, 1993. It is conducted in Kaunas, the second largest city of Lithuania. There are 8 industrial companies from Kaunas city participating in this joint project. The organizers are Environmental Engineering Institute, Kaunas University of Technology; Resource Management Center RENDAN, Denmark; and department of Industrial Environmental Economics, Lund University, Sweden. All the companies participating in the project are representing various branches of industry: textiles, electronics, rubber, food production, etc.

The project is aimed to solve quite a wide range of problems:

- industrialists training in environmental auditing and cleaner production concepts of manufacturing;
- investigation of possible pathways for realization this concept in Lithuanian industry;
- generation of cleaner production examples to show to the manufacturers their possibilities to implement these changes.

The illustration of one specific example follows.

One of the problems of the electronics company participating in the project is enormous losses of heat energy during the heating period. That is because of the inefficient ventilation systems at soldering lines in assembling departments. The approximate amount of ventilated air is 500 $m^3/h$ at a single working place. These losses will be equal to $20,000, if 200 working places are active during the cold season and the outdoors temperature is -5 °C at that time.

The possible solution may be an installation of soldering fumes extractors with vacuum pumps. That will reduce ventilated air flow from 18,000 $m^3/h$ to 80 $m^3/h$ per one assembling line:

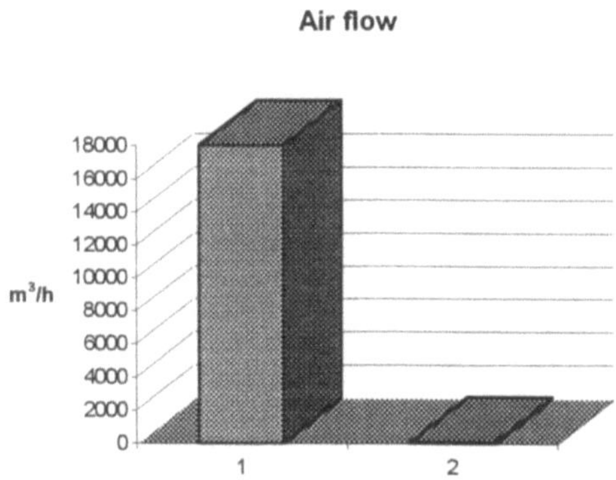


Reduction of electricity consumption is considerable as well, because the capacity of electric drives decreases from 18kW to 1.5kW per single ventilation unit:

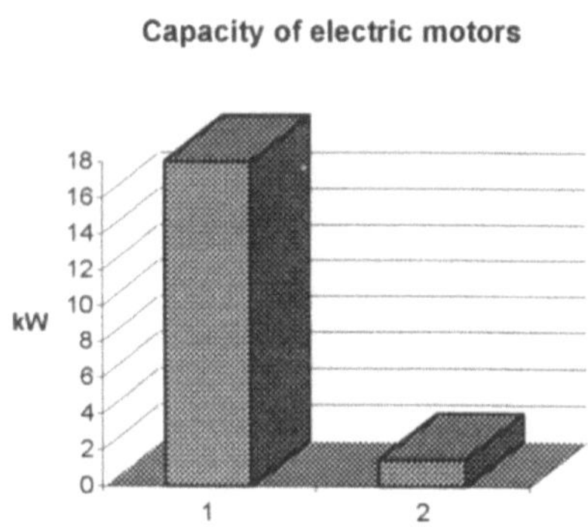


The efficiency is so high due to the location of the fume extractor <u>very close to the pollution source</u> (the thin pipe of the fume extractor is mounted directly on the soldering iron). Heat losses using this equipment will be completely eliminated because the extracted air is passing through filter system and supplying to the workshop afterwards. The efficiency of used ultrafiltration is very high (mechanical filtration of particles 99.995%)

Due to such positive improvement pay back period (PBP) for the mentioned conditions will be approximately 1.1 year. The following graph illustrates that PBP values depends mostly on the average outdoors temperature during the heating period. The number of working places occupied in assembling department has no considerable influence to the mentioned values. It is important that in wide range of conditions PBP does not exceed 2 years:

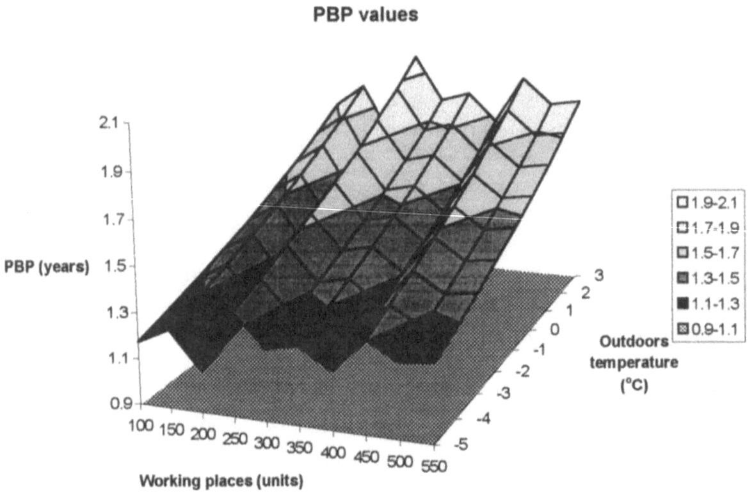


As a conclusion it is possible to say that such situation is not an exception and we believe that ideas of P2 will find a right place in Lithuanian industry.

# The Dutch Ecodesign Program

A. Zweers and H. te Riele

Delft University of Technology
Faculty of Industrial Design Engineering
Section Environmental Product Development,
The Netherlands

## 1. Introduction

Early 1994 we finished our report on the Dutch Ecodesign Program which had been started in 1990. By then it was the first demonstration program together with industry at this scale. At the moment our approach is unique in Europe. We hope that an English translation of our summarizing book ('Ecodesign, 8 examples') will be available early 1995. During the program we learned a lot and developed some tools. Here we would like to share some of our experiences.

## 2. Cleaner Technology, Cleaner Products and Ecodesign

Prevention is an important line within the concept of clean technology. In the Dutch PRISMA project (a large prevention / clean technology program) has been concluded that this line works well for production processes. Products themselves however stay out of reach. So quite naturally, PRISMA has formulated a challenge for focusing on cleaner products. Products should become 'green', 'cleaner' or 'environment friendly'.

In the Ecodesign program we have rejected this approach. We have chosen the product development process within industry as a point of departure. We have had good reasons to do so.

a. The concept of green products is doubtful. There are no absolute standards. The assessment of environmental impact is a difficult one. It works well only in comparing (between products) situation. It is based upon many parameters. Marketability is uncertain.

b. Often the material product in itself is not causing the main problem but the way the product is produced, distributed or discarded. In search for environmental improvement the total product chain (from raw materials via use to waste disposal) should be taken into account as well as the whole product system (a product is serviced, repaired, refilled, co-operating with other products, etc.).

c. Waste prevention concerns waste and emissions. Other problems like resource depletion do not or only indirectly get attention.

NATO ASI Series, Partnership Sub-Series, 2. Environment – Vol. 2
Cleaner Technologies and Cleaner Products
for Sustainable Development
Edited by H. M. Freeman, Z. Puskas, and R. Olbina

d. The negative load of environmental impact can hardly be combined with the necessarily positive drive for product development.

e. The realization of a new product is the essence of entrepreneurship. Products concern the core of a company and are a very delicate matter for discussion.

f. Product improvement needs innovative steps which are in harmony with the economic conditions of the company.

In Ecodesign we have aimed at integration of environmental aspects within the product development process. Some essentials of this approach have been:

lst After commitment of a company's top management Eco-design joined the first or the second phase of the development process (pre-development phase or the actual development phase). We started a dialogue with the product development department (mostly R&D dept.).

2nd Conforming to normal procedures of the companies we should get in contact with other functions of the company, such as marketing, purchasing, strategic management and production.

3rd 'Eco' meant to us both ecologically sound and economically feasible.

4th We worked within the concept of sustainable development.

## 3. Sustainable Development and Ecodesign

Before we go into detail on the program we first clarify our vision on sustainable development.

In the Brundtland report the link is stressed between environmental problems, wealth and economic development for all nations.

When we regard products as need fulfillers for mankind, products are in the center of transformation process (see Figure 1), from resources via products to waste.

The environmental problems have effects of our industrial production that is not smart enough, of our abundant buying and of our carelessly discarding. However we don't want to return to the middle-ages so we choose for growth. Not quantitative but qualitative growth. Environmental quality is just part of the quality of a product.

It is difficult to get hold of the concept of quality of a product. It concerns nature and essence of the object. In arts quality is often expressed in terms of wholeness and harmony of the total of product aspects. In the more pragmatic engineer's approach quality means 'according to the requirements'. Here general standards are pushed forward. They divide product quality into parts and targets.

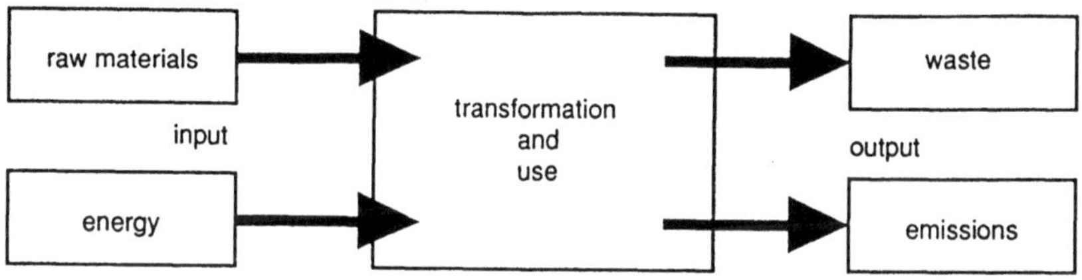


Figure 1. From Resources via Products to Waste

For Ecodesign, on a general level two directions are or value: effectiveness and efficiency. Effectiveness means that the impact of the product - the need fulfillment - is adequate. Efficiency is found in the shortest way of fulfilling and with the least means.

In the Netherlands's the National Environmental Policy Plan this choice is formulated as follows:

- closing of the material cycle in the chain of raw materials → production process → product → waste and the related emissions;

- saving energy together with increasing of efficiency and the use of sustainable energy sources;

- enhancement of quality (more than quantity) of products, production processes, raw materials, waste and environment with respect to a longer use of materials in the economic cycle.

When we translate quality in terms of sustainable development concept we come to the following directions, the so called MET-approach:

a. We must choose for a higher quality and for innovation in the sense that the needs of society are fulfilled more effectively and efficiently with the least input of resources and the least output of hazardous waste. A longer product life is usually more effective and efficient. Higher quality probably also means fulfilling more needs by service products instead of material products.

b. The least input of resources can be obtained by:

- reduction and selectivity of material use;
- closing the material cycle;
- preference for renewable and recycled material;

- saving of energy as much as possible, throughout the product chain.

c. Prevention and minimization of hazardous waste, especially of toxic substances and emissions.

We have put these directions into the MET-matrix model (see Table 1). We have used the model in some projects to help structuring the environmental product analysis, and for the presentation of the Ecodesign results in our final report.

| MET Matrix | Materials | Energy | Toxic Substances |
|---|---|---|---|
| Extraction/Production | | | |
| Distribution | | | |
| Use | | | |
| Discarding | | | |

Table 1. Eco-design MET-matrix

## 4. Ecodesign Set-up

Ecodesign projects have been carried out together with eight companies. In each company a new product was being developed. Its development team was assisted during 6 to 9 months by 2 Ecodesign consultants: an industrial design engineer and an environmental expert. These are professionals, working for consultancies and universities. All together 11 consulting parties have been directly involved during the advising phase of the program.

Although we respected consultants in their own way of acting, and although every company required a different style of intervention, Ecodesign advisors worked according to a five step plan:

<u>Getting started</u>
Agreements are signed in order to regulate aim, planning, work load and publicity rights.

<u>Global environmental analysis of a reference product</u>
Environmental data are obtained. Points of attention are identified. Priorities are set and discussed. Finally environmental aspects are brought into the list of demands for the future product. The reference product can be either the company's current product or a product which is still in its conceptual phase.

<u>Actual design process</u>
Solutions for priority problems are generated. Promising solutions are selected. Models and prototypes are built and tested. Environmental information and data get a very detailed character.

Results definition
Final choices in materials selection and other product details are written down. Future recycling intentions as well.

Environmental comparison between reference product and final product design
In every Ecodesign project the consultants compared the environmental load of the new design to the initial version they met at the start. The Ecodesign final stage (the stage at the moment of our exit) has not necessarily been the design's final stage.

The consultants produced three reports. A chronological report. An extended logically structured report. And one summary for presentation purposes.

## 5. Some results

The companies and products are:

| | |
|---|---|
| Ahrend: | office chair for the professional market |
| Etna: | a four-burner consumer gas hob |
| Olland: | vending machine for hot drinks |
| Focus Veilng: | a breath protection mask |
| Dutch Auctions (VBN): | a transportation tray for potted plants |
| Nedcar: | the dashboard for the successor of the Volvo 400-series |
| Speelhout: | duo-swing to be used in playgrounds |
| Leolux: | chair for a high segment of the consumer market. |

This list shows that Ecodesign has dealt with products having very different product structures, market demands, production amounts, company's cultures and environmental items. The eight case stories in our book reflect this diversity. Here we give an indication of some environmental results of the Ecodesign interventions.

The Ahrend office chair
Materials reduction, materials changes, design for disassembly and agreements with suppliers for take-back, have resulted in a reduction of:
- greenhouse gasses (approx 35%)
- acidification (approx 20%)
- embodied energy (approx 40%)
- disassembly time
- estimated environmental costs in 2000 (from $6.0 to $2.5 per chair).

The Nedcar dashboard
weight optimization, clustering of 'family' - plastics, and electric circuit improvements have resulted in:
- 15% material reduction (180 tons each year)
- better recyclability
- fuel reduction (5.5 million liters each year).

The results of the dashboard case are currently influencing other aspects of the car design. The new car will be introduced within two years.

The VBN transportation tray
Weight optimization, material choice and choice for a mainly re-use system have resulted in reduction of:
- 1000 tons waste per year
- 20% material use
- tray energy content per transportation of 90%
For VBN the new development was rather a logistic system design with a lot of parties involved than a stand alone product design. The new system has been introduced early 1994.

These and other results bring us to the conclusions that:

- Ecodesign works: environmental aspects can be handled by companies and product designers and the results are promising.

- Designers are able to base their operations upon the expert judgement by global environmental chain analyses.

- Ecodesign inside a company's development department tends to influence marketing, purchasing and strategic management.

- With Ecodesign in mind, designers tend to select the technically advanced suppliers. They are the ones who work according to the latest environmental knowledge.

- Ecodesign makes a company self assured again. The company no longer is dependant on external public items coming up unexpectedly. The company gets familiar with its environmental product priorities and can plan its future investments and changes by itself again.

and moreover we found that:

- A dialogue with a company's product development department stimulates the innovation process. It stimulates the integration of environmental aspects in the pool of traditional product design values.

- Many of the environmental design improvements reduce product costs. We found no evidence for the general perception: 'Environmental concerns automatically raise prices.

# Introduction of Modern Methods for Degreasing of Metal Parts.

Gudolf Kjærheim
NATO/CCMS FELLOW
Oestfold Research Foundation
Institute for Cleaner Production
P. O. Box 276
N-1601 Fredrikstad
Norway

## Introduction

Chlorinated organic solvents are beeing used extensively in cleaning and degreasing operation in industries like:

Fabricated metal products industry
Transport equipment manufacture
Printed circuit board manufacture
Defence industries
General mechanical engineering.

Chlorinated solvents have obtained this strong position because their positive properties combine into effective cleaning agents, they evaporate rapidly and are non-inflammatory.
But on the negative side; they contribute to the destruction of the ozone-layer in the stratosphere and are detrimental to human health.
The Vienna Convention in 1985, under the auspices of The United Nations Environment Program, addressed the question of ozone-layer depletion. The decleration of intent from this meeting was followed by signing of the Montreal Protocol in 1987. Norway ratified the agreement on June 24. 1988. The Montreal protocol was renegotiated and strengthened, in 1990 and 1992.

NATO ASI Series, Partnership Sub-Series, 2. Environment – Vol. 2
Cleaner Technologies and Cleaner Products
for Sustainable Development
Edited by H. M. Freeman, Z. Puskas, and R. Olbina

Concrete results include agreements on:
Phase-out of CFC-113, carbon tetrachloride and 1,1,1-trichlorethane by January 1., 1996.
Import restrictions will gradually be imposed on these chemicals, and export to countries that have not ratified the protocol will not be allowed.
Phase-out dates have not been set on other chlorinated solvents, but several are being investigated for having potential carcinogenic properties.
The Montreal Protocol will be revised every fourth year.
The North Sea Declaration supplements the Montreal Protocol in that it states that emissions of toxic material are to be reduced by 50% by 1995. Reference year is 1985.

The reductions apply to the following chlorinated solvents:

- Carbon tetrachloride (tetrachloromethane)
- Chloroform (trichloromethane)
- trichloroethylene (TRI)
- perchloroethylene (PER)
- trichlorobenzene
- 1,2 trichloroethane
- 1,1,1 trichloroethane

The restrictiones that will be imposed on the production, import and use of 1,1,1 trichloroethane in many industrialized countries will force the industry to implement alternative solutions to chlorinated solvents. Switching from 1,1,1 trichloroethane to another chlorinated solvent is most likely a very shortsighted solution.
Many industrial companies are therefore facing similar problems. The demand for a smooth and rapid transition to more environmentally friendly degreasing operations in many industries prompted Oestfold Research Foundation in Fredrikstad, Norway to initiate a two-year program tailored to aid a selected group of industrial companies to find alternative technologies for their cleaning and degreasing operations

## Financing

A program proposal was developed by Oestfold Research Foundation and presented for external organizations in a attempt to obtain financial support. After a period of negotiations we ended up with financial support for the program from:

- Confederation of Norway Business and Industry (Naeringslivets Hovedorganisasjon, NHO): 50%
- Norwegian Environmental Protection Agency 25%
- Participating Comapanies 25%

NHO and SFT were also represented in the steering committee for the program.

SFT (EPA) was interested in all aspects related to implementing environmentally acceptable degreasing and cleaning operations. In addition to the above was NHO interested in improving occupational health and safety conditions.

## Participating companies

In selecting companies to take part in the program we wanted to include a wide range of different industries using chlororganic solvents, chemical suppliers and equipment/process suppliers. We believed that we could obtain better results by having user-companies and suppliers working together towards a common goal.

The participating companies were divided into five groups. Companies within each group were concerned with similar type of problems.

Group 1 Cleaning of transport - and tank equipment
Group 2 Removal of Decor foils

Group 3 Pretreatment before applying lacquer/varnish/paint
Group 4 Substitutes for TRI
Group 5 Treatment of slopwater (oily water from flushing of tanks)

Industrial branches represented include vehicle repair shops, railroad-car manufacture, railroad maintenance, furniture manufacture, electric heater manufacture, electric light armature, fittings production and defence material production.

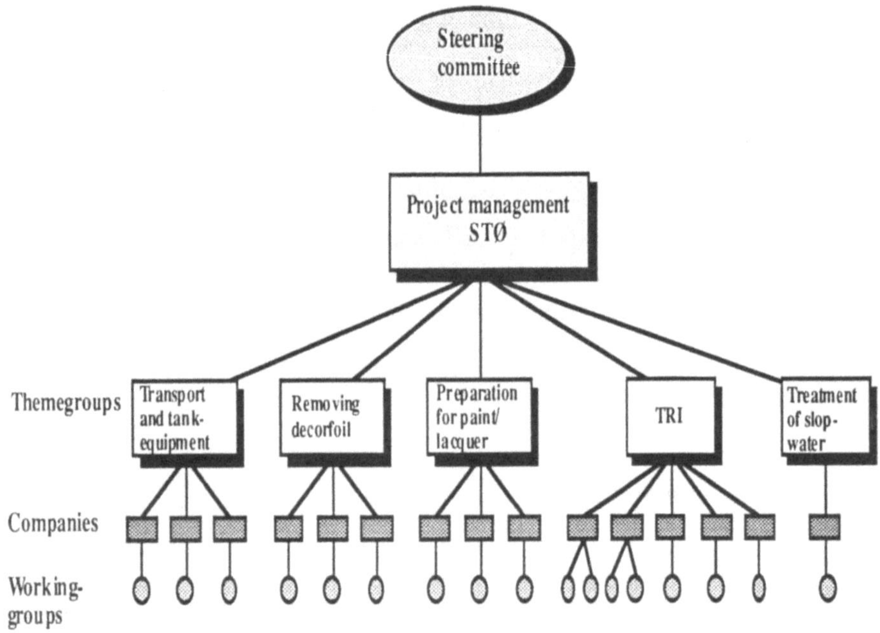


Figure 1 Project organization

The project was organized as shown in Fig.1 with a steering committee, project administration and management and five different theme groups. A project organization was established within each company where most of the technical work was carried out. There were held coordination meetings for exchange of information and experiences on a regular basis throughout the project period. There were also arranged three, two days seminars with all participating companies present, one at startup of the project, one close to the end of the project and one in the middle of the project period.

## Aim

Several general aims were proposed at the start of the program. Each company prepared more specific, prioritized aims for their own plants.

The general aims were:

Give companies an introduction to cleaner production methods so that they would be able to carry out waste minimization opportunity assessments in their own plant.

Motivate people to more resource - and environmentally friendly operation.

Improve working conditions (Occupational safety and Health).

Contribute to increased level of competence and know-how.

Reduce or eliminate the useage of chlorinated organic solvents.

Identify alternatives to present degreasing/cleaning operations

Contribute to reduced costs of chemical, energy and waste disposal

Prepare companies for strickter regulations to come.

## Strategy

The project proceeded along two lines, parallel in time.

1. Conduct a waste mininization opportunity assessment in each company and define own requirements to degreasing and cleaning operationes.

2. Get an overview of what is available on the market in terms of chemicals, processes and relevant practical experience.

## Waste Minimization

These assessments were carried out to get a complete picture of emissions and waste streams from the plant. In addition, this activity was used to motivate people and to encourage to continuous work on waste reduction. We also stressed the importance of using a systems approach in cleaner production efforts. Like in a degreasing operation it is necessary to look both upstream on type and amount of oil applied and downstream on the requirements to a clean surface.

## State-of-the-art

In order to find out what were realistic alternatives to chlorinated organic solvents we:

- Used our contacts to suppliers, both of chemicals and processes.
- Carried out search in data banks and available literature.
- Consulted colleagues in Sweden and Denmark.
- Contacted Solid & Hazardous Waste Education Center in Madison Wisconsin on their course on Pollution Prevention for Solvent Degreasing Operations.
- Contracted Technische Akademie Esslingen Weiterbildungszentrum.

Syposium no.16046/45.222, Reinung mit Kohlerwasserstoffer und Wasser - Verfahrens - Beschreibung fur die Metall - Kunststoff - Elektronik - und Textilbranche.

This resulted in a considerable amount of information which were supplied to the participants in the form of notes, copies of articles or presentations at meetings.

## Degreasing and cleaning operation

Oil and grease are used for corrosion protection of metal parts and components when in storeage.

In a variety of metalworking operationes such as drilling, cutting, grinding, stamping etc., see Fig.2, a fluid is necessary to reduce friction and dissipate heat, thus keeping both tool temperature and workpiece temperature down. The fluid further acts to provide a good finish on the workpiece and to inhibit corrosion. The fluids used include oils, greases and waterbased and synthetic coolants.

However, a clean metal surface is required before surface treatment, plating or coating operations.

Up til now, chlorinated organic solvents have been used extensively in degreasing operations, but to safeguard the environment and protect human health, this project will be looking at alternative degreasing methods.

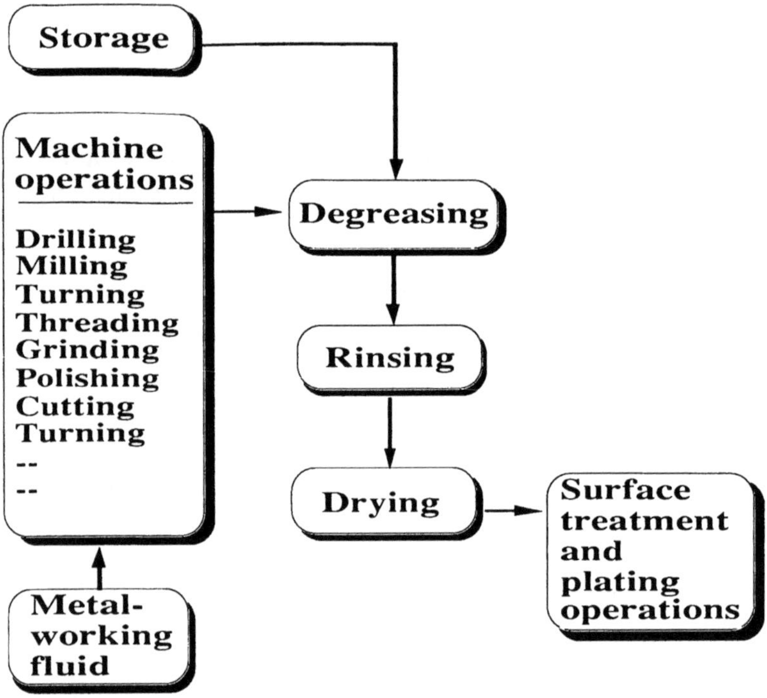


Figure 2 Degreasing and related operations

## Parameters affecting choice of cleaning method

When it comes to selecting an alternative cleaning method, one should not only look at the degreasing operation, but consider the overall system.
It is relevant to ask questions like:

Why do we clean the metal objects?
How clean do they have to be?
How do measure surface cleanliness?

Some of these types of questions will also bring the product design into the picture. Product design should consider degreasing operations and attempt to avoid complicated geometries with hard-to get-at-places.

It is further necessary to know the composition of the contaminants in order to design a suitable cleaning system including sequence of operation. The sequence of operations is particulary important when metal objects are contaminated with several materials. It makes sense, for example to remove a layer of oily contamination by an alkaline cleaner before abrasives are used to remove rust and scale. Cleaners that work well with one metal may not work with another because the reactivity of different metals with alkaline and acid varies. A wrong choice may adversely affect the metal substrate.

The degreasing operation will be simplified if the number of oils that are used can be reduced.

Fig. 3 shows the most important parameters which have to be considered when selecting a proper cleaning method.

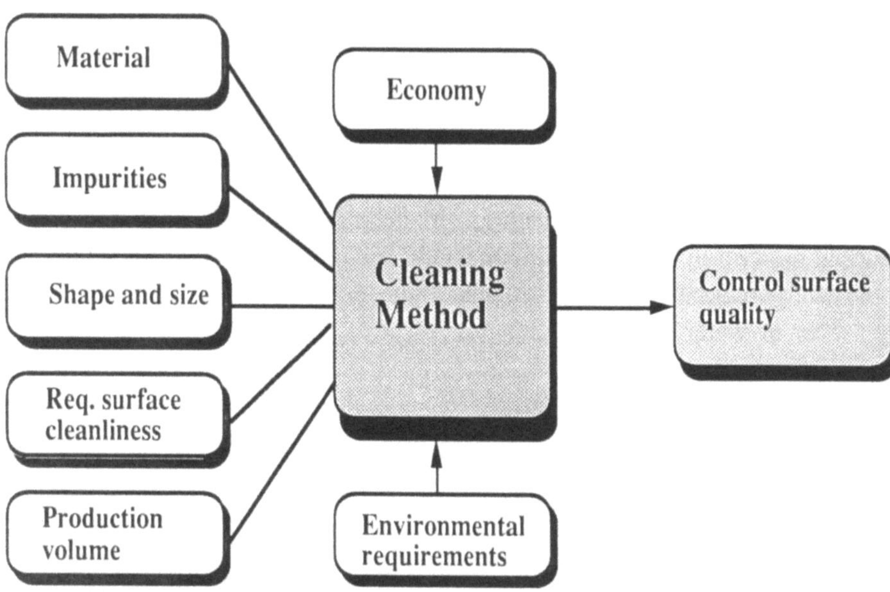


Figure 3 Parameters affecting choice of cleaning method.

## Cleaning effectiveness

Several parameters affect the effectiveness of a given cleaning system. These are:

Physical/chemical effects
Mechanical forces(agitation)
Temperature
Time

These parameters, should, for a given application, be optimized with respect to cleanliness of parts, amount of waste generated and overall economy of the process.

## Cleaning media

In searcing for a new cleaning medium, one should first consider the most environmentally acceptable medium. If the results are not satisfactory, one should proceed to the less environmentally desirable alternative. Generally the following order can be recommended:

- air or water
- abrasives
- aqueous detergent solutions
- acid solutions
- non-chlorinated solvents
- chlorinated solvents

For certain applications water can be used in conjunction with mechanical or ultrasonic agitation. Hot, deionized water may even replace chlorinated solvents successfully, in certain cases.

Abrasives are designed to remove rust and oxides. Common abrasives are sand, aluminium oxide or silicon carbide.

Chemical cleaners may be divided into two main groups; solvent based and aqueous based. See Fig 4.

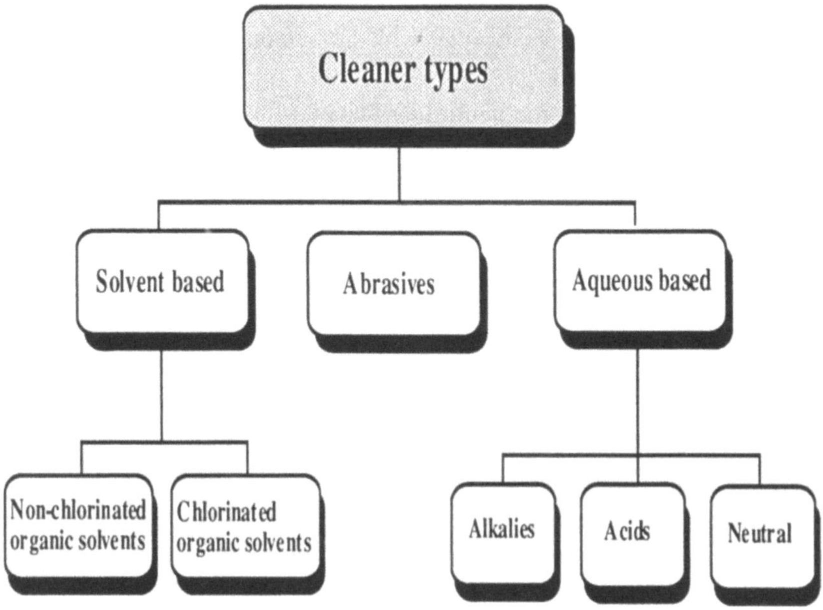


Figure 4 Cleaner types.

Chlorinated organic solvents will be phased out, but other organic solvents can be used effectively in a closed system to remove organic or oil-based contaminants. The low surface tension of the hydrocarbons allows rapid coverage of the surface to be cleaned and penetration under the coating of soil. Heat of evaporation is considerably lower than for water which means lower energy consumption for drying. Hydrocarbons selected for degreasing operations have high flash points. The equipment needed for hydrocarbon cleaning is usually simpler and more compact than for aqueous cleaning systems.

Aqueous cleaners can be dividet into acids, alkalies or neutral agents.

Acidic cleaning solutions are used to remove rust and oxidation scale from metal surfaces. The most common are hydrochloric, sulfuric and nitrie acids, but also organic acids are used together with detergents.
Alcaline cleaners use the reactive chemical power of sodium salts of phosphates, silicates, carbonates together with surfactants.

Neutral cleaners utilize only the neutral surfactants for their cleaning power. Most neutral cleaners, however, contain slightly alkaline corrosion inhibitors. Normally used with mechanical agitation to increase efficiency.

**Aqueous cleaning equipment**

Equipment for aqueous cleaning may be divided into:

- Spray washing equipment
    - Batch spray washer
    - Tunnel washer for continuous production lines
    - High/extra high pressure spray washers

- Agitated immersion cleaning
    - From simple units to multistage, fully automatic systems

- Ultrasonic cleaning
    - Transducers generate ultrasonic waves which cause local "cavities" to occur at the surface of the objects, thereby enhancing the cleaning effectiveness. Frequencies in the range of 35-40 $KH_Z$ are normally used.

- Barelling

  Ratating drum with internal longitudinal fins, may be used for both batch and continuous operation.

- Flood washing
  Cleaning solution is injected into a washing chamber under high pressure through a series of jets. The resulting hydraulic turbulence is more effective than normal agitated immersion cleaning.

## Thermal Cleaning with $CO_2$ pellets

A fairly new method developed in the US: The method may be used for removing paint from vehicles or aircraft, but may also be used for cleaning of objects before surface treatment. The operating principle is based on blowing small particles of dry ice against the surface to be cleaned. Pneumatics provide the driving force. The cleaning power is probably a combination of mechanical interaction when the particles hit the surface and thermal shock, caused by sublimation of the dry ice. The sublimation process cools the surface down to about $50^{\circ}C$. The soily layer becomes brittle, cracks and scales off.

## Separation

To increase bath life or to close continuous cleaning processes, a variety of separation technologies may be installed. The idea is to separate oil and suspended material and at the same time let the cleaning chemicals pass the membrane and return to the bath. Trials may have to be carried out in order to arrive at an optimum solution.

Available technologies include:
- conventional filtration
- sedimentation
- carbon adsorption
- ion exchange
- ultrafiltration
- microfiltration

- reverse osmosis
- cyclones
- destillation
- extraction

The choice of method depends on several parameters like size of particles, density, solubility, ion strength, surface activity etc.

## Biological method

The degreasing is normally done with a weak alkaline solution containing surfactants in a spray washing system.
Oil and other organic material are decomposed by microorganisms, while surfactants are recirculated. The sludge volume generated is low. The surfactants selected should not easily be decomposed biologically.

## Conclusions

The companies involved are in different stages of completing their projects. Most of the companies will have new or modified cleaning systems, without the use of chlorinated organic solvents, installed by summer 1995.

All companies in collaboration with Oestfold Research Foundation have written internal technical reports describing the projects.

Bringing suppliers into the project, together with manufacturing companies have given good results.

There is a need for further training of people in areas like using proper control methods for cleanliness.

Based on the experiences from the project the Norwegian EPA and NHO have decided to give financial support to three other companies that must find alternatives to chlorinated organic solvents. Oestfold Research Foundation is heavily involved also in this project.

## References

Brown, L.M. and Springer, J. Chemical substitution for 1,1,1-trichloroethane and methanol in an industrial cleaning operation, Journal of Hazardous Materials, Elsevier Science Publishers, Vol. 29, p. 179-188, 1992.

Inland Technology Inc. Brochure 1-800-3100,1990,2612. Pacific Hwy. East Tacoma, WA 98424, USA.

JOT'92 Fachtagung uber die industrielle Teilreinigung, JOT, Journal fur Oberflachentechnik, Heinrich Vogel Fackzetschriften GMBH, Munchen, 1992.

MnTAP Membran Filtration: microfiltration, ultrafiltration and reverse osmosis. Fact sheet, The Minnesota office of Waste Management, Mineapolis, 1991a.

MnTAP Evaluation of aqueous cleaners to replace TCA degreasing. Intern Project summary, The Minnesota Office of Waste Management, Mineapolis, 1991b.

Rachlitz, M., Nielsen, M., Jensen, A.H, Chemical Degreasing Operation - Handbook in Technical and Environmenally correct choice of chemical Degreasing Processes in the Iron and Metal Industry (In Danish) Dansk Teknologisk Institutt, 1993

Solids and Hazardous Waste Education Center, Pollution Prevention for Solvent Degreasing Operations. Department of Engineering Professional Development, University of Wisconsin, Madison, 610 Langdon, Madison Wisconsin WI 53703, USA.

Technical Akademie Esslingen Weiterbildungszentrum Reinigung mit Kohlenwassenstoffen und Wassen - Verfahrensbeschreibung für die Metall-, Kunststoff-, Elektronik - und Textilbranche. Symposium no. 16046/45.222, Technische Akademic Esslingen weiterbildungszentrum, Postf. 1265, D-7302, Ostfildern, Germany.

University of California, Waste minimization for the Metal Finishing Industry: Module III. For Alternative Technology Division, California Department of Toxic Substances Control and US EPA.

USEPA, Guides to Pollution Prevention - The Fabricated Metal Products Industry. EPA/625/7-90/006, U.S. EPA, Office of Research and Development, July 1990 d.

UESEPA, Waste Minimization in Metal Parts Cleaning. EPA/530-SW-89-049, U.S. EPA Office of Solid Waste, August 1989.

USEPA, Waste Minimization Opportunity Assessment Manual, EPA/625/7-88/003, U.S. EPA Office of Solid Waste, August 1989.

# Obstacles in Application of Cleaner Production in the Polish Industry

Andrzej Doniec
Faculty of Process and Environmental Engineering
Technical University of Łódź
ul. Wólczańska 175
90-924 Łódź
Poland

Cleaner production (CP) and waste minimization (WM) which is strictly related to it, are the notions determining reduction or elimination of wastes produced by industry. The idea of cleaner production has become an integral part of environmental management and occupies one of the main positions in the ecological policy of many countries in the world. In these countries of highly developed industries, which were pioneers in this field, the propagation of the idea was not spontaneous. As we read in the definition proposed by UNEP, "cleaner production is the conceptual and procedural approach to production", it is necessary to formulate a different approach to production processes. Production is understood as an elimination or reduction of wastes at the source instead of a costly processing of wastes. Hence, we should change the way of thinking and proceeding. This, however, is never an easy task.

Since the beginning of the 70's until May 1989 when UNEP presented the definition of cleaner production, much experience has been gained in applying the principles of environmental pollution prevention. Part of it reflected obstacles in efficient waste minimization in industry, both at the stage of evaluation and project completion. Huising (1989) quotes data presented by Palmer from which it follows that only 10% of them are technical obstacles, while the rest are financial (30%) and political (60%) as the author named them. The second group covers, among others, bureaucratic resistance (20%), human conservatism (10%), public ignorance and misinformation (10%). Thus, socio-psychological obstacles appear to constitute as much as 40%. The above data refer to enterprises functioning continuously in the market system.

A question arises if the same obstacles and difficulties occur in post-communist countries which have just started free-market transformations. There are no complete data on this subject yet, however on the basis of observations of the Polish situation some conclusions can be drawn.

Since the beginning of the 90's various foreign and domestic initiatives to propagate the idea of CP have been started in Poland. Despite undoubted successes they encountered many problems, which became more visible especially during the last two years.

NATO ASI Series, Partnership Sub-Series, 2. Environment – Vol. 2
Cleaner Technologies and Cleaner Products
for Sustainable Development
Edited by H. M. Freeman, Z. Puskas, and R. Olbina

They can be divided into two groups: (i) obstacles in getting industrial plants interested in preventive policy, (ii) problems occurring during development and application of waste minimization projects.

In the first group the most serious obstacle is a complete lack of interest and certain resistance to argumentation. Lack of interest is a result of many factors, the main one being a general financial situation of the firm which has to strive for existence and cannot see any chance for itself in pollution prevention. There is not enough information on cleaner production either. If supplied, however, it meets with rather unfavourable conditions. Schematic thinking causes that seeing "end of pipe" is so strong that it does not admit the fact that "pollution prevention pays". This type of thinking is even multiplied by the old habit originating from the period of centrally planned economy when individuals were released from thinking and responsibility, when prices needed only for statistics values and being tools of policy-making, did not reflect the real value.

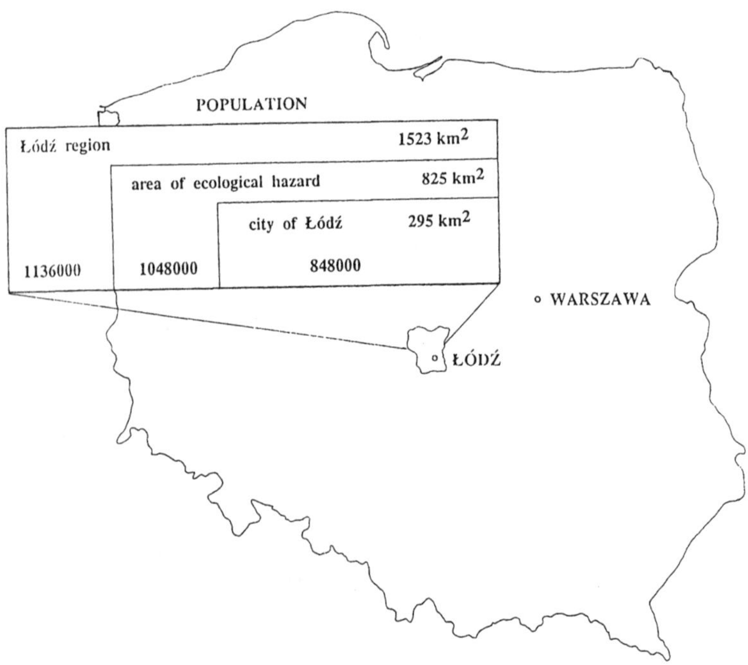


Fig. 1. The Region of Łódź

A good illustration of these obstacles is the situation in the Region of Łódź (1523 $km^2$) situated in Central Poland. 64% of its area is stated a region of ecological hazard. It is a big centre of textile, electrotechnical and chemical industry. There are several hundreds of big and small firms which greatly contribute to the disastrous conditions of the natural environment. On this small area over 520 thousand of cubic meters of wastewater are released. All this wastewater is removed without any treatment to surface water reservoirs. More than 4.5 thousand of cubic meters of solid wastes are produced daily (Stan środowiska..., 1991). Additionally, 182 tonnes of dusts and 257 tonnes of gases are emitted to the atmosphere (Monitoring środowiska..., 1991). In the capital of the Region, Łódź, with 848 thousand inhabitants and the population density 2885 people/$km^2$ these indices are even higher.

It might seem that industry of this region should be particularly interested in quick application of cleaner production principles, especially in low-investment actions which could improve the state of the environment and bring about savings/profits in a relatively short period of time. However, among over 40 participants of the annual International Conference on Environmental Protection (Urząd Wojewódzki..., 1992) organized in 1992 by the Voievodship Office in Łódź there was nobody from industry who took part in the session on cleaner technologies of the Preventive Policy Section. In March 1993 the situation improved. In the seminar entitled "Pollution Prevention Pays" held at the Technical University of Łódź, among 72 participants 36 people coming from industry represented 21 firms from the total number of 108 invited. Unfortunately, next attempts at organizing a School of Cleaner Production within the Norwegian-Polish Programme have failed. There was no firm which would like to have its employees trained in waste minimization.

In such a situation, a programme financed by the Danish government within the assistance to Poland was started with some problems in February this year. A precondition of starting the programme was that 8 to 12 industrial plants apply for taking part in it. Danish experts are to prepare individual programmes of waste reduction for these firms. Some financial support of the investment process was also planned. Despite this it was really difficult to find a sufficient number of plants which otherwise would not be charged in any way for taking part in the programme.

Probable reasons of this situation have been discussed and summarized in Table 1. Nevertheless, it should be added that industry in the Region of Łódź is in a particularly difficult situation. A plant which is not sure about its future is far away from coping with waste reduction problems. On the other hand, plants which are in a better position rather look for sponsors that would cover the costs of an "end-of-pipe" solution. What matters here is also sufficient knowledge and awareness. Industrial plants cannot be blamed when we take into account that for mass media the idea of the above mentioned seminar "Pollution Prevention Pays" appeared not interesting enough.

Table 1. Obstacles in Getting Industrial Plants Interested in Cleaner Production Practice

- poor financial condition of plants fighting for survival
- accustoming to functioning in the "command" system (attachment to "main lines")
- neglecting the relation of pollution reduction and economic efficiency
- schematic thinking – apprehension of "end-of-pipe" problems only
- believing that the only way to improve the state of environment in the plant is to allocate significant funds
- underestimation of organisational solutions
- discrediting of the idea of technical improvement (rationalization) in the period of planned national economy

Table 2. Problems in Completion of Waste Minimisation Projects in Industrial Plants

- accustoming to pretended actions (sketchy goals of the ecological policy, reverse order: first results, then analysis)
- informal relations (favouritism)
- lack of experience in determination of costs and efficiency of enterprises in the market economy
- being unaware of the fact that everything has its price
- regulations discouraging manufacturers from including environmental costs into the final product costs or even into departmental costs
- lack of understanding the necessity for involving the lower-rank personnel into the waste minimisation programme
- belief shared by the personnel that they have no possibility to influence the decisions concerning production processes
- supervising managers are not used to transfer information to the lower-rank personnel
- inadequate form and content of information
- too large gap between the supervisors and workers (the latter ones misunderstand friendly relations)
- underestimation of simple solutions ("housekeeping")

In spite of these obstacles, dozens of waste minimization projects have been developed in Poland, mainly within the Norwegian-Polish Cleaner Production Programme. For two years, World Environment Center (WEC) has been active in this field (World Environment..., 1993). Taking part in WEC's Waste Minimization Impact Project and on the basis of my previous contacts with industry, I have made some observations concerning obstacles encountered during preparation of WM projects. They are resultants of habits and pragmatism (or lack of it?) of the political system which formally came to an end but still exists in people's mentality, concepts, habits and some structures. My observations are presented in Table 2. I think that most severe are, if any, informal relations and connections in which one's own business and loyalty towards "contractors" (favourites) makes it difficult or even impossible to proceed rationally. Of special importance is also lack of understanding of the necessity for gaining all employees to the causes of the undertaken actions and their propagation, as well as original forms of transferring information to all interested parties.

The observations presented above are not backed up by systematic studies which are being prepared now. These are only the observations of a participant in the WM programme. I think, however, that they can be useful for people who deal with practical implementation of CP principles in Central and Eastern European countries, particularly when efficient use of funds coming from international institutions is concerned.

**References**

Huisingh D (1989) Cleaner technologies through process modifications, material substitutions and ecologically based ethical values. Industry and Environment, **12**, 4

Monitoring środowiska. Stan zanieczyszczenia atmosfery w Polsce w 1989 roku. (1991) (*Environmental monitoring. The state of atmospheric pollution in Poland in the year 1989*) Wyd. Instytutu Ochrony Środowiska, Warszawa

Stan środowiska w 1990 r. na terenie woj. łódzkiego. (1991) (*The state of environment in the Łódź Region in 1990*) Nakładem Urzędu Wojewódzkiego w Łodzi, Łódź

Urząd Wojewódzki w Łodzi, Międzynarodowa Konferencja Ochrony Środowiska "Environment 92", 26-27 listopada 1992 (*Voievodship Office in Łódź, International Conference on Environmental Protection "Environment 92", 26-27 November 1992*) Abstracts of papers

World Environment Center, Rezultaty demonstracyjnego programu minimalizacji odpadów u źródła w polskim przemyśle chemicznym (*Results of a demonstrative programme of waste minimization at the source in the Polish chemical industry*) Bydgoszcz, 9 December 1993

# THE PHARE ENVIRONMENTAL PROGRAMME IN POLAND

BY

DR ADAM MIERZWINSKI[1.] MRS LUCYNA DYGAS-CIOLKOWSKA[2.] & PATRICK L O'BRIEN[3.]

*1. Dr Adam Mierzwinski, Deputy Chief Inspector, State Inspectorate of Environmental Protection, Ministry of Environmental Protection, Natural Resources & Forestry, ul Wawelska 52/54, 00-922 Warszawa, POLAND.*

*2. Mrs Lucyna Dygas-Ciolkowska, Senior Specialist, Air Quality Monitoring, State Inspectorate of Environmental Protection, Ministry of Environmental Protection, Natural Resources & Forestry, ul Wawelska 52/54, Warszawa, POLAND.*

*3. Mr Patrick L O'Brien, EU PHARE Adviser, State Inspectorate of Environmental Protection, Warszawa & Project Management Group, Cork, IRELAND.*

NATO ASI Series, Partnership Sub-Series, 2. Environment – Vol. 2
Cleaner Technologies and Cleaner Products
for Sustainable Development
Edited by H. M. Freeman, Z. Puskas, and R. Olbina

# 1. INTRODUCTION

## 1.1 ENVIRONMENTAL ISSUES -

### POLAND AND CENTRAL & EASTERN EUROPE

For nearly fifty years the countries of *Central and Eastern Europe (CEE)* including *Poland* were isolated from the mainstream of European developments, and locked into a system of centralised political and economic control. These countries now seek to become an integral part of Europe's culture and civilisation. They face the enormous challenge of re-building their economies and infrastructure comparable with the changes that have occurred over the last half century in Western Europe. Moreover, these counties have inherited a legacy of *environmental problems* that will require a number of generations to rectify, and at an enormous cost.

The demand for *environmental programmes* in Poland and other CEE countries is increasing yearly. The financial resources available for environmental improvement in these countries will be limited over the next 5-10 years, and the cost of meeting some of the environmental objectives will be very high. In addition to fiscal constraints there is also a lack of institutional capacity within these countries.

*Environmental awareness* unfortunately was something that did not feature as a priority in the agenda of the former Governments of Poland and the other Central & Eastern European Countries ( CEEC's). Environmental problems are by their nature *transnational*. Concerns therefore about environmental pollution problems reflect the need to address these problems on a regional basis. Co-operation not only between Central & Eastern European Countries in resolving environmental problems, but also involving the countries of Western Europe.

Recognising the severe environmental problems, the international community has been responding by providing assistance to Poland and Central & Eastern Europe. The *April 1993 Lucerne Ministerial Conference of Environmental Ministers from East and West,* and in the spirit of the *United Nations Conference on Environment & Development (UNCED)* proposed an *Environmental Action Programme (EAP)* for Poland and the Central & Eastern European Countries.

The European Union has been responding to the environmental requirements of Poland and the other Central & East European Countries by way of the *PHARE PROGRAMME ( Poland Hungary Assistance for Restructuring the Economy )*.
At a meeting of the Council of Ministers of the European Union in December 1989, it was decided to assist Poland and Hungary with fundamental economic, environmental and cultural changes taking place in these countries. As the changes spread throughout

central and eastern Europe, the number of countries covered by PHARE increased from two in 1989 to eleven in 1993.

## 1.2 POLAND - ECONOMIC TRANSITION

Poland has a long established cultural heritage, a rich and highly diversified abundance of flora and fauna. The natural habitats in many regions of Poland have remained in an unimpacted and undisturbed state. However, there are regions in Poland where the environment has, as a result of anthropogenic activities, been severely impacted. This has been attributed to extensive industrialisation and mismanagement of natural resources, throughout the decades of the imposition of an inept communist system with a centrally managed economy, to the detriment of Poland.

The low prices for land and cheap labour and energy overheads, encouraged the establishment of large industrial complexes close to the raw material sources, and in regions with old industrial traditions. Moreover, no thought was given to proper planning and the consequences of the environmental impact of large scale industrial development.

As a result there has been a dramatic deterioration in the living conditions of local populations, along with devastation of the natural environment and artifacts of cultural heritage. Presently at least 30% of the Polish population live in areas where national pollution standards in particular for air are well above the compliance levels. Forests are severely threatened and only about 45% of watercourses carry clean water.

In economic terms, the losses caused by the emission of pollutants to the environment are estimated at several percent of the national income. There has been a decrease in the average lifespan, and increases in infant mortality and diseases, associated with living conditions in industrialised society. This is the price that Polish society has had to pay for years of an inefficient forced economy, dominated by ideological imperatives.

*1989* was the beginning of a new era for Poland, as it started on its way towards a new and economic system, the first country of the former Eastern Bloc to do so. However, the *transition* from a *centrally planned economy* towards a mature *free market western style economy* will be fraught with difficulties requiring significant financial efforts and unfortunately social sacrifices. Poland's immediate task is to establish a target economy model for a democratic Poland.

Moreover, the lack of a transparent strategy in developments presently ongoing could give rise to the much criticised model of the over-consuming society which places additional demands on energy and non-renewable natural resources. The alternative to such an economy is the model of *Sustainable Development* based on maximising efficiency in energy and raw materials consumption and minimising impact on the natural environment. The Sustainable Concept is the basis of the *European Union's Fifth Action*

*Programme on the Environment.*

In the Polish context the sustainable development model is one which can assure Poland of economic development along sound environmental guidelines and that can satisfy legitimate societal needs.

The sustainable development model for Poland was presented in a government paper entitled *National Environmental Policy of Poland* which was approved by the National Parliament in May 1991. It is now reflected in the principles of state economic and social policy for the future, the main features of which are;

* far reaching restructuring of industry guided by the need to satisfy stringent environmental standards for all new and upgraded production technologies;

* *national programme for the improvement of energy efficiency and energy conservation;*

* *rationalisation of transport and mass transit systems in urban areas on a regional and national scale;*

* *restructuring of the agricultural sector to encourage the production of healthy food;*

* *implementation of water conservation and water management programmes;*

* *implementation of programmes for recycling or secondary use of municipal and industrial waste;*

* *overhaul of the current environmental legislative system including environmental enforcement;*

* *designation of protected habitats and establishment of biosphere reserves.*

Financial resources will be the limiting factor towards achieving these goals. The Polish government is committing significant financial resources for environmental protection, but it will also require substantial financial support from the International and European Union, although to date this support only represents *4%-6%* of the domestic expenditure on the environment. The *PHARE Environmental Sector Programme Poland* is supporting Poland in its endeavours to meet its environmental objectives.

## 2. THE PHARE PROGRAMME

### 2.1 WHAT IS PHARE ?

PHARE is the European Union's economic aid programme to support economic restructuring and democratic reform in Central and Eastern Europe. The funding is used to channel technical, economic and infrastructural expertise and assistance to recipient states. The aim of PHARE is to assist these countries achieve market economies based on free enterprise and private initiative.

The term *PHARE - Poland and Hungary Assistance for the Restructuring of the Economy* - initially described international efforts by the 24 industrialised countries (Group of 24) to provide economic support to the emerging Polish and Hungarian democracies, now known as G-24 assistance, co-ordinated by the European Union.

In *December 1989, PHARE* was launched as a specific EU Programme to dovetail with the *G-24* aid, and also works closely with the *World Bank, European Investment Bank and the European Bank of Reconstruction and Development.*

The countries eligible for PHARE funding are *Poland, Hungary, Bulgaria, Czech Republic, Slovakia, Romania, Albania, Latvia, Lithuania, Estonia, and Slovenia.*

Following German unification, PHARE assistance is no longer provided to the former German Democratic Republic. PHARE aid to Yugoslavia has been suspended because of the current military crisis there.

## 2.2 PRINCIPLES GOVERNING PHARE

PHARE provides grants rather than loans to finance the recipient states own efforts to transform their economy. Assistance is "demand driven" as the recipient state decides how the funding is to be used for its own restructuring priorities. These are discussed with the European Commission and the recipient state manages the resulting programmes.

PHARE only funds government reform programmes in the PHARE recipient states. This means *Technical Assistance* in the form of :

* *Preliminary Studies*
* *Action Plans*
* *Policy Advice*
* *Technical Expertise*
* *Training*

*Supply programmes* may provide equipment.

The technical assistance is directed to such sectors as *industry, agriculture, energy, environment.*

*PHARE does not fund private business ventures.* Contracts financed by PHARE are awarded under public procurement procedures by the competent authorities of the recipient countries.

PHARE programmes usually run for a period of two to three years. About 75% of the PHARE funds are allocated to national programmes, problems such as cross border pollution control are resolved using a collaborative approach. Up to 15% of funds are therefore allocated to regional programmes involving several recipient states at a time.

## 2.3 PHARE ASSISTANCE BY SECTOR

In 1990 sector programmes for each PHARE recipient country were influenced by time constraints to deal with *priority issues* by providing *equipment* and *technical assistance* to the sectors eg pollution monitoring equipment & technical assistance to the environmental sector. The emphasis of the sector programmes has been on the strengthening and development of the technical and institutional,infrastructures which are essential for a fully functioning market economy.

## 2.4 HOW DOES PHARE OPERATE ?

The partner or recipient countries reform objectives are identified. Details of individual sectoral programmes with implementation strategies and budget information are worked out within the framework set by the Indicative Programme. Specific measures and projects are identified, analysed and appraised resulting in a Financing Proposal, submitted for approval by the recipient state and the Commission. Formal approval is by way of a signed Financial Memorandum endorsed by recipient state and the Commission.

The recipient states are mainly responsible for establishing Project Management Units to implement the programmes under the supervision of the Commission.

The PHARE budget distribution, funding and budget by sector are given in the text.

# 3.0 PHARE POLAND - ENVIRONMENTAL SECTOR PROGRAMME

## 3.1 OBJECTIVES

The Government of Poland has become increasingly aware of the need to protect the environment, and serious efforts are now being made to improve environmental management. The National Environmental Policy based on:

* *the prevention of pollution at source;*

* *the application of the polluter pays principle;*

* *the introduction of consistent compliance standards & enforcement procedures;*

* *improved environmental monitoring systems;*

* *regionalised environmental management.*

The *Polish Government's strategy* in line with the *National Environmental Policy*, is based on the broader concepts of sustainable development of which the *integration of environmental considerations* in all sectors are an essential feature, rather than on the *narrow pollution control concept*.

*The PHARE Environmental Sector Programme* is endeavouring to resolve institutional constraints such as environmental legislation, environmental management, enforcement & monitoring, and also to improve technical qualifications through the environmental institutions at national and local level.

The PHARE funded environmental programmes have been identified by the *State Inspectorate of Environmental Protection - the Polish National Environmental Regulatory Authority*, and the *Ministry of Environmental Protection, Natural Resources and Forestry*, in agreement with the *Commission of the European Union*.

## 3.2 FUNDING

The present expenditure on environmental protection in Poland under PHARE amounts to seventy five million ECU's. The Environmental Sector Programme comprises sector and sub-sector components which are linked directly to the main immediate priorities of the National Environmental Policy.

The following activities and funding are currently ongoing under PHARE.

*PHARE PHASES I, II, & III 1990, 1991 & 1992 PROGRAMMES*

*PHARE PHASE I 1990*
*Urgent areas requiring immediate attention eg air, water, waste management and nature conservation (22 MECU).*

*PHARE PHASE II COMPRISING PHASE II.1 & II.2*

*PHARE PHASE II.1 1991*
*Integrated programme aimed at decentralisation of effort in environmental protection. Covers seven sectors including Upper Silesia (30 MECU).*

*PHARE PHASE II.2 1991*
*Strengthening of the State Environmental Monitoring System (5 MECU).*

*PHARE PHASE III*

*Assisting with seed capital for investment in environmental protection.*

The implementing authority for the PHARE Environmental Sector Programme is the *Ministry of Environmental Protection, Natural Resources and Forestry* through the *Environmental Coordination & Management Unit - ECMU*.

## 3.3 ENVIRONMENTAL SECTOR PROJECTS

The following is a overview of the PHARE funded environmental sector projects presently ongoing and ranging from institutional strengthening to projects covering air, water and waste.

## A. INSTITUTIONAL STRENGTHENING OF ENVIRONMENTAL MANAGEMENT

### *A.1. Assessment of Proposed and Existing System of Environmental Management*

*A.1.1 Publication in Polish of the EU Environmental Directives*

*A.1.2 Co-financing of the Interministerial Coordinating Unit for the resolution of environmental problems in privatisation*

*A.1.3 Enhancement of the Institutional Structure of Environmental Protection*

### *A.2 Economic and Legal Instruments in Environmental Protection Policy*

*A.2.1 Inventory and Assessments of the Efficiency of the Economic Instruments Applied in Poland and OECD Countries*

*A.2.2 Programme of Tradeable Emission Permits*

*A.2.3 Programme of Compliance Schedules*

*A.2.4 Harmonisation of Polish Environmental Laws with EU Environmental Legislation*

*A.2.5 Design of Draft Legal Provisions on Environmental Impact Assessment Harmonised with EU Legislation*

### *A.3 Definition of Priorities and Costs for Implementation of Medium Term Priorities*

*A.3.1 Development of Cost-Methodologies and Evaluation of Cost-Effective Strategies for Achieving Harmonisation with EU Environmental Standards*

*A.3.2 Designing Liabilities for Past and Present Activities*

## B REGIONAL ENVIRONMENTAL PROGRAMME FOR UPPER SILESIA

### *B.1 Master Plan for Municipal Waste for Upper Silesia*

*B.1.1 Master Plan for Municipal Waste for Katowice*

*B.1.2 Preparation and Implementation of a Pilot Project in Katowice for Municipal Waste Utilisation*

*B.1.3 Training and Equipment Requirements*

### *B.2 Master Plan for Low Level Air Pollution Emission Abatement*

*B.2.1 Study and Design: District Heating Systems and Introduction of Gas as an Energy Source - Equipment/Training & Pilot Project*

**B.3 Optimisation of Water and Waste Water Management**

*B.3.1 Master Plan for Water and Waste Water in Upper Silesia*

*B.3.2 Pilot Project for Optimisation of Water And Waste Water Management in Zawiercie Upper Silesia*

*B.3.3 Supply of Equipment for Elimination of Excessive Water Use*

*B.3.4 Strengthening of Institutional Management and Training operational Staff in Water and Waste Water Treatment based on EU Experience*

**B.4 Study of the Elimination of Food Production in Areas Polluted with Toxic Substances**

*B.4.1 General Study and Pilot Project using GIS on the Elimination of Food Production in areas of Katowice polluted with Toxic Substances*

**C NATIONAL WASTE MANAGEMENT PROGRAMME**

***C.1 HAZARDOUS WASTE***

*C.1.1 Master Plan for Hazardous Waste Management and Site Remediation*

*C.1.2 National Agency for Waste Management - Technical Organisation, Assistance and Training*

**D NATIONAL SYSTEM FOR THE MITIGATION OF ACCIDENTAL THREATS**

***D.1 Master Plan Development for the National System of Accidental Threats***

***D.2 Computer Hardware, Software and Training of Emergency Response Personnel***

**E ENERGY**

***E.1 Alternative Energy Thermal Heating station & Network - Pyrzyce***

**F FRESHWATER SYSTEMS MANAGEMENT**

***F.1 Environmental Protection of the Great Mazurian Lakes Region***

*F.1.1 Strengthening the Existing Natural Resources Data Base and Environmental Management Capabilities with Geographic Information Systems (GIS)*

*F.1.2 Feasibility Study on Fisheries Development*

*F.1.3 Feasibility Study on Pig Slurry Management*

***F.2 Environmental Protection of the Warta River Basin***

***F.2.1 Czestochowa Central WWTP***

***F.2.2 Slesin WWTP***

***F.2.3 Gniezno WWTP***

***F.2.4 Sierakow WWTP***

***F.2.5 Zlotow WWTP***

***F.2.6 Supply of Computer Hardware & Software for the Regional Data Base for the Regional Board of Water Management in Poznan and 18 Regional(Voivodship) Divisions in the Warta River Basin***

**G STRENGTHENING OF THE STATE ENVIRONMENTAL MONITORING SYSTEM**

***G.1 Master Plan for Strengthening of the State Environmental Monitoring System (SEM)***

*G.1.1 Implementation of Key Activities Emerging from the Master Plan*

*G.1.2 Training on Quality Assurance and Quality Control Systems for the Polish State Environmental Inspectorate*

***G.2 Support for the Development of an Environmental Management System - EMS***

*G.2.1 Master Plan and Pilot Project for the Strengthening of the EMS*

*G.2.2 Implementation of Key Activities Emerging from the EMS Study*

*G.2.3 Computer Hardware and Software Equipment for EMS*

***G.3 SUPPLY OF EQUIPMENT TO KEY FACILITIES - LABORATORIES AND ENVIRONMENTAL MONITORING STATIONS***

*G.3.1 Automatic Surface Water Quality Monitoring Equipment*

*G.3.2 Ambient Air Quality Monitoring Equipment*

*G.3.3 Equipment for Operational & Organisational Infrastructure in the State Inspectorate of Environmental Protection*

*G.3.4 Equipment for Upgrading Air, Water and Waste Monitoring*

**H ESTABLISHMENT OF A NATIONAL ENVIRONMENTAL EDUCATION CENTRE**

***H.1 Establishment of National and Regional Education Centres***

*H.1.1 Programme Development and Equipment Requirements*

All the above PHARE funded projects have a Technical Assistance element associated with them. Technical experts from the European Union provide the technical assistance input, working alongside Polish experts, in order to ensure the smooth running and management of the various projects.

## 4. CONCLUSIONS

Since its inception the *PHARE Environmental Programme* has made a significant impact on the amelioration of Poland's environmental problems. This impact has resulted in the *improvement of infrastructure facilities* for water and waste water treatment, atmospheric emissions monitoring ( including mobile air quality monitoring vehicles), soil and groundwater contamination.

Moreover, the *Polish Government* is playing an active role in effecting protection of the environment, by *promoting public awareness* and perception of indigenous environmental issues. Environmental information available in the public domain will increase peoples awareness to national and international environmental issues.

Despite the range of obstacles to be overcome, in what are very much the initial stages of economic transformation, the response of Poland to PHARE support for the development of environmental capabilities and policies, has been, and continues to be, undeniably positive.

*Poland* is aspiring towards *full membership* of the *European Union* in the fullness of time. Now that *Poland* has acquired *Associate Membership* is indeed an important achievement for Poland in realising its desires to become part of a greater Europe.

**REFERENCES**

European Commission Phare Information Office Directorate General for External Economic Relations, DG1, May 1994.

# The Evaluation of Textile Aids and Chemicals Used by The Textile Industry in order to Reduce Environmental Impact

Jerzy Rouba
Environmental Protection Department
Textile Research Centre
ul. Piotrkowska 270
90-361 Łódź
Poland

## 1. Introduction

The textile industry processes natural and synthetic fibers into fabrics for domestic and industrial use. The majority of textile mills employs full processing of textile goods being a combination of pretreatment, bleaching, dyeing, printing and functional finishing. As a result of these so called "wet" processes textile effluents carry away the impurities and pollutants from the fibers as well as used up the textile chemicals into water courses, directly or indirectly. Therefore, the textile industry, to be exact, the textile wet processing sector is seen by water authorities at local and national level, environmental organizations and public opinion as a user of significant amounts of good quality water and a heavy polluter of water courses. The textile effluents usually have a bad effects on aquatic systems leading in turn to penalties, charges for disposal and severe restrictions upon the textile processors. Additionally, the textile effluents, with regard to their specific constituents, make a lot of troubles for sewage treatment plant operators.

NATO ASI Series, Partnership Sub-Series, 2. Environment – Vol. 2
Cleaner Technologies and Cleaner Products
for Sustainable Development
Edited by H. M. Freeman, Z. Puskas, and R. Olbina

## 2. The textile industry - wet processing sector as a large processor and polluter of water courses

To assess properly the impact of the textile industry on the environment we must know very specific circumstances under which it is working. The textile industry in its business activity depends upon many factors. It depends on:

- deliveries of huge amounts of the textile raw materials (natural and synthetic fibers) from all over the world,
- supply of a great number of the textile chemicals (dyestuffs, chemicals, auxiliaries),
- supply of large quantities of water and energy,
- meeting very strict quality standards respective to the requirements of the customers,
- very changeable trends in fashion which affect dyeing, finishing and quality of goods,
- increasingly tighter environmental regulations.

All these factors are reflecting in the volume and composition of the textile wastewater. The total volume of wastewater discharged from the textile mill comes mainly from wet processing of goods and cooling systems. The volume of textile effluent in one textile mill is almost impossible to compare to the other due to the differences in processing equipment, sort of processed fibers or blends, requirements of the handled orders, amounts of cooling water and in fact is very changeable. The same refers to the composition of the textile wastewater which depends upon the sort of the textile raw materials, the textile chemicals and typical processes used in the particular textile mill. A great deal of pollutants and impurities both organic as well as inorganic may be present in various combinations and varying concentrations in the textile wastewater.

Classification of the textile chemicals, dyestuffs and auxiliaries occurring in textile effluent with regard to their difficulty of treatment and pollution category (1-5) (1 - the least and 5 - the most harmful) is shown in tab. 1 [Cooper P. - 1992]. One can see from this table how many different and very toxic chemicals the textile industry is employing in the course of the treatment of goods. These substances introduced almost entirely into the textile effluents exhibit high pollution potential dangerous to the aquatic environment. To reduce their environmental impact is becoming, nowadays of greatest importance for textile industry.

## 3. The pollution load of the textile effluent

Almost all the pollution load in the textile industry comes from the wet processing sector but the quantities vary considerably depending upon the process. In the course of wet processing the textile goods we can distinguish several stages which are the basic sources of pollution the textile effluent.

First basic source of water contamination is scouring. This process removes dirt, processing oils, metal salts and undesirable components of natural and synthetic fibers. The chemicals used include caustic soda, soda ash, acetic acid, phosphates, chelating agents, surfactants. The textile raw materials are usually imported from very distant countries of its origin and there are also stored for a long time before they will be processed into fabrics. It has been proved that natural fibers as cotton, wool, jute, before reaching the textile mills are already seriously contaminated. It results from the fact that cotton plants producers use up to 15 herbicides during the period of growth, similarly the jute producers, while the wool contains several insecticides.

On the other hand the synthetic fibers are also contaminated though less because contain the rest of monomers, catalysts used during the production. Furthermore, during the transport and storing some chemicals are applied to preserve the quality of textile matter. The growth of mildew, mold, fungus or rot is inhibited by application of biocides. The used compounds contain chlorinated phenols like pentachlorophenol - PCP, DDE, DDD, DDT,

PCB, Heptachlor and metallic salts (zinc, copper, mercury). The tributyltin compounds identified in textile effluents have been widely used in antifungicidal treatment and are the basic component of pesticides [Laing I.G. - 1991, Hazel B.G. - 1991].

The textile chemicals used during the scouring contribute to the wastewater a great load of organic and inorganic matter adding some impurities associated with these compounds, for instance: caustic soda introduces mercury, zinc compounds cadmium etc. All these pollutants and impurities are washed off and enter the aquatic environment by direct discharges from textile mill. According to EC Directive 76/464/EEC most of these pollutants and impurities is put on the Black List (I) of chemicals which are prohibited in the countries of European Union as dangerous for the environment. Unfortunately, environmental regulations concerned with the use these chemicals are not so tight and strict in the countries supplying raw materials as are in the countries processing fibers, mainly of European Union.

The second basic source of the textile water contamination is desizing operation used in the course of cotton processing. In the result of desizing large quantities of size, mainly starch with some chemicals are washed off into water stream increasing largely total load of dissolved organic matter. It is worth to keep in mind that in the textile effluent more than half the organic matter comes from the scouring and desizing operations [Park J., Shore I. - 1984].

The third basic source of the textile water pollution is the rest of wet finishing processes like: bleaching, dyeing, printing, functional finishing. Although, this part of pollution load is not very large (load of BOD is not very high) but due to the presence of dyes, heavy metals, dyebath additives such as carriers and some constituents of special finishing (mothproofers, rotproofers, flame retardants) which are known for their toxicity, mutagenity or persistency this source of pollution exhibits very serious danger for aquatic environment.

Another sources of pollution in the textile effluents are often the water treatment plant, cooling towers and boiler plant if they are located at the premises of the textile mill. Presence or lack such facilities may considerably affect the pollution load of textile wastewater mainly due to the high load of electrolytes discharged into the wastewater stream. The same refers to the blow down from cooling towers and boilers which contain biocides and corrosion inhibitors - phosphates.

## 4. The specific constituents of textile effluents

Analyzing the total pollution load in the textile wastewater a great attention must be paid to some very specific constituents of this pollution which affect on the aquatic environment very seriously. Of particular importance in the textile effluents are:
- colour,
- heavy metals,
- chlorinated organic compounds,
- surfactants,
- phosphates.

The colour in the textile wastewater originates from the unfixed dyes and is easily detectable by ordinary people in the wastestreams as well as in the watercourses. At present, colour discharges are becoming increasingly unacceptable by sewage plants operators, water authorities and public opinion.

The heavy metals may enter the textile effluent from the fibers, yarns or fabrics, common textile chemicals used in large quantities, several classes of dyes, dye oxidants (dichromate, copper), sizing preservatives (zinc), latex (zinc) catalysts for resin treatment, acid fulling (zinc), the processing water and the corrosion of the machines and pipelines.

The chlorinated organic compounds are introduced into the textile wastewater as pesticides with natural fibers and textile goods in the course of processing, as carriers during the dyeing and the finishing compounds during special treatment of textile goods as well as biocides in blow down from boilers, cooling towers etc.

The surfactants are broadly used in many textile processes. Most of the textile auxiliaries belongs to the surfactants which exhibit anionic or nonionic character.

The phosphates are recognized as a significant pollutant responsible for the unwanted growth of algae in surface watercourses similarly as nitrates (toxic effects of nitrogen compounds). The phosphates are broadly used in manufacturing of surfactants as builders.

A variety of organic and inorganic chemicals as well as auxiliaries employed for purposes of wetting, levelling, retarding, buffering, carrying, softening, sequestering, defoaming etc. in the course of textile processing form very complex mixture in the bulk of the wastewater. It results in many undesirable side-reactions leading to still unknown intermediates and final compounds being even more toxic for environment than the substrates. Formation of organohalogens (AOX) in the bleaching effluent is a very good example for it. Due to the application of sodium hypochlorite for bleaching textile goods it has been observed occurring in the textile effluent very toxic substances, chlorinated organic compounds so called organohalogens (AOX).

Another example is formation of heavy metal complexes with auxiliary chemicals which can act as complexing agents. Ligands such as acetates, ammonia salts, pyrophosphates, perborates make the heavy metals less susceptible to chemical or biological treatment. It makes possible the heavy metal complexes pass unchanged through the wastewater treatment system and release the heavy metals in aquatic environment
[Norman P.I., Seddon R. - 1991].

# 5. The in-plant measures taken to reduce pollution load at the textile mill

## 5.1. The minimization of water use

To reduce pollution load at the textile wastewater first of all should be decreased water consumption at textile mill. As one knows, the use of chemicals depends upon liquor ratio and it is also well known that % unfixed dyes increases by increasing liquor ratio. Therefore the basic measures taken at the textile mill level include - water use reduction (saving water),
- water reuse.

The first step in saving water at the textile mill is to measure the amounts actually used, to know exact volumes of water used to various processes. Water use reduction is usually carried out by elimination of unnecessary water consumption through conventional processing at lower liquor ratio, improved dyeing techniques which require less water, avoiding excessive amounts of rinsing water, by preferring continuous processes as well as good housekeeping practices.

Water reuse is done by using the same water in more than one process. The possibilities exist for combining 2-3 separate processes into one bath, for reuse of uncontaminated cooling water in processes requiring hot water, reuse of rinsing waters, reuse of wash water from bleaching in scouring, reuse of process liquors, reuse of scouring rinses for desizing, etc. But it must be mentioned that specific technical and economic factors affect the application of most of these in-plant measures at a given mill. Therefore, each implementation must be considered under its own mill specific requirements. Additionally, all these measures are closely linked with flow monitoring equipment, storage facilities and proper rearrangement in water piping systems.

## 5.2. Minimization of the textile chemicals use

Further efforts directed towards the reduction of pollution load must take into consideration the textile chemicals. Considering the pollution load of the textile effluent we must remember that the part of this load coming from textile raw materials and chemicals (impurities) cannot be altered very much at the textile mill level. Any decrease of the pollution load must therefore come from the changes in the dyestuffs, chemicals and auxiliaries employed at particular textile mill. Of great concern are the textile chemicals and auxiliaries. It results from the fact that while the dyes are nearly 90% exhausted onto the fabrics then wetting agents (auxiliaries) for example are about 100% rinsed out to the discharge. After closely examining the textile chemicals consumption at the works one can find that the dosages of chemicals and auxiliaries used in wet processing are in many cases excessive for the requirements. It is very important from the environmental aspect to keep to a minimum the quantities of the textile chemicals having high pollution potential or toxic properties, particularly those which have to be employed in considerable amounts. The textile mills laboratories should perform regularly tests to determine minimally effective dosages of textile chemicals. To study the list of chemicals and auxiliaries used at the textile mill to consider whether any of them can be reduced in quantity should be permanent duty of finishing staff.

The reduction to a necessary minimum dyestuffs, chemicals and auxiliaries should be done by very precise dispensing of chemical substances, mainly in the liquid state, with use of automation and computer controls of dosages based on measurements of conductivity, redox potential, pH or colour.

The decrease of chemicals use can be also obtained through simultaneous enhancement of physical agents action i.e. elevated temperature, pressure, electromagnetic fields, ultrasonics, infrared, low temperature plasma. All these physical agents can activate the course of reactions between the chemicals and fibre resulting in the decreasing of dosed chemicals. Particularly interesting thing in this matter is application of electromagnetic fields of high frequency to activate such reactions. Their action causes activation of dipolar moments of chemicals, water and fibers what enables the diffusion processes on the surface and the inside of the fibers [Brzeziński S. - 1991].

The minimization of use of the textile chemicals, their substitution or elimination at last are generally regarded as the most effective in-plant measures to reduce pollution load of textile wastewater.

## 5.3. The substitution and elimination of textile chemicals

The standard method of looking at pollution problems must be adopted by all the textile finishers. The basic approach in this field is either to substitute (replace) the pollutant with an alternative less- or non polluting compound or eliminate it at last. Studying the list of chemicals and auxiliaries used at the textile mill in order to substitute them with substances introducing less impurities or more biodegradable and even eliminate should be also the duty of finishers of every textile mill and kept in view all times. But also should be kept in mind that when dangerous substance is replaced by less- or non-dangerous alternative, the latter must be no less effective in the textile process in which is used and no greater quantity should be required.

Usually, the cost to substitute other chemical substances for those containing toxic pollutants is much less than the cost to remove the pollutants from a mill's discharge. The substitution must cover also such substances which though itself undergoes easy biodegradation but their metabolites exhibit toxic properties to the environment. Such example are alkylphenol ethoxylates - surfactants used in the textile processing which metabolites (phenols) are very toxic to fish.

They are substituted by linear ethoxylated fatty alcohols. Another very specific example of substitution is replacement EDTA, DTPA sequestrants by NTP chelating agent forming less stable complexes with heavy metals in order to easier split off heavy metal in the course of textile effluent treatment.

Some examples of the substitution of the textile chemicals are given in tab. 2. It seems that there is a considerable scope for environmental thinking for finishers in dealing with effluent problems. One must also consider if the pesticides, mothproofing agents etc., are really essential, necessary and worth all the problems they cause? If any of the finishing processes at the textile mill cause serious environmental problems can this process be modified, carried out at a more appropriate location or eliminated totally, perhaps. The answer for such questions belongs to the finishing staff. It is of prime importance for the finishers at the textile mill to know from which process a particular form of pollution is arising and how affect the environment. At present, nobody from the finishing staff knows which of the process chemicals employed has the highest COD or BOD, what is the concentration of surfactants in spent process liquors, what is the concentration of auxiliaries in the spent dyebaths etc. Up to now the finishers did not need to acquire such knowledge. But now it seems to be very important and necessary for them to join finishing skills with environmental awareness.

There is much work to do in the field of textile engineering to change processes and chemicals. Very promising area of concern for finishers in order to reduce environmental impact of textile industry is replacement of chemical technologies applying aggressive chemicals by biotechnologies with use of various enzymes. The scouring of textile goods has a very bad effect on environment, therefore enzymatic treatment of cotton fabrics carried out at pH ~7 and temp. 40oC make possible the course of the process without any chemicals and in milder conditions. Similarly, technology of enzymatic hydrolysis can replace process of wool carbonization reducing significantly the impact of textile effluents on the environment.

The second process to be used to treatment of textiles with high potential of application seems to be low temperature plasma treatment. Chlorination of wool can be substituted by plasma treatment eliminating the use of water and chemicals and making the textile processes more friendly for environment [Brzeziński S. - 1994].

Another area of search to reduce the environmental impact are works on new preparatory treatment of cotton fibers so called cationisation to increase the reactive dyes fixation, form more stable links with fibers and also use other classes of dyes as well as developing new brands of reactive dyes applicable at very low of salts concentrations. [Brzeziński S. - 1994]

Furthermore, the existing textile wet processes must be regularly reviewed in accordance with pollution prevention principles. The best opportunity for such activities present regular environment auditing at the textile mills.

## 5.4. The substitution of auxiliaries - case study in Poland

In the search of alternative solution to reduce pollution load in the textile effluent, at the Textile Research Centre in Łódź (Poland) has been done the study. The purpose of this study was examination the auxiliaries most commonly used in Poland by textile industry in order to know their pollution potential with regard to dissolved organic matter and toxicity and then choose the best auxiliaries in each category of application. The study dealt with 50 auxiliaries used by the textile industry, mainly abroad origin.

Environmental evaluation of these auxiliaries included following parameters:

- load of dissolved organic matter introduced to the wastestream by unit weight of the product expressed as

BOD5, COD,
- biodegradability (%),
- lethal concentration CL50 (mg/dm3),
- load of surface active agents (anionic or nonionic) (mg/dm3),
- load of oils, fats (ether extract) (mg/dm3).

All these data were collected and worked out in the form of catalogue which was thought as the aid to the textile finishers at mill in the selection of auxiliaries with possibly the least load of organic matter and toxicity for aquatic systems. Following the study recommendations the finishers at several textile mills could through careful selection of auxiliaries to reduce the total load of dissolved organic matter expressed as BOD5 in the effluent 20-25% on average. It was relatively simple and very effective action decreasing environmental impact of the industry "at source" as a matter of fact without any investment costs.

Unfortunately, this approach to pollution problems in Polish textile industry as yet is still not common practice. It seems, as long as the strict execution of environmental regulations in relation to industry will be poor as long the implementation of cleaner production principles will be not sufficient as well.

## 6. The recovery and recycling some auxiliaries and chemicals

The next step to reduce environmental impact of the textile aids and chemicals used by the textile wet processing is their recovery and recycling. The basic aim to implement such practices is to reduce processing costs while the reduction of pollution load seems to be a secondary benefit.

One area that has already received a great deal of attention in this matter and is applicable for large scale is desizing. The effluent at this stage is highly concentrated, creates the greatest BOD5 and might be more easily treated in this form rather than by just adding it to general effluent to be diluted.

Therefore, specialized treatment at the machine point of source could be possibly the best solution. The substitution of starch by other recoverable and recyclable synthetic sizes as PVA, CMC, Acrylates makes possible to reduce environmental impact of the textile chemicals on a very large scale. The next typical application of the recovery and recycling are concerned with dyeing operations - some classes of dyes (indigo) and preparatory operations - some surfactants. It is very important to point out again that in every case such recovery and recycling installations should be located as closely as possible the source of the pollution. That means, having determined the origin of the pollutant it is easier to treat in its most concentrated form - the recovery and recycling are the most efficient. Therefore, the segregation of waste streams containing different materials is in the service of easier recovery and recycling.

When we consider the recovery and recycling of valuable components of textile wastestreams we should also take into consideration that we get fairly clean water which also is recyclable. Both in the first and in the second case there is a need for storing the recovered constituent as well as the water and maintaining service piping system entirely separate from the original process water supply. Distribution lines must be kept separate and when the amounts of recycled water increase this become more and more difficult and expensive. It would be most profitable and convenient to foresee similar solutions and include such systems at the stage of plant designing to have both the water supply and drainage systems easily accessible and capable of being changed readily to allow for many variations in their use.

## 7. The textile effluents treatment

Nowadays, in the light of new environmental regulations is becoming more and more probable that municipal sewage works in the care of its own treatment system may refuse to deal with textile effluent which exhibit excessive colour, content of heavy metals or toxic matter. Water authorities are obliged to accept domestic sewage but as for discharging industrial effluent the matter is becoming more troublesome and difficult for textile finishing industry.

One can expect that the all textile wet processors in the nearest future will be forced to treat their effluents on site according to the level required by local water authorities. In the case of very severe requirements the final effluents would be much less contaminated than the current effluents. In the consequence it could lead to the almost totally closed water system at the textile mill. Only in such situation environmental impact of the textile mill could be reduced as far as its is possible. But we must kept in mind that only when the main components of pollution have been substituted, removed and recycled will the effluent be merged for treatment. If the treatment is to be effective must be applied to most concentrated wastestreams "at source" what is linked with separate sewerage system in the works. It can be expected that the wastewater treatment plant will be distributed within the works according to the main sources at pollution and the pollution control equipment will consist of small pieces of plant at selected, various stages throughout the mill [Hazel B.G.- 1991].

## 8. Concluding remarks and recommendations

To summarize the considerations on the reduction of environmental impact caused by the chemicals and textile aids (auxiliaries) in the textile effluents it would be suggested to take following steps in the proper sequence as effective in-plant measures:

1) Operate with more and more clean the textile raw materials and the textile chemicals.
2) Decrease the total volume of water at the textile mill.
3) Minimize the use of the textile chemicals.

4) Substitute the textile chemicals by more sound for environment alternatives.
5) Recover and recycle some valuable constitutens of textile effluent like sizes, dyes, auxiliaries as well as some very toxic pollutants ("at source").
6) Treat an unavoidable textile effluent in order to make closed water system at the works.

It is important to stress that all these steps should be not only "one day" action but permanent process inherent in business activity of the textile mill.

Apart from effective in-plant measures also more general actions should be undertaken in this matter. General measures taken in order to reduce environmental impact of the textile aids and chemicals ought to include following issues requiring multiple oriented and systematic approach by the whole textile industry.

1) The textile wet processors should be provided with ample and reliable information on impurities and dangerous substances associated with the textile raw materials. There is an urgent need of close cooperation between the textile wet processors and the suppliers of textile raw materials in this matter. The application of biocides and preservatives should be limited to a necessary minimum or replaced by other chemicals sound for environment.

2) The manufacturers of the textile chemicals, usually large, multinational chemical companies should be obliged to provide the users of their products with full information about quality and possible effects on human and environment taking into account short and long term risk.

3) The improvement of quality of textile raw materials, the textile chemicals, the textile goods in relation to their effects on human and environment should be under permanent control of the textile industry resulting in better pollution prevention. That means the research and industry will have to continue their efforts to develop alternative and environmentally safe methods for finishing. It will require partnership in research by the textile chemicals manufacturers, machinery manufacturers, researchers and wet processing industry.

4) It would be desired to establish pollution prevention centre promoting sound wet processing technologies and products for needs of the textile industry to reduce its environmental impact.

5) To implement successfully pollution prevention principles is necessary to educate and train the finishing staff at all levels to enhance their environmental awareness.

## 9. References:

Brzeziński, S. (1994) Ecological and certification problems of textile goods. Proceedings of International Conference Łódź-Sulejów (in Polish)

Cooper, P. (1992) The consequences of new environmental legislation on the UK textile industry. Text. Horizons International 10:31-38

Hazel, B.G. (1991) Current problems and future restrictions in dyehouse operations. JSDC 11:392-394

Laing, I.G. (1991) The impact of effluent regulations on the dyeing industry 21:56-71

Norman, P.I., Seddon, R. (1991) Pollution control in the textile industry - the chemical auxiliary manufacturer's role. JSDC 4:150-152

Park, J., Shore, J. (1984) Water for the dyehouse: supply, consumption, recovery and disposal. JSDC 12:383-397

Table 1. Classification of the textile chemicals, dyestuffs, auxiliaries with regard to their difficulty of treatment and pollution category [Cooper P. - 1992]

| *General Chemical Type* | *Difficulty of Treatment* | *Pollution Category* |
|---|---|---|
| Alkalis<br>Mineral acids<br>Natural salts<br>Oxidizing agents | Relatively harmless<br>Inorganic pollutants | 1 |
| Starch sizes<br>Vegetable oils, fats & waxes<br>Biodegradable surfactants<br>Organic acids<br>Reducing agents | Readily biodegradable | 2 |
| Dyes & fluorescent brighteners<br>Fibers & polymeric impurities<br>Polyacrylate sizes<br>Synthetic polymer finishes<br>Silicones | Dyes and polymers difficult to biodegrade | 3 |
| Wool grease<br>PVA sizes<br>Starch ethers and esters<br>Mineral oil<br>Surfactants resistant to biodegradation<br>Anionic and non-ionic softeners | Difficult to biodegrade<br>Moderate BOD | 4 |
| Formaldehyde and N-methylol reactants<br>Chlorinated solvents & carriers<br>Cationic retarders & softeners<br>Biocides<br>Sequestering agents<br>Heavy metal salts | Unsuitable for conventional biological treatment<br>Negligible BOD | 5 |

Table 2. The substitution of textile chemical to reduce pollution load of textile effluent [Norman P.I., Seddon R.- 1991]

| *Textile chemicals* | *Substitutes* | *Processes* | *Purpose of substitution* |
|---|---|---|---|
| natural size-starch | synthetic sizes (PVA, CMC, PAA) | weaving | lower BOD of effluent |
| surfactants alkylphenol ethoxylates<br>R<br>\\<br>/ \\<br>l l<br>\\ /<br>$O(CH_2CH_2)_xH$ | linear ethoxylated fatty alcohols<br>$R-O(CH_2CH_2)_xH$ | prepara-tion bleaching | lower toxicity of effluent |
| sodium hypochlorite | hydrogen peroxide | bleaching | avoid formation of AOX |
| acetic acid | mineral acid | | lower BOD of effluent |
| metal containing dyes | other type of dyes | dyeing | lower content of heavy metals |
| sequestering agents, chelates EDTA, DTPA | NTA polymers based on acrylic acid | | lower stability of heavy metal complexes |
| phosphates, phosphonates | acrylic polymers, acrylic copolymers | | lower content of nutrient matter |

# An Overview of Pollution Prevention in Portugal

Ana Gonçalves, Vitor Limpo and Filomena Bartolomeu
Institute of Technology and Innovation for Entrepreneurial Modernisation (ITIME)
Environmental Technology Demonstration Unit (UDTA)
Azinhaga dos Lameiros à Estrada do Paço do Lumiar 1699 Lisboa Codex
Portugal

## *Abstract*

We present a summary about the situation of portuguese industry and economy as well as their consequences in the environment.
Next we draw a general view of the legislation evolution referring to the Environment and its connection whit industrial policy that exists in Portugal.
Finally, we describe some works developed in the area of pollution minimization.

## *Portugal: economy, industry and environment*

Portugal is the most occidental country in Europe, located in the Iberian Peninsula. It's a country whit several asymmetries and the river Tagus is a physical boarder of climacterics and topographic between north and south.

These differences were important for the social and economic development of this country. The big urban and industrial centres are locate along the coast between Braga and Setubal. The population density is 241 habitant/$Km^2$ in the coast having the interior of the country only 34 habitant/$Km^2$ and the average for the country is 110 habitant/$Km^2$ .

After becoming a member of European Community (EC) in 1986, Portugal had a significative growth in his annual GDP and from 1985 to 1990 grew by 4,6%, that is 1,5% above the EC average and most economic indicators have showed a favourable evolution.

Due to Community Programs over the last years, it has been possible to improve the once traditional scenario, since great investments have been made, specially in the area of transportation and traffic networks.

A big effort was made to give the national territory basic structures and support mechanisms for productive investment were established. This economic growth started a new economic structure where the agriculture lost importance and the industrial sector stabilise and there was a growth of employment in the services sector.

NATO ASI Series, Partnership Sub-Series, 2. Environment – Vol. 2
Cleaner Technologies and Cleaner Products
for Sustainable Development
Edited by H. M. Freeman, Z. Puskas, and R. Olbina

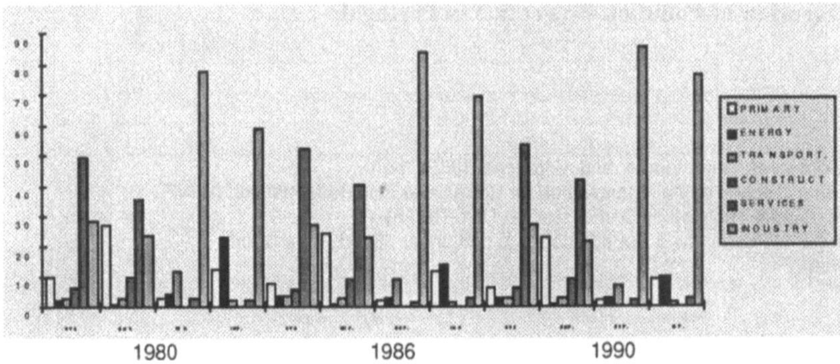


*Fig. 1: Portuguese economy structure in percentage (source: INE)*

The industrial sector is one of the most important in national economy and employes 22% of active population, producing 27% of GPV in comparison with agriculture that employes 23% of active population and produces only 6% of GPV .

This growth is associated with structural changes and with regional disparities ,and generates several types of pressure on the environment:natural resource use,pollution and the restructuring of land use. The creation of new industrial activities and the emigration from interior to the coast city create new environmental problems.

During the last decade,in Portugal a new thought regarding the environmental problems was developed by the authorities and the industrials.

In 1987 the Parliament passed the Framework Law on the Environment and in 1990 the Ministry of the Environment and Natural Resources was created.

Most of the legislation produced has direct implications on the industry, having as goal to avoid the environmental consequences resulting from this activity.

So, the beginning of a new environmental legislation and the investments in the protection of several environmental sectors are, in a certain way, the resulting effort of adaptation of our country to the new market.

EC aid for the environment, under the Community support programme of 1989-1983,has mainly taken the form of financial aid for investment in measures to control pollution.

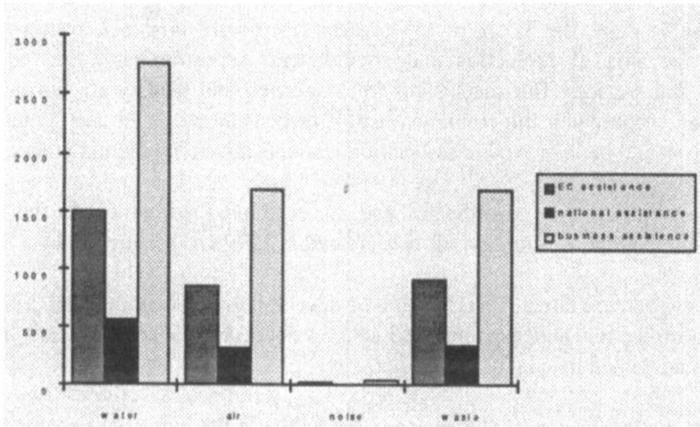


*Fig 2: Distribution by environmental media funding source, 1991 (Source: GEPAT)*

With the implementation, by the Ministers of Industry and Energy, of the PEDIP - Strategic Program of Dinamization and Modernisation of Portuguese Industry, support systems to projects of investment and environmental protection were created, that ascribe as lost funds the acquisition of equipment destined to protect the environment in the industrial units.

## *Evolution of the Environmental Legislation*

During the eighties a great effort was developed in Portugal in the way to restructure the existing legislation creating therefore the base of the environmental law.

After 1986 much of the existing legislation had to be modified. In certain areas it was completely not adequate to reality for example the "Water Law", existing since 1916. In others sectors legislative holes were noted.

In a short period of time, a large amount of legislative and regulatory texts were created, making possible the building of a solid corpus of environmental law.

Although the existence of a great amount of laws, there is not accuracy in their application. This situation exists because the recent environmental legislation is mostly a result of the transposition from Community Directives to Portugal appearing in some cases difficulties in its application. These difficulties are due to the adoption of a group of laws, conceived and harmonised by countries whit different realities of ours, where these questions were studied during many years.

Despite the strict legislation, a decision not to punish violators of Portugal's new pollution control laws was made. Giving the industrial enterprises a period of adaptation to the legal pollution levels.

Its appropriate to refer the effort of the present restructure in which concerns the State Organism in the area of protection and environment inspection with the redefinition of competencies and services. But mechanims for inspection and penalty are not yet operating efficiently, thus encouraging the continued use of technically outdated and sometimes illegal practices, they should be improved in association with incentives or voluntary measures.

The national institution that co-ordinates and inspects the Environment is the Ministry of Environment and Natural Resources that was created in 1990 and restructured in 1993.

The general environment director has the role of developing the environmental policy, through the study, co-ordination and execution of actions necessary to preserve and improve the Environment and defend its quality.

It is necessary the integration of the environment policy in the industrial policy creating legal instruments to ordinate and control the appearing of new industrial installations.

Since 1990 Portugal has carried out environmental impact assessments for all projects covered by the EC directives.

In summary, the environment questions posed to industry are mainly related with the quality of water, the quality of the air, residues and noise.

By analysing the national legislation evolution, we can verify that until the seventies existed a hole at the legislative level that allowed the appearance of zones with pollution problems on several areas of the country, mainly in great urban and industrial areas. Since the eighties this hole has been gradually reduced.

Therefore, from rare or even non-existent environmental legislation on a near past and facing the environmental problems created, the present legislation tends to control and impose limited values of emission, becoming therefore, mandatory the most immediate correction measures (end-of-pipe treatments).

It is necessary to make the legislation incentive the development of new industrial production technology based mainly on the reduction of raw materials consume and

energy, originating therefore less volume of industrial residuals. This strategy has economical advantages not only on the costs of raw materials and energy but also on the costs of despoliation that industry has to pay.

In conclusion, on the level of legislation, there was a gradual evolution of attitudes: in the beginning passive, at the moment controlative.

We predict that in a near future, under the EC directives, the attitude to have will be volunteer, with the industry itself applying internal measures to reduce pollution.
The pollution should be optimised. To reach this goal there is not legislation one hundred per cent efficient. It is necessary mainly a gradual and progressive change of the way to face and respects the environment that surrounds us.

This goal will be, only, achieved with integrated strategies including the technological modernisation and the environmental education and training.

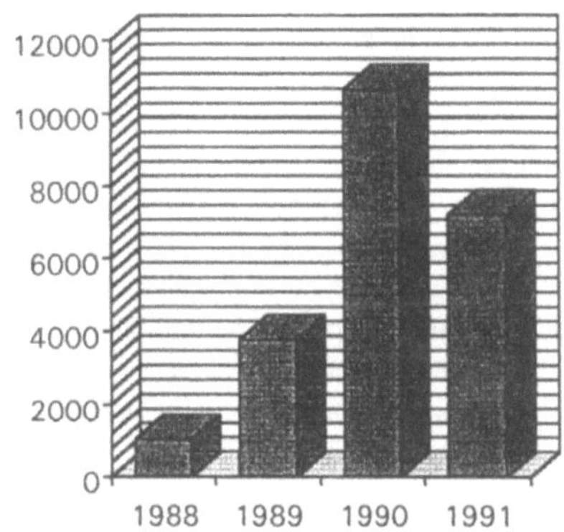


*Fig. 3: Investment by enterprises to combat pollution (million escudos) (Source: GEPAT)*

## *The minimization of pollution in Portugal*

The portuguese industrials know the environmental problems caused by their activity but only in the nineties the investment in the environment was relevant. The most part of it was in the end-of-pipe treatments.

There doesn't exist actually specific regulation or a global or regional programs to encourage and promote national pollution prevention. The industry doesn't have yet incentives to the adoption of pollution prevention strategies and most of the solutions to industrial effluent problems were mostly cheap and simple "end-of-pipe" technologies.

There are some isolated efforts by the industry relating to the adoption of techniques of waste minimisation. For example, the portuguese Hovione (pharmaceutical industry) had made investments in source reduction, recycling and reducing raw materials cost by 50%.

Several portuguese research institutions have been developing some research in this area but don't obey to an integrated effort.
The Environmental Technologies Demonstration Unit (UDTA) carries out integrated studies of pollution reduction by industrial sectors or unit operations to make pollution prevention a permanent cost-effective part of industrial processes.

At the moment the UDTA is performing a research project with financial support of JNICT, a national government department for supporting research, entitled "Implementation of a waste minimisation program in a ceramic industry plant".

This project pretends to implement a programme to reduce quantitative and qualitatively (volume and toxicity) the wastes generated in a ceramic factory to the extend that is economically practical.

The results of this project allow to improve the understanding of pollution prevention option for resolving environmental problems and demonstrate innovative pollution prevention approaches and technologies.

In Portugal there are many small and medium companies belonging to the ceramic sector. These companies are concentrated near the coast and several of these units are installed in habitation areas with high population levels so, although not very pollutant it's important to know its environmental effect.

Our project is still in an initial phase of gaseous effluents, waste waters and solid waste characterisation.

We have detected some classical possibilities of pollution minimisation, mainly in some process operations, equipment cleaning and handling, storing and raw materials transfer.

It's relevant the application of this kind of projects to some sectors of portuguese traditional industry. This is the case of the cork sector, where there is still very little information concerning pollution data. This makes it much more difficult, for those industries, to be able to obtain the ecological label.
The UDTA is developing a project for pollution characterisation of the cork industry.

In conclusion, Portugal is a country that still has problems in environmental pollution, but we hope that the "lessons learned" from other countries' problems and successes can be directly applied to Portugal's environmental problems. It is necessary training and information efforts in Portugal because communication along the administration, companies and the public is still occasionally insufficient, as result of habits inherited from the past.

*References*

Eduardo L. Rodrigues, "A política de ambiente na EC e a acção da DGI", presented to Seminar: Novas tecnologias ambientais e estratégia empresarial", April 1993.

Facility pollution prevention guide, U.S. EPA, January 1992.

Facts about portuguese industry, Gabinete de Estudo e Planeamento, 1991/1992

Glen McLeod and John O'Hara, "EC proposals for integrated pollution prevention and control", in *Chemistry and Industry*, November 1993.

Harry Freeman, Teresa Harten, Johnny Springer, Paul Randall, Mary Ann Curran and Kenneth Stone, "Industrial Pollution Prevention: A Critical Review", in *J. Air Waste Manage. Assoc.*, Vol.42, N° 5, May 1992.

Krish R. Krishnan, Beth Leamer, "Effective strategies for integrated hazardous substances management", presented to Environmental Conference Lisbon Portugal, June 26-30, 1994.

Luis Mira Amaral, "A vertente ambiental do desenvolvimento económico - o desafio dos anos 90", in *Ingenium*, Maio 1993.

L.W. Baas, M. Van der Belt, D. Huisingh and F. Neumann, "What some governments are doing and what all governments can do to promote sustainability", in *European Water Pollution Control*, Vol 2, Number 1, January 1992

Manuel C. Pássaro, "A indústria portuguesa face ao ambiente no contexto comunitário", in *Pequena e Média Empresa,* 1993.

OECD .Environmental Performance Reviews: Portugal, Paris 1993.

Teresa Gouveia -Ministra do Ambiente, "Ambiente -Factor de inovação tecnica e melhoria da competitividade " -in *Industria,* Dezembro 1993

"Waste reduction guide", U.S. Postal Service, AS 552, February 1992.

"Waste minimization opportunity assessment manual", U.S. EPA, July 1988.

# Considerations on Sustainable Development and Its Perspective in Romania

Corneliu A. L. Negulescu and Viorel Harceag
Environmental Research & Engineering Institute
Splaiul Independentei 294, Sector 6
P.C.77703 Bucharest 78
Romania

## *ABSTRACT*

Like in the most countries, especially in Eastern Europe, present pollution control systems into Romania are based mainly on "end of pipe" approach. Because of the large number of different waste sources involving a multitude of contaminants potentially present, pollution control is generally both ineffective in solving all environmental problems, and financially expensive requiring significant capital in terms of investment and operating costs.

The "pollution prevention at the source" is a better approach of environmental protection and it can result in economic benefits through reduced waste treatment costs and reduced costs for material input.

The paper describes some basic considerations about paradigm shift from pollution control to pollution prevention, the state of sustainable development and its perspective in Romania.

NATO ASI Series, Partnership Sub-Series, 2. Environment – Vol. 2
Cleaner Technologies and Cleaner Products
for Sustainable Development
Edited by H. M. Freeman, Z. Puskas, and R. Olbina

## *SUSTAINABILITY*

Sustainable development was defined by the WORLD COMMISSION ON ENVIRONMENT AND DEVELOPMENT in 1987 as "development that meets the needs of the present without compromising the ability of the future generations to meet their own needs". In a sustainable process system no products will be manufactured which are not either biodegradable or can serve as raw materials for other production system (Moser F. 1992). No toxic substances must be produced. Chemical products are to be manufactured, as far as possible, from natural raw materials on a bio-technological basis and not on a petro-chemical basis.

Pollution prevention includes strategies that encourage production process changes, in order to reduce both the amount of hazardous waste released to the environment and the use of resources, including raw materials and energy. Waste minimisation involves the following steps (R. Qassim 1992):

- source reduction (raw materials substitution, product alteration, change of process operating conditions, modification of operational procedures);
- recycling (substitution of fresh materials by the waste generated ones);
- recovery (transforming the waste from one industrial process into useful raw material for other industrial process);
- treatment and incineration;
- storage and disposal.

At present, there is certainly an enormous potential in alternative materials and processes that makes winners on each side, the ecology and economy (Niederl K.

1993). Many of those new technologies already exist on the market and just have to be adapted to the specific situation.

In order to make the change from pollution control to pollution prevention an expanded set of guiding principles is necessary. These principles include a series of activities (Moser A.,Narodoslawsky M. 1993):

- to use mainly solar energy in all options in future, this "technology mix" includes the use of the following techniques: photovoltaic, solar, thermal and geothermal power stations, wind and water power stations, biomass, hydrogen storage;
- to use renewable raw materials in their high diversity adapted to climates regional and social conditions:

biomasses from agriculture and forestry for industrial bulk production will replace fossil non-renewable materials;

- to replace toxic heavy metals by renewable (bio)catalysts;
- to produce (bio)degradable products;
- to recycle non-degradable materials like minerals;
- to close the cycles of biocompatible materials in the nearest environment possible by embedding them into natural cycles with their assimilation capacity.

There is an increasing volume of literature reporting case studies of waste minimisation projects in various industrial sectors. These case studies demonstrate not only environmental attractiveness of waste minimisation projects, but substantial economic benefits which are usually involved.

## *ROMANIAN PERSPECTIVE*

Romania, one of the post-communist East European countries, is going through a period of transition to the market economy, involving great difficulties in economic, financial, social and environmental protection fields. Environmental problems are, in a fundamental sense, due to present economic and technological inefficiency. That is why Romanian pollution control systems consist usually in "end of pipe" technologies.

The change from pollution control to pollution prevention results in environmental improvements. Some Romanian achievements in this field are presented bellow.

Adoption of appropriate legislation s the first decisive step to be taken and this should gradually change the relation towards environment on all levels of society. The new Romanian environmental protection law has drawn up and now is waiting for adoption in the Parliament. The basic goals of this new law are:

- to prevent the pollution of the environment;
- to re-establish the balanced natural background, based on its regeneration capacities;
- to restore the environment damages caused in the past;
- to reduce natural resources exploitation;
- to preserve the renewing natural resources;
- to improve the environmental pollution control technologies and equipment.

There are also several settlements drew up by Trade and Industries Ministries related to recycling systems of packages, in order to minimize the use of the raw materials and the amount of solid waste.

In the last years many ecological high schools and universities have been set up in Romania. In their syllabi there are foreseen pollution prevention courses, in order to educate people who will work as technicians in industry in this environmental protection approaches.

Presently, within many Romanian industrial units, technological improvements and refurbishment take place (this often based on modern imported technologies and equipment), and will result in the decrease of the emission of pollutants.

Some results of Environmental Research and Engineering Institute, already implemented in industrial plants (C.A.L. Negulescu 1993), constitute other steps in the sustainable development field. They consist mainly in useful materials recovery by pre-treatment at sources and recycling them into technological processes. Main examples are related to:

- cellulosic fibres recovery in pulp and paper and wood processing industries by pre-treatment of waste waters using static screens;
- dye-stuffs recovery from waste waters in textile industry by membrane processes (ultrafiltration), using Romanian membranes;
- oil recovery from waste waters in metal processing industry by ultrafiltration;
- processing of some ore deads for noble metals recovery (Romanian-Australian co-operation).

Other proposals for industrial process improvements at Chloro-sodic Products Company Govora are relevant:

- The use of high performance filters, instead of the existing -Solvay type- allows for the separation of the lye from $NaHCO_3$ wash water. In this way, the volume of discharged waste water and its chloride load are reduced;

- to improve the limestone and coke dosing system resulting in reducing of the amount of solid waste from the lime kilns.

## *CONCLUSIONS*

The pollution prevention concept has begun to be better known in Romania. At different conferences and symposia, it has been presented conclusions drawn up at the Graz 1993 N A T O Symposium, and many specialists have manifested a great interest in this topic.

The possibility for works related to environmental protection to be profitable is extremely attractive, especially in the transitional counties.

## *REFERENCES*

Moser A. , Narodoslawsky M. (1993), Embedding technologies into the natural cycles of biosphere, EPA/NATO workshop Pollution Prevention Techniques and Technologies, Graz, p.173

Moser F. (1992), Basic considerations pertaining to the development of circular processes for a sustainable world, ENVIROTECH, Viena, p.89

Negulescu C.A.L. (1993) Pollution prevention research: The Romanian experience,EPA/NATO workshop Pollution Prevention Techniques and Technologies, Graz, p.201

Niederl K. , Schnitzer H. (1993), Ecoprofit Graz, EPA/NATO workshop Pollution Prevention Techniques and Technologies, Graz, p.13

Qassim R. (1992), Waste minimization in the batch process industries, ENVIROTECH, Viena, p.104

# Major Constraints to Successful Implementation of Cleaner Technologies to Slovak Industry[1]

Miroslav Chodák
Centre for Environmental Impact Assessment
at the Department of Landscape Ecology
Comenius University in Bratislava
Mlynská dolina B-2
842 15 Bratislava
Slovakia

## Introduction

The model of sustainable development implies an approach, in which economic development is combined with environmental protection. This means that progress in environmental protection is achieved along with improvements in economic performance. Resources for realizing both goals may be generated from savings in investment funds that become available due to more efficient use of energy and materials as well as the less intensive use of waste treatment and waste disposal services.

With this respect, the concept of cleaner technologies offers a considerable opportunity for industry in Slovakia to combine improved economic performance with reduced impact on the environment through source reduction and on-site recycling. The implementation of cleaner technologies to Slovak industry can generate a starting point for sustainable development.

---

[1] Disclaimer: The opinions expresses in this paper is the professional opinion of the author and does not represent the official position of the Comenius University, Bratislava, Slovakia.

NATO ASI Series, Partnership Sub-Series, 2. Environment – Vol. 2
Cleaner Technologies and Cleaner Products
for Sustainable Development
Edited by H. M. Freeman, Z. Puskas, and R. Olbina

This paper identifies and discusses problems and barriers related to successful implementation of cleaner technologies to industry of Slovakia. Presented results have been identified within the first part of the project "Opportunities for Cleaner Production in Transition Economies". The research has been carried out by the Centre for EIA of the Comenius University in Bratislava within its Pollution Prevention Program and is funded by the Research Support Scheme of the Central European University in Prague.

Information for the paper is based on published data, analysis of project reports of the Slovak/Norwegian and the Czech/Norwegian Cleaner Production Project, questionnaires, visits to industrial companies, and interviews and consultations conducted with a number of cleaner production experts and governmental officials.

## Definition of Cleaner Production

For the purpose of this paper, cleaner technologies as well as pollution prevention, waste minimization and cleaner production, will refer to: redesign and/or reformulation of products; process modification; changes in plant equipment; raw material substitution; and simple operating changes and good-housekeeping processes which contribute to meeting environmental quality objectives and increasing productivity (Fénérol, 1993). This definition excludes end-of-pipe treatment and control technology, but focuses on measures which prevent or eliminate pollution at source.

## Problems and barriers to implementation of cleaner technologies

Up to now, more than 50 barriers to successful implementation of cleaner technologies to Slovak industry have been identified within the project "Opportunities for Cleaner Production in Transition Economies". Despite a wide array of contributing factors to the impediment of cleaner production progress, it is suggested that the identified barriers can be summarized under six key areas. These are: present situation; human resources; governments; economy transformation process; policy instruments; and enterprises.

In the following text, each of the key areas is discussed in details. As well, a general framework for the key areas is proposed and the mutual implications are examined with respect to the implementation of cleaner technologies.

Present situation

The "present situation", together with the "economy transformation process", is the origin of constraints to successful implementation of cleaner technologies which are exclusively specific for the post-communist countries, Slovakia in particular. It includes issues such as geographical, social, historical, political, and economical characteristics of the country.

Within this key area, the major constraints to cleaner technologies are:

1. *Structure of industry*. Slovakia is heavy industry oriented. In terms of shares of overall industrial output, in 1992 the leading sectors were metalworking (15.2%) and foodstuff and beverages (14.5%), followed by machinery, chemicals, oil refining, and power, water and gas supply which all exceeded 5% (Busik et al., 1993). Apart from the fact that such a production is highly energy and raw materials demanding, the present structure of Slovak industry is based on productions where there are excessive capacities in the world (Zamery..., 1994).

2. *Previous dependance on export markets and sources of raw materials in Eastern Europe*. This factor is a direct result of the Slovak postwar industrial development (Busik et al., 1993).

3. *State ownership*. The fact that many larger industrial companies are still state owned enables them to behave irrationally. It is the heritage from past times, where goods are produced without any sales secured, expensive and not profitable facilities are maintained despite the bad economic situation of a company, etc. (Borovsky, 1994). The state ownership has another implication, which is the absence of competitive business environment in Slovakia (Ako vidite..., 1993).

4. *Energy and raw materials consumption patterns*. In past, enterprises tended to establish such manufacturing processes where all semi-products were manufactured by themselves. This led to an inefficient production due to the ineffective use of production facilities directly affecting their profitability (Borovsky, 1994).

Other important factors include: *geographical location, historical aspects (including the split of the Czech and Slovak Federative Republic), social pressures (unemployment), outdated infrastructure,* etc.

Economy transformation process

Economic reforms were introduced to Slovakia in 1990. The reform scenario established three goals: privatization, foreign trade and price liberalization, and restrictive monetary, fiscal and income policies (Busik et al., 1993). The major obstacles of the economy transformation process to implementation of cleaner technologies include:

1. *Privatization*. The prospect of privatization certainly spurred some of the monolithic state industries to clean up their production methods. On the other hand, there was considerable unwillingness on the part of prospective buyers to accept financial responsibility for decontaminating the site, while the state owners could often not afford to do so. In other cases, investors simply leased the site from the state or bought the clean modern part, leaving the dirty part in the hands of the state. The competition between foundering businesses to

attract foreign investors enabled prospective buyers to choose which parts of factories they wanted, potentially leaving the state treasury or municipality with "nothing but clean-up obligations" (Manser, 1994). Another issue is that during the privatization process a company is "freezed", it is not allowed to make any decision affecting its assets or production process. It is clear that, on one hand, no cleaner production project can be carried out during this period, and, on the other hand, if the privatization process takes too much time, it can have a significantly negative impact on the company's economic performance. I some cases, a conscious mismanagement of industrial companies in order to reduce its price in the privatization process has been experiences as well.

2. *Economic recession.* The economic reforms have been accompanied by falls of over 20% in GDP and industrial production (Busik et al., 1993). The overall recession is attributable to problems associated with the transformation maneuver and to other external factors: the disintegration of the socialist market system, and the stagnation and recession of the world's economy.

Other factors which directly or indirectly impede the cleaner technologies progress in relation to the on-going economy transformation process in Slovakia include: *uncompetitive business environment, not solved ownership relationships,* and *insufficient in-country market demand.*

Human resources

It is generally acknowledged that recognizing and solving barriers to human resource development is an important element for achieving sustainable development.

As far as human resources are concerned, the major obstacles to implementation of cleaner technologies to Slovak industry include:

1. *Resistance of thinking*, as the most serious constraint, is resulting in maintaining standard

business practices which strongly influence the effectiveness of environmental management.

2. *Missing managerial skills* are an additional element in Slovakia, however, is necessary for creating and guiding the methodological capacity required for cleaner technologies.

3. *Absence of training and education*. Besides better skilled managers, successful cleaner production programs require qualified staff who are capable of performing at the standards set by the demands of existing challenges.

Moreover, when implementing cleaner production projects, the *lack of creativity*, and the *unwillingness to undertake personal initiatives* and *to accept the responsibility* by the staff members have been experienced as well.

## Governments

The governments at all levels are seen as the only actor among all the key areas which is able to generate conditions for the countrywide implementation of cleaner technologies on a sustainable basis, and thus initiate the process of transition towards sustainable development in Slovakia.

However, there are some very serious obstacles which will have to be overcome. They include:

1. *Political and economical instability*. The continuously changing political and economical preferences is an important constraint not only to the implementation of cleaner technologies, but generally to the whole restructuring of Slovak economy (Ako vidite..., 1993).

2. *Unskilled government*. Slovakia does not have enough of skilled professional politicians. The endless personnel changes in government are impeding the transformation process and increasing chances for major mistakes in governing the country.

3. *Approach to cleaner technologies and defining the concept of cleaner production.* The concept of cleaner production has been found, in most cases, unknown to governmental officials at all levels. They often mix it with best available technologies, or no-waste technologies. Most essential information about cleaner technologies are missing. This all contribute to the fact that there is no cleaner production policy neither a capacity to develop it.

4. *Specific environmental media orientation.* Most of environmental programs are organized along specific environmental media lines (e.g. air, water or soil) and thus the multidisciplinary approach necessary for cleaner technologies does not exist in Slovakia (REC, 1994).

5. *Bureaucracy and administration.* Cleaner production programs need to be integrated into the environmental as well as economic policy at the national or local level. Such integration is currently limited in Slovakia partly by the endless administration procedures and also by the insufficient level of cooperation between government offices and ministries. Sometimes, the official policy of the Ministry of Economy and Finance was reported in direct conflict with the one of the Ministry of Environment (REC, 1994).

6. *State budget deficit* will become the major obstacle to impede the establishment of a financial support scheme for cleaner technologies once the cleaner production policy is implemented.

7. *Funds from the State Environmental Fund.* There are two problems related to the State Environmental Fund. First, companies are not able to pay, and the amount of fines and charges officially imposed is much higher (ca. 30 times) than the amount actually collected and deposited in the State Environmental Fund which redistributes the money through a grants program (REC, 1994). Second, the money collected are not redistributed back to improve the performance of the charged companies, but used in other sectors according to a priority scheme where cleaner technologies are missing.

Policy instruments

Major obstacles to implementation of cleaner technologies in this key area include:

1. *Incompatible system of legislation.* In the transforming process from centrally-planned economy to a market one, the whole legislation system of Slovakia has to be reconsidered and new acts to be adopted. During this period, there are vast opportunities to misuse present gaps in the legislation and avoid punishment.

2. *Existing environmental legislation* in Slovakia is strictly media (air, water, soil, waste, etc.) oriented, and does not enforce neither stimulate implementation of cleaner technologies.

3. *Absence of methodological guidance.* Amendments to the environmental regulations do not contain any methodological guidance which most industrial companies agreed to be helpful.

Enterprises

This is the key area where all the activities related to implementation of cleaner technologies are taking place. Therefore, to recognize the barriers and problems to this implementation is extremely crucial. These include:

1. *Outdated technology.* Outdated technology works both ways, if too old, there is no other solution than the replacement. On the other hand, outdated technology offers numerous opportunities for source reduction (much more than modern technology).

2. *Lack of capital.* This has been recognized as the most serious constraint in Slovak industry to the implementation of other than no-investment cleaner production options. Unfortunately, many Slovak industrial companies do not have any money for investment actions which is,

however, an essential precondition for their further growth and development. On the other side, the amount of available investments is depending on the industry sector and may vary from company to company according to other important factors (e.g. the form of ownership, the share of foreign capital, etc.)

3. *Company performance.* A company with weak economic performance is unlikely to adopt cleaner production principles and to implement cleaner technologies.

4. *Management system and division of responsibilities.* The inflexible management and the tough hierarchical structure in a company has a negative impact on the implementation of new ideas.

5. *Position of environmental department* within the organizational structure of a company. Often, environmental departments in Slovak industrial companies are in a position of a watch-dog with almost no rights and many duties. It is isolated and perceived as useless or disturbing.

6. *Discouraged management* is a serious obstacle. The reasons are often related to the privatization process where representatives of the top management are reluctant to adopt new strategies because: a) they are restricted by the privatization law, and/or b) they are not certain how their recent strategic decisions would be assessed by the new owners.

7. *Remuneration system.* Especially, low salaries have a negative impact on the motivation of workers. In some cases, the cleaner production efforts are not desirable because their implementation would cut-off the extra bonus in the workers' salary for working in risky conditions. As well, the lack of rewards for successful waste reduction should be considered as an obstacle.

8. *Accounting system* which does not allocate the total environmental costs. The manner in which a company recognizes costs is critical to decision-making on whether or not to implement new waste reduction opportunities.

9. *Quality concerns.* Generally, there is a fear that modifications leading to cleaner technologies have a deterious effect on product quality.

10. *Belief in required pollution control.* The belief that legally required pollution control is enough is also critical.

Other important factors are: *incomplete data on the exact sources and amounts of waste, administrative barriers, technology limitations, relations with state authorities and suppliers,* etc.

General framework for the key areas

In practice, all the above mentioned constraints are very closely interlinked and mutually reinforcing both inside and outside their key areas. Dividing the constraints into the six key areas enables us to generate a general framework where the mutual implications can be easier recognized and examined. Figure 1 presents the framework of the constraints to implementation of cleaner technologies which is assumed to be valid not only in Slovakia, but in all Central and Eastern European countries of economic transition.

The distinguished key areas are placed within three layers outlined by three circles of a different size. The smallest circle (No. 1) includes key areas which have the most direct and immediate impact on implementation of cleaner technologies to Slovak industry, namely: governments, policy instruments, and enterprises. In practice, those are the enterprises themselves and the governments using various policy instruments who can stimulate directly the implementation of cleaner technologies. On the other hand, experience show that this mechanism alone would not ensure the implementation of cleaner technologies on a sustainable basis.

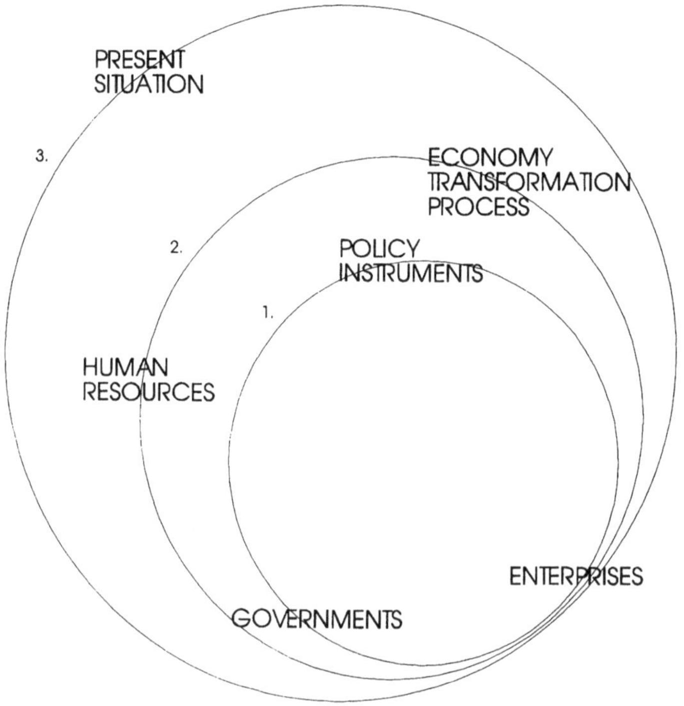


Figure 1: General framework of the constraints to cleaner technologies

Sustainable implementation can be achieved only with the support resulting from developing the human resources for cleaner production. As well, the smooth economy transformation process helps to generate conditions for establishing a capacity for cleaner technologies implementation in a long run. Therefore, the circle No. 2 is important for achieving the sustainability, although resolving the barriers under this layer would affect the immediate implementation of cleaner technologies less directly.

The circle No. 3, represented by the "present situation" key area, provides a milieu for all other key areas. This layer determines the total potential for cleaner technologies of the country, however, is most difficult to influence.

## Conclusion

From the above, it can be concluded that resolving the major constraints to successful implementation of cleaner technologies to Slovak industry, summarized under the six key areas (present situation, human resources, governments, policy instruments, economy transformation process, and enterprises) can affect the implementation process in terms of its and sustainability.

The Figure 1 clearly showed that enterprises, governments and policy instruments are the key areas which have the biggest potential for immediate improvements of the cleaner technologies implementation. In order to continue in these improvements on a sustainable basis, the human resources for cleaner production have to be developed. However, completing the on-going transformation of the Slovak economy is an essential requirement for influencing the present situation and establishing sustainable development.

## References

Borovsky, J. (1994): Zvaz priemyslu Slovenskej republiky. In: Profit 31/94, Bratislava.

Busik, J., et al. (1993): The Slovak Republic, Country Report. Bank Austria AG, Vienna.

Fénérol, C. (1993): Supply side policies to augment government support for promoting cleaner technologies. OECD, Paris.

Manser, R. (1994): Going West, Market reforms and the environment in Eastern Europe: The first three years. In: The Ecologist 24/1.

REC (1994): Strategic environmental issues in Central and Eastern Europe, Environmental needs assessment in ten countries. Vol.2, Regional Environmental Center, Budapest.

(1994): Zamery hospodarkej politiky vlady Slovenskej republiky na rok 1995. In: TREND, Bratislava.

(1993): Ako vidite ekonomicke perspektivy Slovenskej republiky? In: TREND, Bratislava.

# WASTE MINIMIZATION IN A DYE HOUSE

Janez Petek, Peter Glavič
University of Maribor
Department of Chemical Engineering
Smetanova 17
62 000 Maribor
Slovenia

## *INTRODUCTION*

Waste minimisation in process industries consists of source reduction and recycling. It is possible to save money by reducing waste treatment and disposal costs, to improve the quality of products, to reduce potential environmental liabilities, to protect public and worker's health, to increase safety and to protect the environment. Waste minimization by source reduction and recycling is the cheapest method for waste reduction. It can be adopted by all process industries: textile, pulp and paper, leather, food, chemical, oil, ceramic, pharmaceutical, metallurgical.

In Slovenia, we are staying behind the developed countries. Therefore, it is urgent to start solving the problems. Ecological dumping behind export to developed countries will not be allowed as waste minimization increases costs and reduces competitiveness in the European markets. As our enterprises export without profit they will not be able to compete with modern technologies. They will have to invest a great deal or shut down the plants. They have to get ready in time for the changes that are coming.

Together with the Department of Chemical Engineering at the Faculty of Technical Sciences in Graz, we have started to minimize waste in textile industry. Our endeavour is sponsored by the Research Fond of Maribor, and the Austrian and Slovenian government within the project PREPARE (Preventive Environmental Protection Approaches in Europe) which is a part of EUREKA/EUROENVIRON program. We have performed the waste minimization assessment in a dye house.

Textile industry is a large consumer of process water, energy, chemicals, textile auxiliary agents and dyes. It produces a lot of effluents (waste water, waste heat,

NATO ASI Series, Partnership Sub-Series, 2. Environment – Vol. 2
Cleaner Technologies and Cleaner Products
for Sustainable Development
Edited by H. M. Freeman, Z. Puskas, and R. Olbina

hazardous and toxic wastes, exhaust gases from dryers, gas emissions from boiler plant), which are mostly released to the environment.

There are four types of waste which are produced by textile wet processing (preparation, dyeing, printing, finishing and other processes) and are most amiable to source reduction:

1. hard to treat (dyes, metals, phenols, toxic organic compounds, phosphates);
2. dispersible (print paste, lint, waste from coating operations, waste solvents, batch dumps, unused finished mixes);
3. hazardous or toxic (metals, chlorinated solvents, non-degradable surfactants);
4. high volume wastes (waste water, alkaline waste from preparation, bath dye waste).

## *1. PROCESS AND PRODUCTION DATA*

The textile factory produces all kinds of yarn, woven and knitted goods of wool, cotton, cellulose, synthetics and blends. The factory employs about 600 workers. Table 1 shows the production in the dyeing and finishing mill.

Table 1: Average production and capacity of dyeing and finishing mill.

| **Sort of the textiles** | **average production (t/a)** | **capacity of the mill (t/a)** |
|---|---|---|
| yarn, warp beans, cross-wound cones | 400 | 800 |
| woven and knitted goods | 800 | 1 000 |

Average production of waste water is 726 000 $m^3/a$. 91.7 % of the waste water is process waste water, 6.9 % is municipal waste water and 1.4 % is rain-water. The process waste water consists of waste water from dyeing and finishing processes (24 %), washing (34 %) operations, rinsing operations, spillage, leaks (38 %), loss of condensate and cooling water (4 %). There is no treatment plant. After screening, to separate lint, the effluent flows directly into the river. The quality of waste water released into the environment is found in Table 2.

Table 2: Quality of the waste water.

| | Actual value | allowed value |
|---|---|---|
| color | dark | * |
| temperature (°C) | 25 | 33 |
| pH | 10 | 6.5-9.0 |
| electric conductivity (μS/cm) | 1 440 | - |
| TOC $\rho(C)/(mg/l)$ | 76 | 17 |
| anionic tenzides (mg/l) | 2,9 | - |
| non-ionic tenzides (mg/l) | 13,7 | - |
| COD $\rho(O_2)/(mg/l)$ | 400 | 160 |
| $BOD_5$ $\rho(O_2)/(mg/l)$ | 120 | 30 |

The production process in the dye house consists of the following operations:
- filling of the dyeing machine,
- washing,
- preparation process,
- rinsing,
- dyeing,
- rinsing,
- second dyeing (in the case of dyeing blends),
- rinsing (in the case of dyeing blends),
- finishing process,
- rinsing.

Depending on the textile material, kind of blend, dye, and dyeing machine the dyeing process can be carried out in one or two baths with a short or long dyeing bath. Table 3 contains types and capacities of monthly production in the dyeing house. Table 4 presents the consumption of dyes, textile auxiliary agents, chemicals, steam and water in the two dyeing mills and in the whole factory.

* smell, color and turbidness of waste water may not change the river.

## 3. ASSESSMENT PHASE AND GENERATION OF WASTE MINIMIZATION OPTIONS

We began with the assessment phase in the machine dyeing mill. In the first step of the experimental work we planned to reduce production disturbances due to incorrect flow of warm water and to reduce a need for additional soft water in the case of large production load. In the second step we worked out material and heat balances for single dyeing apparatuses as well as for the whole dye house; we evaluated textile production, calculated costs and offered four alternatives for waste minimization. We arrived at a conclusion that it is possible to lower wastes in the machine dyeing mill by reducing the usage of water and energy. In such a manner we could lower costs of dyes, auxiliary textile resources, steam and water.

Table 3: Type and capacity of dyeing apparatuses with their monthly production.

| Type of the apparatus | bath capacity (kg) | monthly production (kg/m) | No. of charges |
|---|---|---|---|
| **machine dyeing mill** | | | |
| THIES ECO-BLOC M/P 900LB | 150 | 5,060 | 34 |
| THIES ECO-BLOC M/P 900LB | 150 | 5,060 | 34 |
| THIES P DUO MAT 800LB/KB | 100 | 5,900 | 59 |
| THIES ECO-BLOC S II 900LB | 150 | 8,900 | 59 |
| THIES ECO-BLOC S II 600LB | 50 | 2,970 | 59 |
| PROLETER RL 250 | 270 | 6,300 | 34 |
| PROLETER RL 250 | 270 | 6,300 | 33 |
| **piece dyeing mill** | | | |
| THIES SOFT STREAM SIII 140 | 60 | 1,000 | 50 |
| THIES SOFT STREAM 140 SV | 450 | 5,280 | 27 |
| THIES SOFT STREAM 140 SV | 450 | 5,700 | 27 |
| THIES ROTO STREAM 140 | 480 | 6,200 | 28 |
| THIES ROTO STREAM 140 | 480 | 6,100 | 28 |
| WINCH (8 apparatuses) | 400 m | 1,000 m | 50 |

The next phase of the research was to look for possibilities of waste minimization in a piece dying mill, where we offered seven alternatives for waste minimization. We concluded that it was possible to lower wastes in the piece dyeing mill by reducing the usage of water for 50 % and energy for 30 % by improving housekeeping and maintenance.

During our work we searched for production changes that would demand no investment costs. Some practical improvements were introduced directly in the production. For other investments we asked for quotations from equipment producers and visited one of them.

Table 4: Consumption of dyes, chemicals, textile auxiliary agents, steam and process water in the dye house.

| | Machine dyeing mill | piece dyeing mill | the factory |
|---|---|---|---|
| dyes (t/a) | 14.4 | 16.1 | 30.5 |
| textile auxiliary agents and chemicals (t/a) | 210.0 | 180.0 | 516.5 |
| steam (t/a) | 1,500.0 | 5,300.0 | 36,760.0 |
| process water ($m^3$/a) | 130,000.0 | 217,000.0 | 726,000.0 |

*Options for implementation in the machine dyeing mill:*

1. Good housekeeping using computer control instead of a pneumatic one on both Thies apparatuses.
2. Change all washing out with combination of shorter prewashing and afterwashing under pressure.
3. Heat recovery from waste water.
4. Installation of a new dye kitchen.

*Options for implementation in the piece dyeing mill:*

1. Good housekeeping and regular maintenance.
2. Dyeing of knitted goods in hank dyeing machine without prewashing out.
3. Recycling of condensate.
4. Re-using cooling water.
5. Reducing cost of water and electricity by simultaneous cooling and washing out.
6. Heat isolation of the dyeing machines.
7. Preparing optimal volume of process solutions for pad-batch finishing.

Table 5 shows estimated capital investments and possible savings.

Potential savings are not stable because capacity depends on various factors such as orders, quality of raw materials and workers awareness. We expect that maximal savings could reach 70 % of the total savings found (620 000 USD/a).

## *4. CONCLUSION*

It is possible to reduce wastes, consumption of water and energy in the two dye houses. It is also possible to minimize waste by use of dyeing baths, ultra filtration of exhausted dyeing baths, and by substitution with biodegradable chemicals, dyes and auxiliary agents. Table 6 shows the reduction of pollution load after the planned implementation of the proposed waste minimization options.

Table 5: Capital investment for the planned waste minimization alternatives and savings.

| Option | investment costs USD | savings USD/a | payback period a |
|---|---|---|---|
| 1.1. | 73 800 | 116 000 | 0.68 |
| 1.2. | - | 10 200 | - |
| 1.3. | 52 000 | 196 000 | 0.26 |
| 1.4. | 28 000 | 42 000 | 0.67 |
| **total** | **153 800** | **364 200** | **0.42** |
| 2.1 | 2 000 | 313 000 | 0.006 |
| 2.2. | 9 000 | 18 000 | 0.50 |
| 2.3. | - | 75 000 | - |
| 2.4. | 75 000 | 100 600 | 1.34 |
| 2.5. | - | 1 000 | - |
| 2.6. | 7 000 | 7 000 | 1.00 |
| 2.7. | - | 5 400 | - |
| **total** | **93 000** | **520 000** | **0.18** |

In the future we shall carry out heat and energy balances of the plant and optimise its heat exchangers network. In the first step we will locate steam users, reproduce network structure and roughly evaluate consumption and losses. The analysis will comprise the boiler and steam users: heat exchangers for room heating, dyeing machines in the mill, dryers, sizing machine and evaporators. In the second step we shall carry out a rigorous analysis of the network and utility consumption to propose possible solutions.

Table 6: Planned reduction of pollution load after implementation of proposed options.

| | | |
|---|---|---|
| waste water | 153 000 $m^3/a$ | 21 % |
| condensate | 3 060 t/a | 10 % |
| dyes | 7 300 kg/a | 24 % |
| textile auxiliaries and chemicals | 73 800 kg/a | 14 % |
| waste heat | 1 928 000 MJ/a | 3 % |

## *7. LITERATURE*

EPA, Waste minimization Opportunity Assessment Manual, Hazardous Waste Engineering Research Laboratory, U. S. Environmental Protection Agency, Cincinnati, Ohio, 1988.

Petek J., Glavič P., Waste Minimisation in a Dye House, Faculty of Technical Sciences, Department of Chemical Sciences, 1992.

Petek J., Glavič P., Waste Minimisation in a Piece Dyeing Mill, Faculty of Technical Sciences, Department of Chemical Sciences, 1993.

Smith B., Pollution Prevention by Source Reduction in Textile Wet Processing, Pollution Prevention Pays Program, Raleigh, NC, 1988.

# Cleaner Energy Production Technologies For Fossil Fuels

Aysel T. Atımtay and Aysegul Aksoy
Environmental Engineering Department
Middle East Technical University
Ankara, 06531 Turkey.

## 1. Introduction

In recent years, the development of new technologies to generate clean energy has been a great concern for *pollution prevention*. There is an increasing demand in the world for energy due to industrialization. This increasing demand forces the scientists to investigate *new technologies for production of electricity in a more efficient way*. Development of more efficient processes for energy production also aims the process to be cleaner from environmental point of view.

There exist huge feedstocks of coal available in the world. This huge feedstocks make the coal an important source of energy for the future. Also, the stability of coal prices make the power plants using coal economically attractive. Electrical energy is one of the most favourable clean energies which is generally produced by conventional systems known as pulverized coal-fired or stocker-fired boilers with steam turbine and generator systems. In these systems, the energy conversion efficiency is around 30-35 %. This means that only about one third of the heat generated can be converted to the electrical energy and two third is wasted.

Various studies have been conducted in order to increase the energy conversion efficiency while taking into consideration that the emissions should be kept below the standards. For

NATO ASI Series, Partnership Sub-Series, 2. Environment – Vol. 2
Cleaner Technologies and Cleaner Products
for Sustainable Development
Edited by H. M. Freeman, Z. Puskas, and R. Olbina

this purpose new technologies have been developed in the United States of America and other industrialized countries in order to generate electrical energy more efficiently from coal. Among the new technologies developed Integrated Gasification Combined Cycle (IGCC) system is the one of the most promising one. When coal is gasified, the sulfur in the coal reacts with steam to form hydrogen sulfide and small amounts of other sulfur-containing compounds. Hydrogen sulfide is a toxic gas and it contributes to the formation of acid rain when it is oxidized to $SO_2$ and/or $SO_3$. It is, therefore, necessary to remove as much of the hydrogen sulfide as possible from the coal gasification stream prior to release to the atmosphere. Additionally, turbines and related equipment need to be protected from corrosive action of sulfurous compounds in the coal gas.

This paper summarizes the research findings for the development of supported sorbents for use in hot gas cleanup system of the IGCC. The sorbents should have high capacity and sulfidation efficiency for hydrogen sulfide removal, high attrition resistance for suitability in fluidized bed applications and good regenerability.

## 2. New Energy Conversion Systems

Considerable amount of work has been carried out to increase the efficiency of generating electrical energy in power plants and new technologies are being developed in the United States and other industrialized countries. Another very important issue is to decrease the amount of pollution due to energy production. The advanced energy conversion sysytems have been identified as having significant efficiency and economic advantages over conventional pulverized coal-fired and stoker-fired boilers to produce electricity. New energy conversion systems under development for power production are as follows (MRTC/DOE, 1987):

* Pressurized fluidized bed combustion combined cycle system
* Integrated gasification combined cycle system
* Gasification molten carbonate fuel cell
* Direct coal fueled turbine

**Pressurized Fluidized Bed Combustion Combined Cycle System (PFBC)**

Figure 1 shows a schematic diagram of a PFBC system. The combustion of coal takes place in a fluidized bed system at a temperature of 815 - 920°C and at a pressure of 7 -17 atm. Limestone or dolomite particles are also fed into the bed as sorbent. High temperature - high pressure (HTHP) combustion gases exit the combustor and are expanded in a gas turbine to produce electricity. The heat recovered from the gases leaving the turbine are used to preheat the boiler feed water and this water is sent into the heat exchanger in the fluidized bed to extract the heat released during combustion. HTHP steam is produced in heat exchangers. This steam is used to drive a steam turbine which generates 65 to 75 % of the total power output of the system. The power output of the gas turbine is 25 to 35 % of the total. Altogether, the estimated PFBC cmbined cycle conversion efficiency reaches to 40 % which is much greater than the conventional systems.

**Integrated Gasification Combined Cycle System (IGCC)**

Among the new technologies, "Integrated Gasification Combined Cycle (IGCC)" system is one of the most promising ones. A schematic diagram of an IGCC system is given in Figure 2. In this system coal is gasified under high temperature and pressure. The coal gas exiting the gasifier is cleaned from the particulates and sulfurous gases. The clean gas is then sent through the gas turbines. The excess heat in the gas exiting the gas turbines is recovered in a heat recovery system to produce steam which in turn is used to run the steam turbines. Gas turbines are used together with the

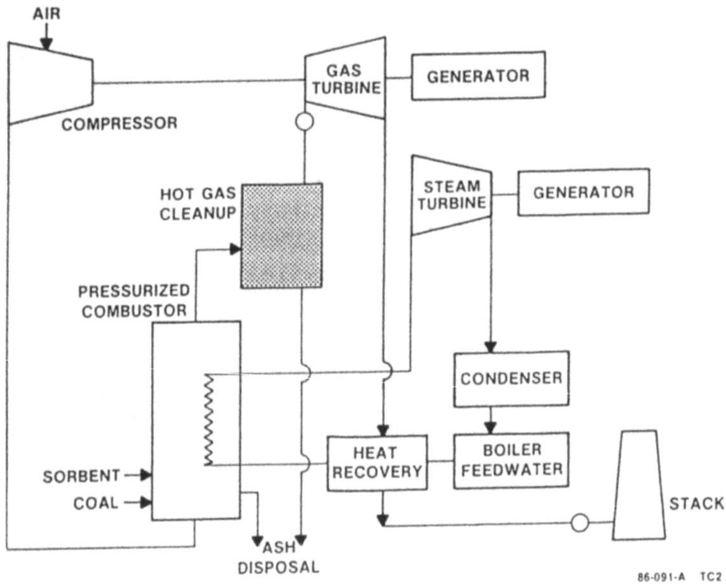


FIGURE 1 PFBC STEAM-COOLED SYSTEM

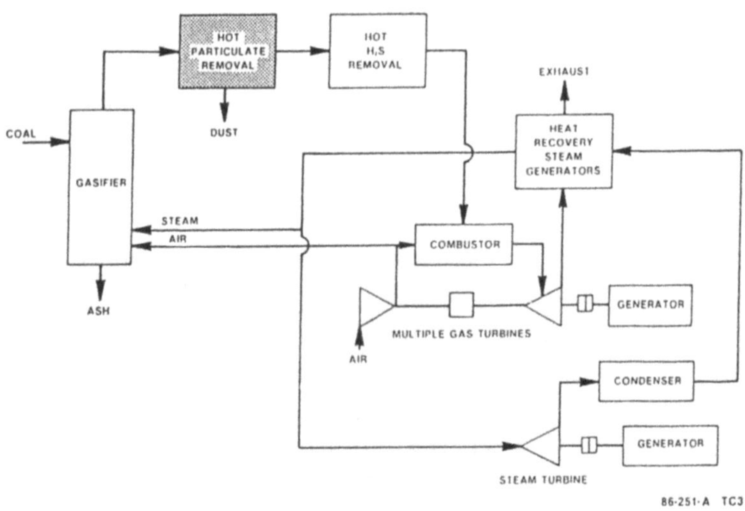


FIGURE 2 INTEGRATED GASIFICATION COMBINED CYCLE

steam turbines in this technology and that is why the name of "Combined Cycle" is given to the system. A high energy conversion efficiency reaching up to 50 % can be obtained with this combined cycle system.

In IGCC, the $NO_x$ and $SO_x$ emissions are much less than the conventional systems. But emission of $H_2S$, that is generated at the gasification unit, creates problems such as corrosion at the mechanical parts of the system and environmental pollution. In order to apply IGCC in commercial scale, $H_2S$ emission from the gasifier unit must be decreased to a tolerable limit. From a typical gasifier unit, the $H_2S$ concentration is around 5000 ppm. The tolerable limit for the system is around 150 ppm (Atımtay, et al., 1993). For the removal of $H_2S$ from coal gas several sorbents are under development. The research results of some of the sorbents reported in the literature are given here.

## 3. Experimental Section

The sorbent investigated in this study is zinc oxide and iron oxide on a support material, and it is prepared by wet impregnation. LZ-Y82 zeolite (1/8" extrudates, Union Carbide) is used as the support. LZ-Y82 has the following characteristics : $SiO_2$ = 65.6 % by weight, surface area = 625 $m^2/g$, crush strength = 1.77 kg/mm, pore diameter = 5.4 °A. The wet impregnation procedure includes : 1) Impregnation of the zeolite with a solution of the metal salt that we want to use in a rotary vacuum evaporator, 2) Dehydration of the sorbent in a vacuum oven at 70°C and 0.1 inch Hg, 3) Calcination of the sorbent in a muffle furnace at 750°C for 7 hours. Wet impregnation procedure can be applied to the zeolite for several times in order to provide the absorption of the desired amount of metal oxide on the zeolite. The procedure was repeated for three times in this study. The iron and zinc contents of the sorbent were 9.33 % and 4.88 % by weight, respectively. The BET surface areas and the crush strengths of the sorbents were also determined before and after the experiments.

A schematic diagram of the experimental system is shown in Figure 3. The reactor made of quarz tube was placed vertically in a tubular furnace which was connected to a temperature control unit. Reaction gases entered from the bottom of the reactor. A balance was used to follow up the weight change of the sorbent sample in the reactor with time. Sorbent pellets to be tested were placed in a pan made of platinum. The balance was connected to a computer system through a RS-232 interphase. With a special software written for our system, the weight of the sorbent in the reactor was measured and this measurement was transferred to the computer memory simultaneously. The data collected this way was stored in the computer in desired time intervals. Spreadsheet and graphics programs were used for data processing.

After the gases were mixed in a manifold, they were fed into the reactor. The gas mixture consisted of $H_2S$, $H_2$, and $N_2$. The $H_2S$ concentration was changed between 5000 and 15000 ppm. The temperatures tested were 500, 600 and 650°C.

The experiments consisted of a sequence of sulfidation and regeneration runs. The reactor temperature was held constant during the sulfidation and regeneration experiments. Reactor pressure was atmospheric. Regeneration of the sulfided sorbent was conducted with air and with a 6 % and 12 % oxygen/air (by volume) mixtures. After the sulfidation run was stopped, the reactor was purged with nitrogen for at least 15 minutes. The sorbents were subjected to five sulfidation/regeneration cycles to establish sulfur sorption capacity and regenerability. Sorbent samples were analyzed for total weight, metal content, crush strength, surface area before and after reaction.

## 4. Results and Discussion

The sulfidation efficiency of the supported zinc ferrite sorbent at 600°C and 5000 ppm $H_2S$ has been shown in Figure 4. Sulfidation efficiency reaches to 88.2 % and at 650°C the conversion efficiency is even higher (95 %). When sulfidation efficiency of zinc ferrite is compared with the individual oxides (zinc oxide and iron oxide separately supported on zeolite), zinc ferrite shows the highest value. The conversion also increases with temperature.

The BET surface area of the zinc ferrite sorbent has been determined by using a monosorb equipment of model "Quantachrome". The surface area of the unreacted zinc ferrite sorbent has been found to be 300 $m^2$/gr, however after the sulfidation reaction this area has decreased to 288 $m^2$/gr at 600°C. Sintering can also result in a decrease in the BET surface area, but sintering is not expected at the temperature range studied. So, this is probably due to the higher specific volume of ZnS when compared with ZnO.

The crush strength analysis on the sorbents tested at 600°C and $H_2S$ concentration of 1 % by volume revealed that sulfidation resulted in a decrease in the crush strength of the fresh sorbent. Zinc ferrite has a higher crush strength (2.04 kg/mm) when compared with the original crush strength of the zeolite (1.7 kg/mm) after the sulfidation reaction. Before the sulfidation reaction the crush strength of the fresh sorbent was 2.87 kg/mm.

Regeneration tests were conducted with partially sulfided zinc ferrite. Zinc ferrite was sulfided by 70 % before the regeneration test. Regeneration tests were run at three different $O_2$ concentrations of 6, 12 and 21 % by volume at two different temperatures of 600 and 650°C. The general regeneration process is the reaction of a metal sulfide with oxygen to form metal oxide and $SO_2$. At low temperatures and high oxygen concentrations sulfate formation is also possible for zinc ferrite. The effect of oxygen concentration on the

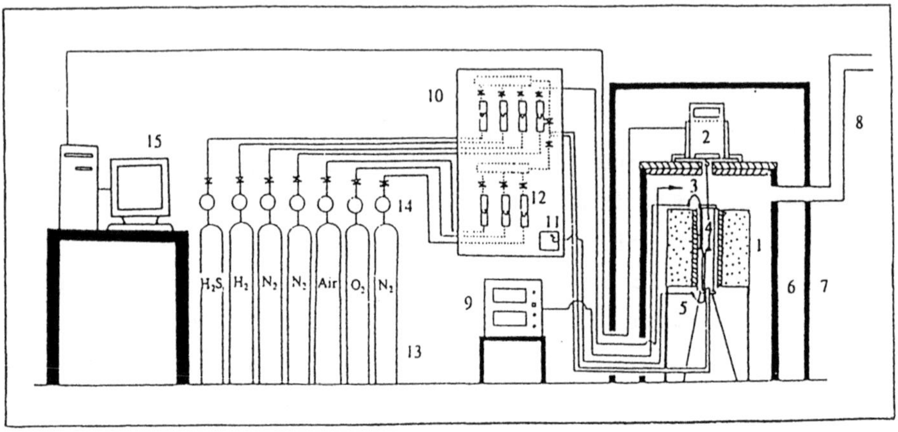


Figure 3 Schematic Diagram of the Experimental Setup

1. Furnace
2. Balance
3. TC for Furnace
4. Quartz Reactor
5. TC for Reactor
6. Inner Cabin
7. Outer Cabin
8. Gas Suction Pipe
9. Furnace Control Unit
10. Control Panel
11. Temperature Indicator
12. Rotameters
13. Gas Cylinders
14. Regulators
15. Computer

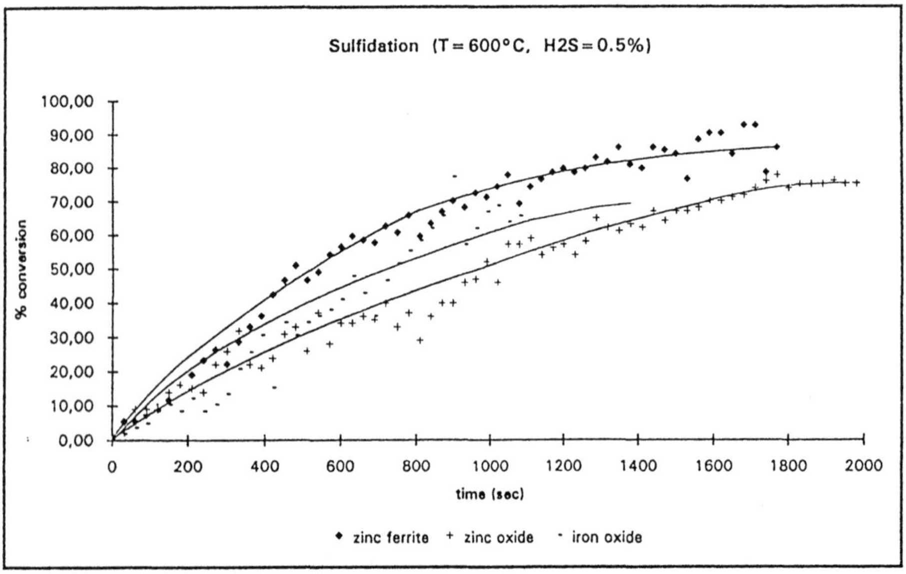


Figure 4 Sulfidation Efficiency Curves of Zinc Ferrite, Zinc Oxide and Iron Oxide at 600°C and 0.5% $H_2S$ by volume

conversion due to regeneration is given in Figure 5. Complete stoichiometric sulfidation for the zinc ferrite sorbents, partially sulfided to be used in the regeneration tests, produced a final dimensionless weight of 1.02 (for 70 % sulfidation). This dimensionless weight value is the initial dimensionless weight of the sample to be regenerated. Complete regeneration should produce a final value of 1.0. Values greater than 1.0 may reflect incomplete sulfide removal or especially the formation of zinc sulfate (Focht, 1989). For all oxygen concentrations the final dimensionless weight value reached were above 1.0. This is probably due to sulfate formation. In a study by Silaban (1991) some sulfate formation was observed at 600°C in bulk zinc ferrite. The presence of oxides of iron and zinc together in the composition of zinc ferrite decreases the sulfate formation when compared with the individual zinc oxide.

In order to test the multicycle sulfidation-regeneration usage of the supported zinc ferrite, a test of 5 1/2 cycles was made, where a cycle consisted of a sulfidation and a regeneration reaction. The runs were made at 650°C and with $H_2S$ concentration of 1 % by volume. For regeneration 6 % oxygen was used. The results obtained from the multicycle test is given in Figure 6. As can be seen from this figure, multicycle test shows good regenerability. Sulfidation efficiency decreases after the first cycle, but it is more or less the same for the following cycles. The behaviour in the first cycle can be attributed to the stabilization of the sorbent structure. After 5 1/2 cycle of testing, no negative effect was observed on the BET surface area. A decrease in the crush strength was observed, but it was still higher (2.25 kg/mm) than the crush strength of the original support zeolite (1.7 kg/mm).

## 5. Conclusions

Zinc ferrite (combination of zinc oxide and iron oxide) was supported on a zeolite in order to increase the structural

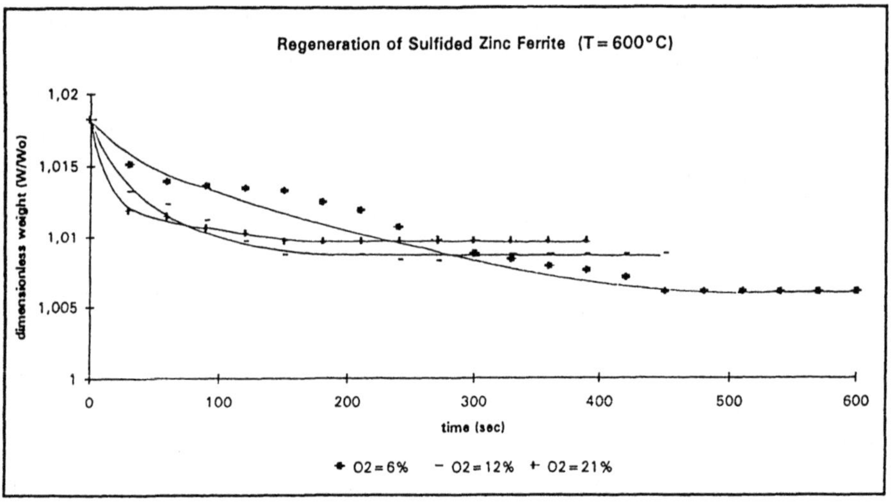


Figure 5 Effect of $O_2$ Concentration on the Regeneration of Sulfided Zinc Ferrite (T=600°C)

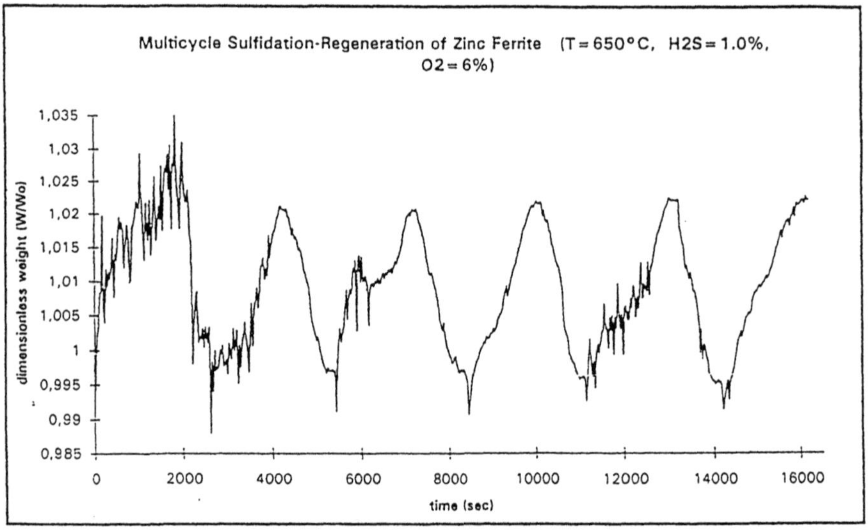


Figure 6 Multicycle Sulfidation-Regeneration of Zinc Ferrite

stability of the sorbent. The results of sulfidation tests showed that presence of iron oxide and zinc oxide together in the formulation of zinc ferrite resulted in better sulfidation efficiencies when compared with these metal oxides individually. The sulfidation efficiencies at 600 and 650°C were the highest for zinc ferrite when compared with the sulfidation efficiencies of zinc oxide and iron oxide. In conclusion, the results obtained for supported zinc ferrite are quite satisfactory. Zinc ferrite combines the easy regenerability of the iron oxide and high sulfur absorption capacity of the zinc oxide.

## References

METC (1987) Gas stream cleanup (Technology Status Report), Report no: DOE/METC-87/0225,3-7.

Atimtay AT, Gasper-Galvin LD, Poston JA (1993) Novel supported sorbent for hot gas desulfurization, Environ. Sci. Technol. 27(7): 1295-1303.

Focht GD, Ranade PV, Harrison DP (1989) High temperature desulfurization using zinc ferrite regeneration kinetics and multicycle testing, Chem. Eng. Sci. 43(11): 2919-2926.

Silaban A, Harrison DP, Berggren MH, Jha MC (1991) The reactivity and durability of zinc ferrite high temperature desulfurization sorbents, Chem. Eng. Comm. 107:55-71.

## Acknowledgment

The financial support provided by the Turkish Scientific and Technical Research Council for this project is deeply appreciated.

# Prevention of Air Pollution Problems in Thermal Power Plants in Turkey

Aysel T. Atımtay and Günal Özenirler
Environmental Engineering Department
Middle East Technical University
Ankara, 06531 Turkey

## Introduction

Energy is a very essential element for the industrial development of a country. Therefore, when a country is developing it needs power plants for the production of energy. The energy requriment in Turkey has shown an exponential increase throughout the years as it can be seen from Figure 1. The energy requriment can be met by hydraulic or thermal power plants not considering the nuclear ones. According to the figures published in 1991, out of Turkey's total capacity of energy production (17,200 MW(e)), 41 % is produced by hydraulic and 59 % is produced by thermal power plants.

The fuel used in thermal power plants in Turkey are mostly coal and specifically lignites. Usually, Turkish lignites are high in sulphur and ash contents. The emission of $SO_2$ and particulates create environmental problems. Therefore, it is a big issue in Turkey to be able to make these coal fired power plants environmentally friendly and control their emissions according to the Turkish Air Quality Protection Regulations.

NATO ASI Series, Partnership Sub-Series, 2. Environment – Vol. 2
Cleaner Technologies and Cleaner Products
for Sustainable Development
Edited by H. M. Freeman, Z. Puskas, and R. Olbina

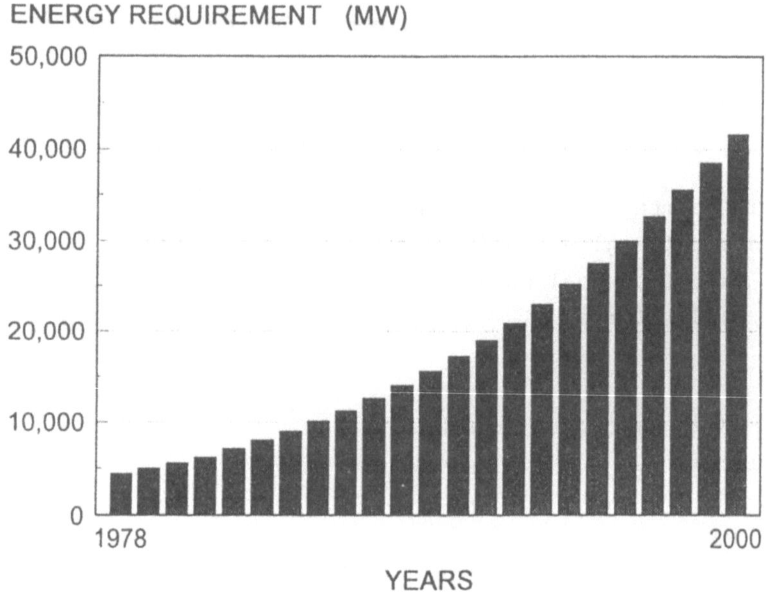


**Figure 1.** Energy Requirement of Turkey

**Thermal Power Plants in Turkey**

This paper will summarize the status of power plants and their air pollution control. The total capacity of thermal power plants in Turkey by January 1993 is 9078 MW(e). The names, status and fuel types used in thermal power plants are listed in Table 1. Most of the thermal power plants are located in the western part of Turkey due to the fact that large coal reserves suitable for power production are found in that part of the country. Two of those plants, Hamidabat and Ambarlı Power Plants (both on Combined Cycle System), are located along the natural gas pipeline. This pipeline distributes the natural gas purchased from Russia.

**Table 1.** Names and Status of Thermal Power Plants

| Name of Plant | Installed Capacity, MW(e) | Fuel | Status |
|---|---|---|---|
| Afşin-Elbistan | 4 * 340 | Lignite | Operation |
| Çayırhan | 2 * 150 | Lignite | Operation |
| Kangal | 2 * 150 | Lignite | Operation |
| Kemerköy | 3 * 210 | Lignite | Construction |
| Orhaneli | 1 * 210 | Lignite | Operation |
| Seyitömer | 4 * 150 | Lignite | Operation |
| Soma-A | 2 * 22 | Lignite | Operation |
| Soma-B | 6 * 165 | Lignite | Operation |
| Tunçbilek-A | 64 + 65 | Lignite | Operation |
| Tunçbilek-B | 2 * 150 | Lignite | Operation |
| Yatağan | 3 * 210 | Lignite | Operation |
| Yeniköy | 2 * 210 | Lignite | Operation |
| Ambarlı | 630 | Fuel Oil | Operation |
| Aliağa CC | 180 | Fuel Oil | Operation |
| Hamidabat CC | 1200 | Natural Gas | Operation |
| Ambarlı CC | 1350 | Natural Gas | Operation |
| Denizli | 15 | Geothermal | Operation |
| Çatalağzı-B | 2 * 150 | Hard Coal | Operation |

Ambarlı CC thermal power plant has been awarded by "Power International" as the largest power plant on combined cycle system (CC), for being energy efficient (51.4 %) and environmentally friendly .

The characteristics of Turkish lignites are given in Table 2. As can be seen from this table, the S and ash contents of these coals are high. Coal reserves in Turkey are rich in lignite with an estimated amount of 8,100 million tons. The reserves of various other coals in Turkey are given in Table 3.

**Table 2.** Characteristics of Turkish Lignites

| | Range of Values | Average Values |
|---|---|---|
| Moisture, % wt. | 12 - 42 | 32 |
| Ash, % wt. | 14 - 42 | 22 |
| Volatile Matter, % wt. | 16 - 38 | 26 |
| Fixed Carbon, % wt. | 11 - 39 | 20 |
| Total Sulphur, % wt. | 0.8 - 8.5 | 1.8 |
| Calory Value, Kcal/kg | 950 - 4500 | 1800 |

**Table 3.** Coal Reserves in Turkey

| Reserve Type | Amount (million tons) |
|---|---|
| Lignites | 8,100 |
| Hard Coal | 1,300 |
| Asphaltite | 75 |
| Bituminous Material | 810 |

**Emission Limits For Power Plants**

The Air Quality Protection Regulation which has become effective on November 2, 1986 is the regulation that all power plants in the country should comply. This regulation was prepared in line with the purpose and principles envisaged in different articles of the Environmental Law passed on August 9, 1983.

The emission limit values are given in this regulation according to the fuel types and the remaining operating hours of the power plants. Table 4 gives the $SO_2$ limits for different size plants. Table 5 summarizes the limits for Dust and $NO_x$.

**Table 4.** $SO_2$ Emission Limit Values

| Combustion Plants | $SO_2$ Emissions ($mg/Nm^3$) | | | | | |
|---|---|---|---|---|---|---|
| | < 300 MWt | | New Plants | > 300 MWt | | |
| | Remaining Operating Hours | | | Remaining Operating Hours | | |
| | < 20,000 | > 20,000 | | < 20,000 | 20,000 - 50,000 | > 50,000 New Plants |
| Solid Fuel Fired Plants | - | (*) 3200 | (*) 2000 | - | (*) 3200 | (*) 1000 |
| Liquid Fuel Fired Plants | - | (*) 3200 | (*) 1700 | - | (*) 1700 | (*) 800 |
| Gas Fired Plants | (**) 60 | (**) 60 | (**) 60 | 60 | 60 | 60 |

(*): If limits are exceeded, $SO_2$ emission degree have to be reduced to 10% in the plants with a thermal capacity of up to 300 MW(t) and to 5% for those with a thermal capacity of higher than 300 MW(t).

(**): This limit is 100 mg/Nm$^3$ for the plants having a fuel thermal capacity below 100 MW(t).

**Table 5.** Dust and $NO_x$ Emissions

| Combustion Plants | Dust Emission, (mg/Nm$^3$) | | $NO_x$ Emission, (mg/Nm$^3$) | |
|---|---|---|---|---|
| | Old Plants | New Plants | Old Plants | New Plants |
| Solid Fuel Fired Plants | (***) 250 | (***) 150 | 1000 | 800 |
| Liquid Fuel Fired Plants | 110 | 110 | 1000 | 800 |
| Gas Fired Plants | 10 | 10 | 500 | 500 |

(***): In case of lignites having an original ash content of over 18 % this limit can be increased by 200 % with the permission of General Management of Environment.

## $SO_2$ and $NO_x$ Emission Values in Turkey

In a master thesis carried out by G.G. Palaogullarından in the Environmental Engineering Department of the Middle East Technical University, in 1991, the emissions from 132 individual plants (cement, sugar, iron and steel, paint and varnish, fertilizer, soap and detergents, sulphuric acid, aluminum, copper smelters, sulphur recovery, refineries and power plants)were calculated by using the "emission factors" compiled by EPA. In this study, suspended particulates, $NO_x$ and $SO_2$ emissions from each plant were estimated and results from individual plants were added up to determine emission inventory of the whole country.

Annual emissions of $NO_x$ and $SO_2$ in Turkey according to the above mentioned study are given in Table 6 in tons/year.

**Table 6.** $NO_x$ and $SO_2$ Emissions in Turkey (1991)

| Emissions | Amount (ton) |
|---|---|
| $NO_x$ | 240,000 |
| $SO_2$ | 1,300,000 |

The percentages of $SO_2$ and $NO_x$ emissions for thermal power plants are 69 % and 75 % of the total emissions, respectively. Thermal power plants are main sources of $SO_2$ emissions from the industries because of the high S content lignite coals used. $SO_2$ emissions are usually above the limits given in Turkish regulation. Therefore, flue gas desulphurization (FGD) systems are required to lessen the $SO_2$ emissions. For this purpose, the first FGD plant in Turkey was installed in Çayırhan Thermal Power Plant.

Since Turkish coals are low in nitrogen the amount of $NO_x$ emitted is below the standards, as far as the particulate emission is concerned, all power plants in Turkey are equipped with Electrostatic Precipitators for the control of dust emission.

## The Design Characteristics of the Çayırhan Thermal Power Plant

Çayırhan Thermal Power plant is located in the central part of Turkey, about 120 km northwest of Ankara. The plant consists of two units, each having a capacity of 150 MW (e). The fuel used in the plant is Beypazarı lignite, which is mined in a near by location. The characteristics of the lignite coal which has been taken as basis for the design of the combustion chamber are as follows;

| | | |
|---|---|---|
| Moisture | : | 27.4 % by wt. |
| Ash (dry basis) | : | 41.3 % by wt. |
| Total Sulphur | : | 4.65 % by wt. |
| Combustible Sulphur | : | 4.15 % by wt. |
| Lower Heating Value | : | 2700 + 100 Kcal/kg |

The waste materials produced from the combustion chambers are fly ash, bottom ash, and the flue gas. The bottom ash is disposed at the Ash Dam 1 km away from the plant. In order to prevent the dusting of the ash during transportation and in deposition, the ash is wetted with water. Since the soil in the Çayırhan region is of impermeable character, there is no danger of ground water pollution. The fly ash is collected at the Electrostatic Precipitators which work with an efficiency of about 99.5 %. The amount of fly ash emitted from the stack can be lowered to 137 mg/Nm$^3$ at full load.

**Installation of The FGD System in Çayırhan Thermal Power Plant**

Since the combustible sulphur content of the lignite used at Çayırhan Thermal Power Plant is as high as 4.15 % by weight, the concentration of the $SO_2$ in flue gas can go up as high as 21,000 mg/Nm$^3$. If two generating units work at full load, it has been calculated that 30 tons of $SO_2$ will be emitted to the atmosphere each hour. Therefore, Turkish Electricity Board has decided to install a Flue Gas Desulphurization system to the Çayırhan Thermal Power Plant to prevent the air pollution problem.

An international tender was opened for the FGD plant in 1985 and the contract for the construction of the plant was awarded on July 1987 to a consortium consisting of Bischoff-Klockner-Lentjes of Germany and their Turkish partner Gama on turn-key basis. The trial runs for the FGD System began on November 1991. The total cost of the FGD plant was one third of that for the thermal power plant. Temporary acceptance of the plant was made at the beginning of 1992. Today, the plant is working without any major problems.

The process of the FGD System is based on the wet absorption principle using $CaCO_3$ as the absorbent additive to reduce $SO_2$ emissions. Limestone is obtained from nearby quarries. Recoverable gypsum is produced as a by-product.

**Table 7.** Design Values for Full Capacity

| | |
|---|---|
| Flow rate of the incoming gas | 600,000 $Nm^3/hr$ |
| $SO_2$ concentration at the inlet | 17,600 $mg/Nm^3$ |
| $SO_2$ concentration at the outlet | 800 $mg/Nm^3$ |
| Overall $SO_2$ removal efficiency | 95 % |
| Particle concentration at the inlet | 400 $mg/Nm^3$ |
| Particle concentration at the outlet | 100 $mg/Nm^3$ |
| Temperature of the gas at the inlet | 160°C |
| Temperature of the gas at the outlet | 60°C |
| Water consumption (for each unit) | 77.7 $m^3/hr$ |
| Limestone consumption (for each unit, as $CaCO_3$) | 17.9 ton/hr |
| Energy consumption (for each unit) | 9,565 kW-hr |

There are three main sections in the FGD system:

1. Limestone handling and absorbent preparation unit
2. Sulphur oxides scrubbing system
3. Gypsum dewatering system

**Power Plants in the Program for an FGD System**

According to information obtained from the authorities of the Turkish Electricity Board and papers published by the Directory of Environmental Division, the plants which have been included in a program to be installed with FGD's are as follows;

**Table 8.** FGD Systems Planned for Future

| Power Plants | Capacity MW(e) | Process |
|---|---|---|
| Orhaneli | 1 * 210 | Limestone scrubbing |
| Kemerköy | 3 * 210 | Limestone scrubbing |
| Yatağan | 3 * 210 | Limestone scrubbing |
| Yeniköy | 2 * 210 | Limestone scrubbing |
| Soma | 6 * 165 | Limestone scrubbing |

**Other Refurbishment Studies**

There is a "coal pollution abatement project" started with the World Bank.
That project includes:

1. Improvement of ESP in the Soma-B Power Plant, Manisa
2. Conversion of the conventional pulverized coal combustion system in Soma-A Power Plant to the fluidized bed combustion system.

Besides, World Bank will support the installation of the continuous $SO_2$, $NO_x$ and particulate monitoring instruments in the power plants with a budget of 1,000,000 US$.

**Conclusion**

As described above, FGD systems are quite expensive and their construction as well as their operation is a burden for developing countries. However, for the prevention of global air pollution problems, abatement programs for $SO_2$ emission are essential, though not adequate.

In this regard, the studies mentioned above are important steps to decrease the $SO_2$ emissions from thermal power plants of Turkey.

**References**

The Air Quality Protection Regulation, Official Gazette #19269, 2.11.1986.
Gülen Sibel, 1991, "Çayırhan Thermal Power Plant with the FGD System", Summer Practice Report.
Kayhan, B. A., Esmerliğil, L. N., 1978, "Nuclear Fuel Conversion and Related Nuclear Types"; Turkey Third General Energy Conference, Vol. 1. pg. 616, Ankara.
"Evaluation of Pollution Control Technology for Coal Combustion" Status Report on the Control of Air Pollution from Fuel Combustion in Turkey prepared for NATO/CCMS, June 1993, York, U.K.
Palaoğullarından G. G., 1991, "Emission Inventory Studies in Turkey", Master Thesis, METU, Ankara.
Turkish Electricity Board Publications.

# A Life Cycle Study for Common Construction Materials

**Aysen Müezzinoğlu and Hikmet Toprak**
**Department of Environmental Engineering**
**Dokuz Eylül University**
**Bornova 35100, İzmir**
**Turkey**

## 1. INTRODUCTION

Environmental importance of block housings to answer the tremendous demand of dwelling units of the increasing population in Turkey as a symbol of the developing world is studied in this presentation. With this purpose a housing complex named "Mavişehir" presently being established in the city of İzmir has been taken as a case.

The method of investigation is to study the life cycle steps of common construction materials with the concept of "from-cradle-to -grave". Important life cycle stages are demonstrated in the form of flow schemes. Key steps are studied for material and energy balances and some figures to relate the resource utilization with the "Mavişehir" buildings are established.

## 2. IMPORTANCE OF CONSTRUCTION SECTOR AS AN UNSUSTAINABLE ACTIVITY

There is a permanent gap between demand and supply of new dwellings in Turkey probably like many other parts of the world. This shortage of supply is due to high costs involved in building new houses in constrast to the high growth rate of the population in urban centers. It is expected that similar trends exist in other developing countries, too. This means that more and more natural resources including the land for construction must be used for new houses all over the world.

Construction business is the leading sector in the Turkish

NATO ASI Series, Partnership Sub-Series, 2. Environment – Vol. 2
Cleaner Technologies and Cleaner Products
for Sustainable Development
Edited by H. M. Freeman, Z. Puskas, and R. Olbina

economy. However, availability of enough planned zones in city plans for future growth with acceptable infrastructure facilities is always behind the actual demand and therefore building unplanned squatter housings is a striking reality especially in big cities. Dependable statistical information pertaining to those unplanned regions is missing and is always an important point of dispute. That is why in the following case study only the planned and (therefore) licenced dwellings are taken into consideration. It is roughly estimated that the source utilization rate for squatter housings is at the same order of magnitude as the planned residential buildings that are dealt with here.

In Turkey gross national income was calculated as 135 billion USD at year 1990 out of which only 6% was related with the construction activities. However, the national budget was 15 billion USD and construction expenditures compare with 33% of this amount. In the construction sector 62% belonged to private and governmental dwellings, which means that a sum of about the size of 20% of the Turkish budget has been spent for licenced houses.

In Turkey annual urban population increase rate is 3.38% (clearly above the overall population growth rate of slightly less than 2% per year after 1990). Turkey has 814,578 $km^2$ of real land area out of which approximately 11,000 $km^2$ must be reduced for lakes and reservoirs. Out of the remaining "land" if %80 is assumed to be "dwellable" by excluding the high mountains, rocks, rivers, marshlands, etc. 642,950 $km^2$ of land is really open for buildings.

Estimated trends in the growth of urban areas for the next few centuries is a striking one. It is recorded that about 44 million $m^2$ per year of new <u>usable dwelling floor space</u> has been licenced in 1990 in Turkey. If we assume that the real figure was twice as much by the addition of the unrecorded squatter housings, it is roughly estimated that 88 million $m^2$ of net dwelling floor space was made available at this year. If this figure is assumed to be built on single flat houses only, it means that 0.015% of new dwelling land is consumed each year out of the 642,950 $km^2$ of "dwellable" national territory. Starting from a % 10 urban land coverage at present and assuming that the urbanization takes place at the fixed rate of year 1990, urban land will eat out all of the territorial land of Turkey in about 65 centuries.

On the other hand, assuming a ratio between net usable dwelling area and gross land utilization for construction is about 15% to allow for enough green areas. Presently this figure is 25% in Turkey causing concrete covered land surfaces over the cities. With 15% assumption the period of time for gross dwelling land use is calculated to be 9.75 centuries. If we assume that 35% of the urban land is dedicated to residential blocks, the period of time until all Turkey will convert into a huge city is only 3.41 centuries, or 341 years. This means that after approximately 5 generations all of the Turkish population will start living in a huge single city (probably to be named Istanbul as this is the fastest growing city presently with already more than 10 million population). Similar calculations may be made for many countries of the developing world. In this extreme situation there will be no room for any agriculture, any large scale industry, any rural land, or an appreciable coverage of forests, meadows, beaches, recreational empty space etc.

This means that the present trend in urbanization is clearly unsustainable with respect to the land use aspects within a short period of time. However, unjustly land use is not the only environmental cost of urban growth. In such an important activity sector, huge amounts of materials, energy and efforts are spent, too. Also to avoid the ever increasing costs of planned urban land to be prepared for construction, city administrators have to permit high rise buildings which is another serious cause of environmental conflicts. As it is known, high rise buildings create secondary adverse effects on the urban environmental quality.

This serious unsustainability in urban growth is the reason why possible total environmental costs of construction activities are being treated in this paper. With this in mind a planned (rather luxury) type of a housing complex being constructed in Izmir is taken as a case to study the **direct** (or primary) environmental costs of housing constructions.

## 2. DESCRIPTION OF THE STUDY METHOD AND THE SELECTED CASE

Because life cycle analysis is a good method to test the sustainability of a good or a service, environmental costs of structural materials of sand, gravel, water, cement and steel which are the main elements of any construction work are established by means of this type of analysis in this study.

In Izmir (the third crowdest city of Turkey) urbanization growth rate figure is 3.30% which is about the same as the country average. Therefore any conclusion to be withdrawn from İzmir urbanization rates is quite representative of all of Turkey. That is one of the reasons why a large construction zone called "Mavişehir" in İzmir has been chosen as a case study in this presentation. This is a residential complex covering a large area and it consists of high rise buildings. The housing lot consists of blocks of 16, 17, 19 and 20 floors. There are 2,872 individual apartment flats in this complex.

### 2.1. IMPORTANT ENVIRONMENTAL CONFLICT AREAS IN THE STUDIED CASE

Following discussion is an incomplete effort taking some material costs that can be studied in more detail. However, we must note that several other repercussions may also exist and jeopardize the sustainability of urban life. Important impact factors for which some environmental cost figures must be affiliated are:

**Land use disturbances:** "Mavişehir" housing complex is a large part of the city land use plan and it is located at a delicate coastal environment. This land is a recently dried salt marshland formed by an old delta of Gediz River which is flowing today at further north. About a century ago lower portion of the river has been diverted in order to protect the İzmir Bay from filling with sedimentary materials. So this old delta has been freed from alluvional assault and stayed as a salt marshland, a kind of delicate wetland, in the further north of which a very valuable bird's sanctuary exists. In between "Mavişehir" and the sanctuary huge salt pans and a very large salt manufacturing facility takes place. Another large portion of the marshland will soon be converted into a series of large stabilization ponds of the natural wastewater treatment category. Environmental impacts of this system has been studied in detail.

Presently there is no widely acceptable method to internalize the environmental costs of use up of delicate ecological land such as this one. We can safely say that there will be left very little natural ecological balance, soon, perhaps not only due to this project but also the rate of overall developments in and around İzmir. However, it is almost impossible to evaluate the environmental cost of the lost ecological balance due to the urban developments.

**Emissions due to fuel use in material transport and work machinery:** "Mavişehir" buildings are being made by means of a "panel construction system" functioning with what is known as "sliding casts" which is a rather fast method of construction. High strength concrete type "B225" (concrete to withstand 225 $kg/cm^2$ pressure for 7 days) is being used in this construction. According to "Turkish Concrete Standards", 1 $m^3$ of "B225" consists of 800 kg sand, 1,278 kg gravel and 350 kg of Portland cement. Total amounts of these constituents used in "Mavişehir" concrete elements are 106,250 tons of sand, 134,100 tons of gravel, 119,000 tons of cement (Fig.1).

Sand: Sand has been dredged with excavators at the upstream section of Gediz river bed near "Salihli", a town 127 km far away from the "Mavişehir" and transported by means of trucks with carrying capacities of 30 tons each. Properties of the excavator used for dredging sand in Salihli are as follows:

- Loading capacity : 1.22 tons
- Loading time : 1 minute
- Diesel consumption : 18.0 kg/hour

Using these figures total diesel consumption for on-site production of sand was calculated to be 26.1 tons. For transportation of sand number of vehicles on route was 3,542 and the total vehicle kilometers was 899,668 km. Considering unit diesel consumption of a truck which is $2.64 \times 10^{-4}$ ton/km, total diesel consumption for transportation of all of the sand was found to be 237.5 tons.

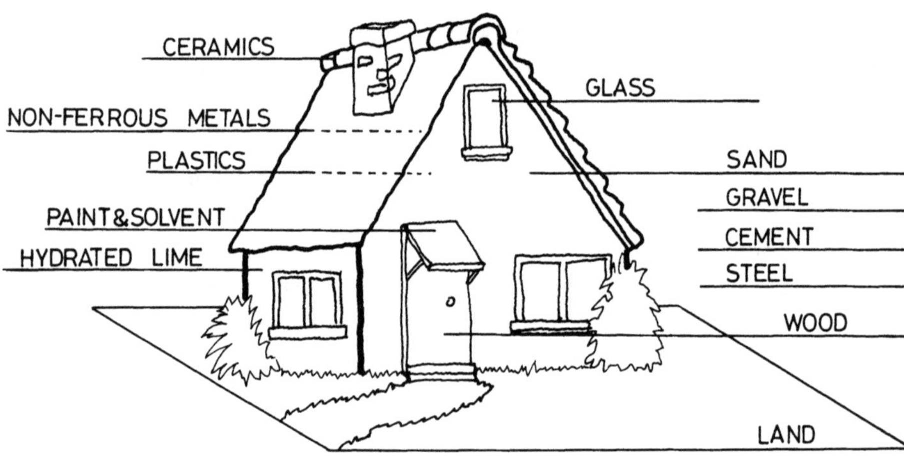


Fig.1. The types of materials being used in a dwelling

Gravel: Gravel and cement have been purchased from "Batı Anadolu Cement Plant" located at 27 km distance to the construction area. Gravel has been produced in the stone quarries by blasting and crushed at the site. Unit diesel fuel consumption for blasting and crushing are 1.92 and 0.53 kg/ton, respectively. Total diesel consumption for on-site production of gravel was determined to be 502.1 tons. Number of total vehicles on route were calculated to be 4,470 and total vehicle kilometers was found to be 241,380 km. Considering the trucks used for gravel transportation properties of which are the same as sand trucks, total diesel consumption for transportation of gravel was calculated to be 63.8 tons.

Cement and Concrete: Cement, has been brought from "Batı Anadolu Cement Plant" by means of 30 ton capacity container trucks. Number of trucks on route were 5,950 and total distance travelled was 309,400 km. Total diesel consumption was found to be 81.6 tons for this transportation.

Concrete has been prepared in the construction site by diesel engine-driven mixers. According to "Turkish Unit Cost Analysis for Contractors", for producing of 1 $m^3$ concrete 2.11 kg diesel fuel is to be utilized. Considering the total amount of concrete used in "Mavişehir" was 340,000 $m^3$, total diesel consumption was found to be 717.4 tons.

Steel: Steel has been provided from an iron and steel mill complex which is located at the city of İskenderun 1,045 km far from "Mavişehir" at the Southeastern Mediterranean coast. Steel has been transported by trailers of 50 ton capacity. Number of trailers on route between İskenderun and "Mavişehir" was calculated to be 540 and total transportation distance travelled was 1,128,600 km. Unit diesel consumption of a trailer is $5.29 \times 10^{-4}$ ton/km, therefore the total diesel consumption was calculated to be 596.9 tons.

Calculated overall diesel-fuel consumption for both on-site production and transportation sector related to sand, gravel, cement and steel that are treated in this section is summed in Table 1. In order to evaluate the air pollution potential of this much use emission factors of a diesel driven vehicle are given in Table 2.

Table 1. Diesel consumption in "Mavişehir" for the main raw materials of the construction (tons).

| Constituent | On-site production | Transportation | Total |
|---|---|---|---|
| Sand | 26.1 | 237.5 | 263.6 |
| Gravel | 502.1 | 63.8 | 565.9 |
| Concrete | 717.4 | - | 717.4 |
| Cement | - | 81.6 | 81.6 |
| Steel | - | 596.9 | 596.9 |
| Total | 1245.6 | 979.8 | 2225.4 |

Table 2. Air pollutant emission factors for diesel-engined machines/vehicles (g/kg diesel).

| Substance | Emission factor |
|---|---|
| CO | 82.8 |
| Hc | 12.0 |
| $NO_x$ | 38.5 |
| Dust | 4.3 |

From Table 2 total diesel fuel consumption on the construction of "Mavişehir" was calculated as 2,225.4 tons. Emissions corresponding to this much use of diesel fuel calculated from the factors listed in Table 2 are given in Table 3.

Table 3. Diesel vehicle emissions for "Mavişehir" (g).

| Substance | Emission factor |
|---|---|
| CO | $1.8x10^8$ |
| Hc | $2.7x10^7$ |
| $NO_x$ | $8.6x10^7$ |
| Dust | $9.6x10^6$ |

## 2.2. EMISSIONS EMBEDDED IN CONSTRUCTION MATERIALS

According to Figure 1, it should be remembered that not only the used costruction materials but also their raw materials are polluting matter. Their regular stack emissions and effluents are of well known characteristics and their production, storage, transport and use at construction create a high quantity of **fugitive emissions** which are of extreme importance to **Pollution Prevention** studies. Figures 2 to 10 shows the block diagrams showing the processes for manufacturing of common construction materials beginning with cement as the main material and its related raw materials and intermediate products and continuing with construction steel, cement, glass, paints, varnishes, wood and wood products, lime, sand and gravel, ceramics, plastics.

Cement: 1 ton clinker (the intermediate product in cement manufacture) is blended with 5 % additives by weight converts into 1.05 ton cement. Basic flow diagram of cement manufacture and flow diagrams of raw and auxiliary material manufacturing processes are shown diagrammatically in Figure 2.

Pollutant emission quantities coming from the rotary kiln stacks are summarized in Table 4. This table does not include fugitive dust emissions and are corrected for high efficiency electrostatic precipitators for dust control in the cement manufacture.

It is calculated that during the construction of "Mavişehir" complex 119,000 tons of cement has been utilized. To produce this amount of cement 113,333.3 tons of clinker has been processed. Cement manufacture emissions are shown in Table 5. According to the literature for the production of 1 ton of cement 147 kWh of electricity is used (Buonicore, 1993). In Turkey this comes from the national interconnected network. For production of 1,000 kWh of electricity 66.82 kg $SO_2$, 22.1 kg dust and 1,224 kg carbon gases (carbon dioxide and carbon monoxide as carbon equivalent) have been emitted from the power plants (Müezzinoğlu, 1994).

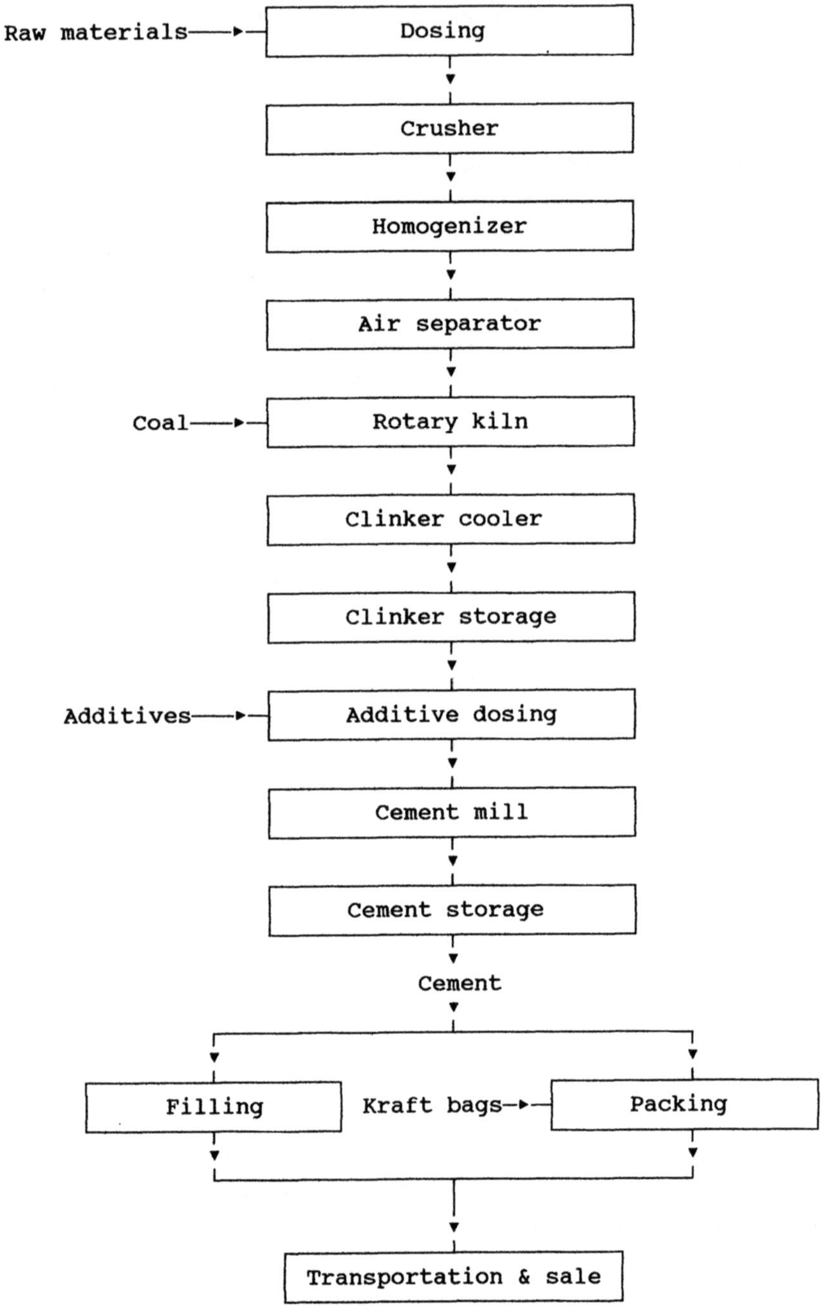


Fig.2. Flow scheme of cement manufacture.

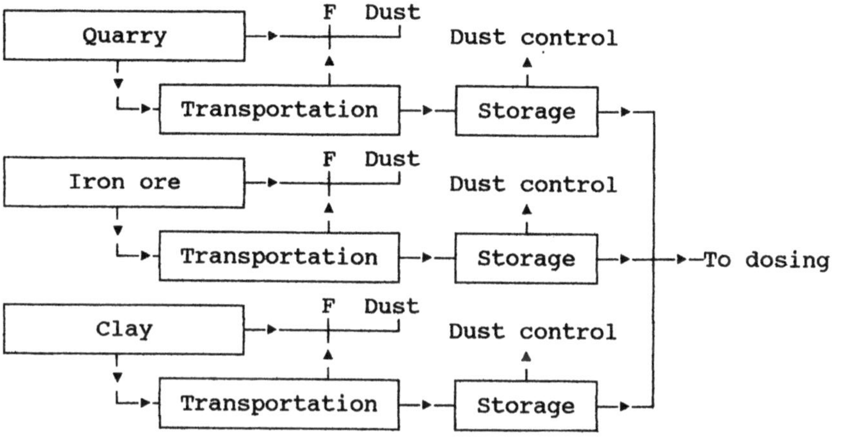


Fig.2.a. Raw material preparation for cement manufacture.

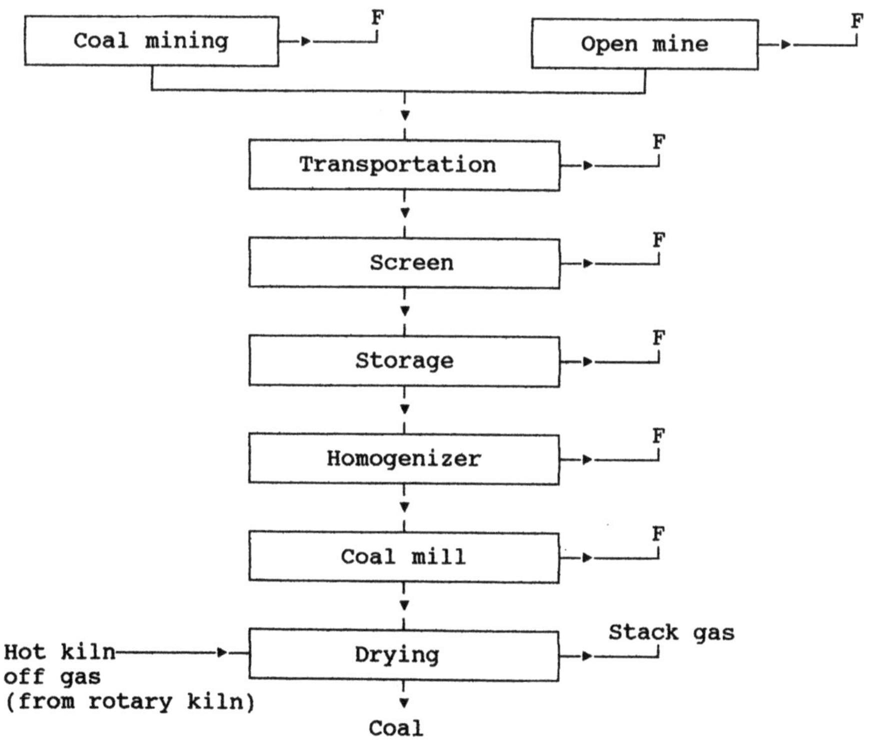


Fig.2.b. Coal preparation for cement manufacture.

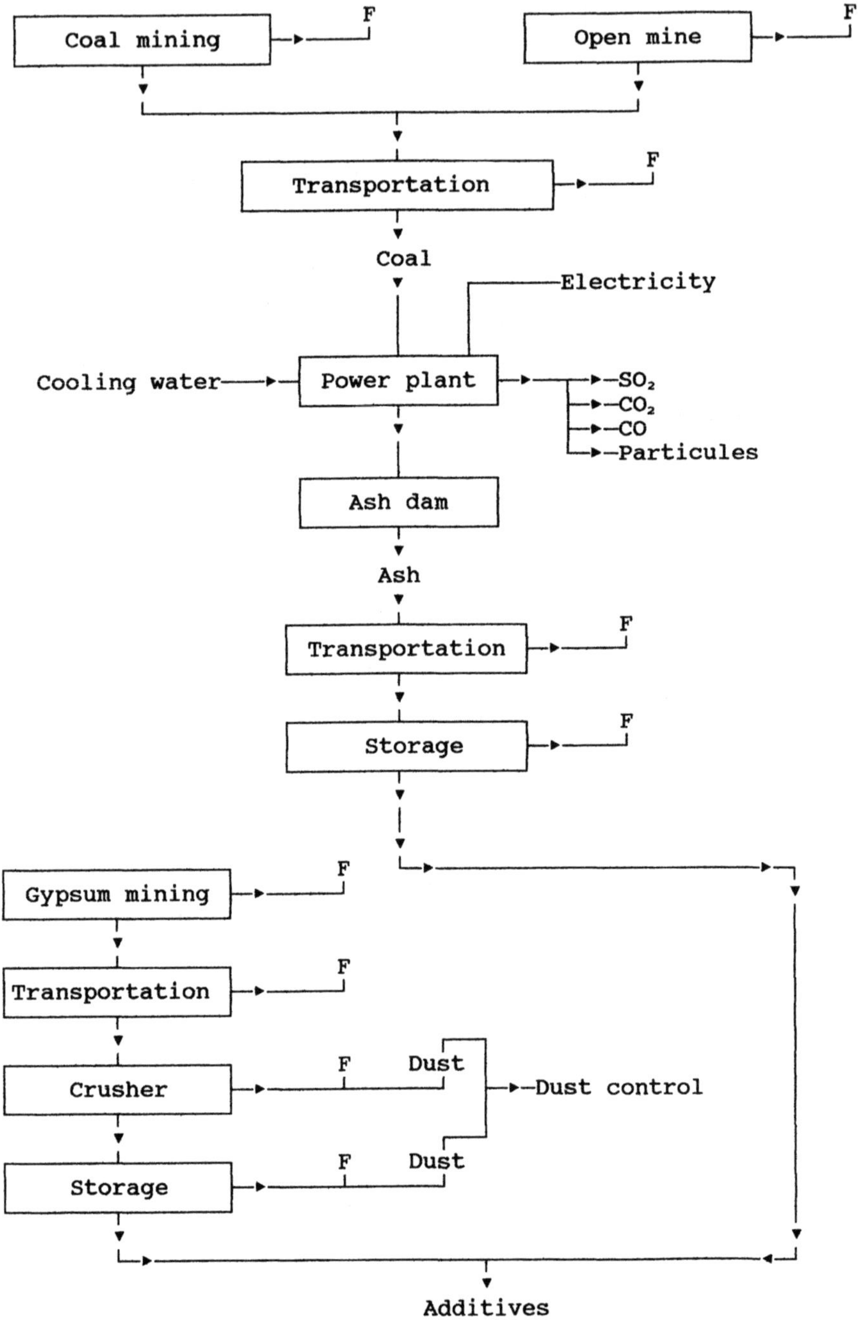


Fig.2.c. Additive preparation for cement manufacture.

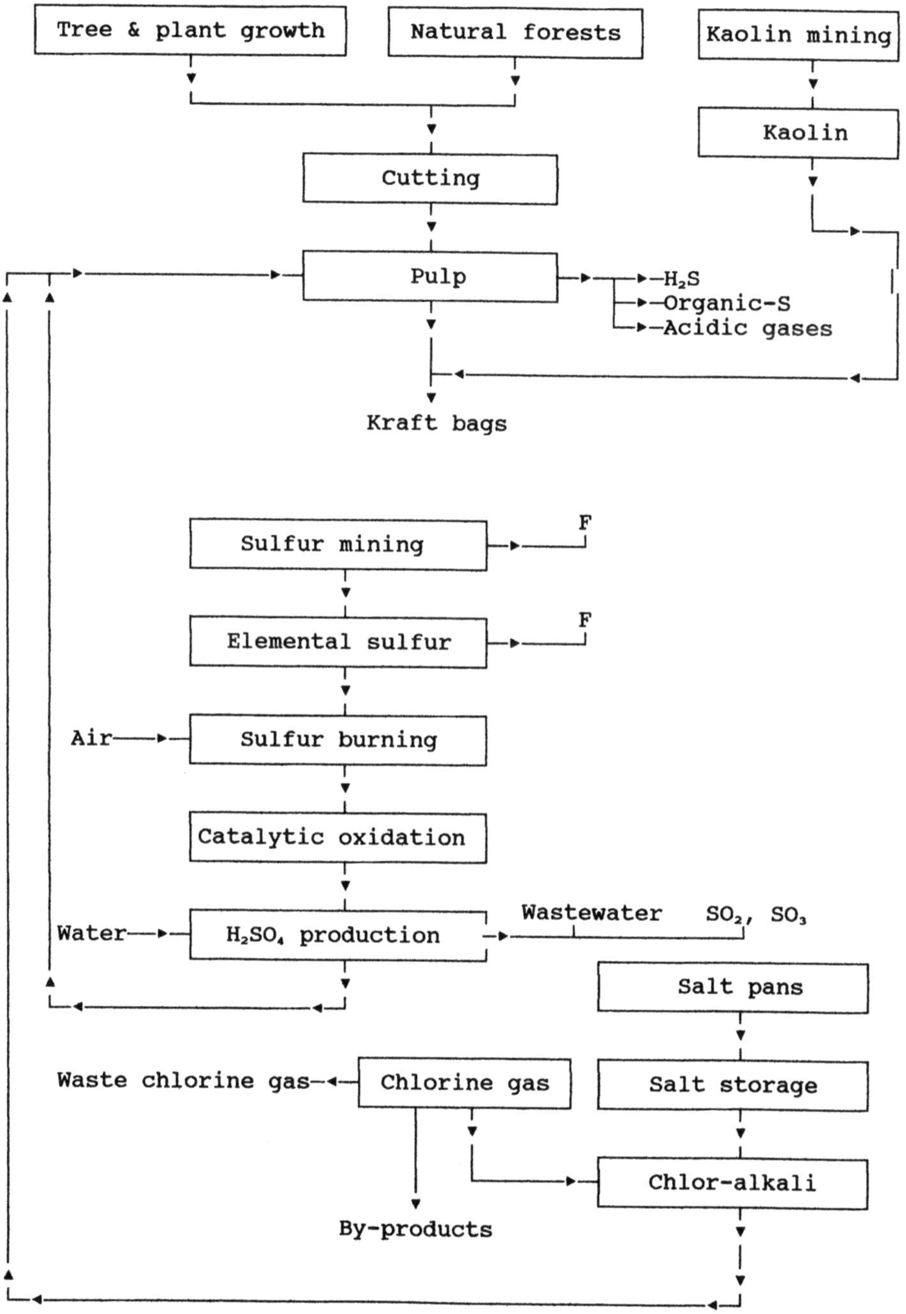


Fig.2.d. Kraft bag production for cement manufacture.

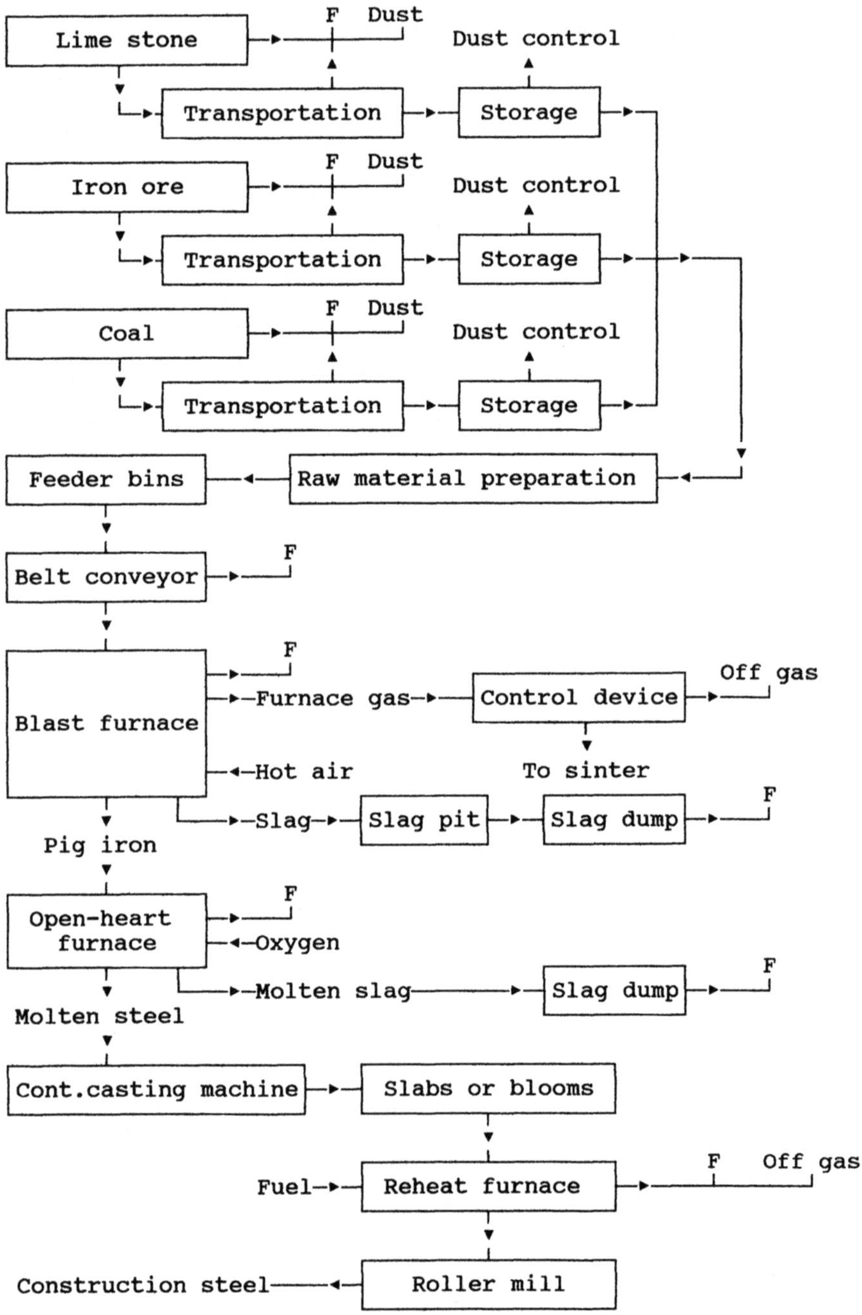


Fig.3. Flow scheme of iron and steel manufacture.

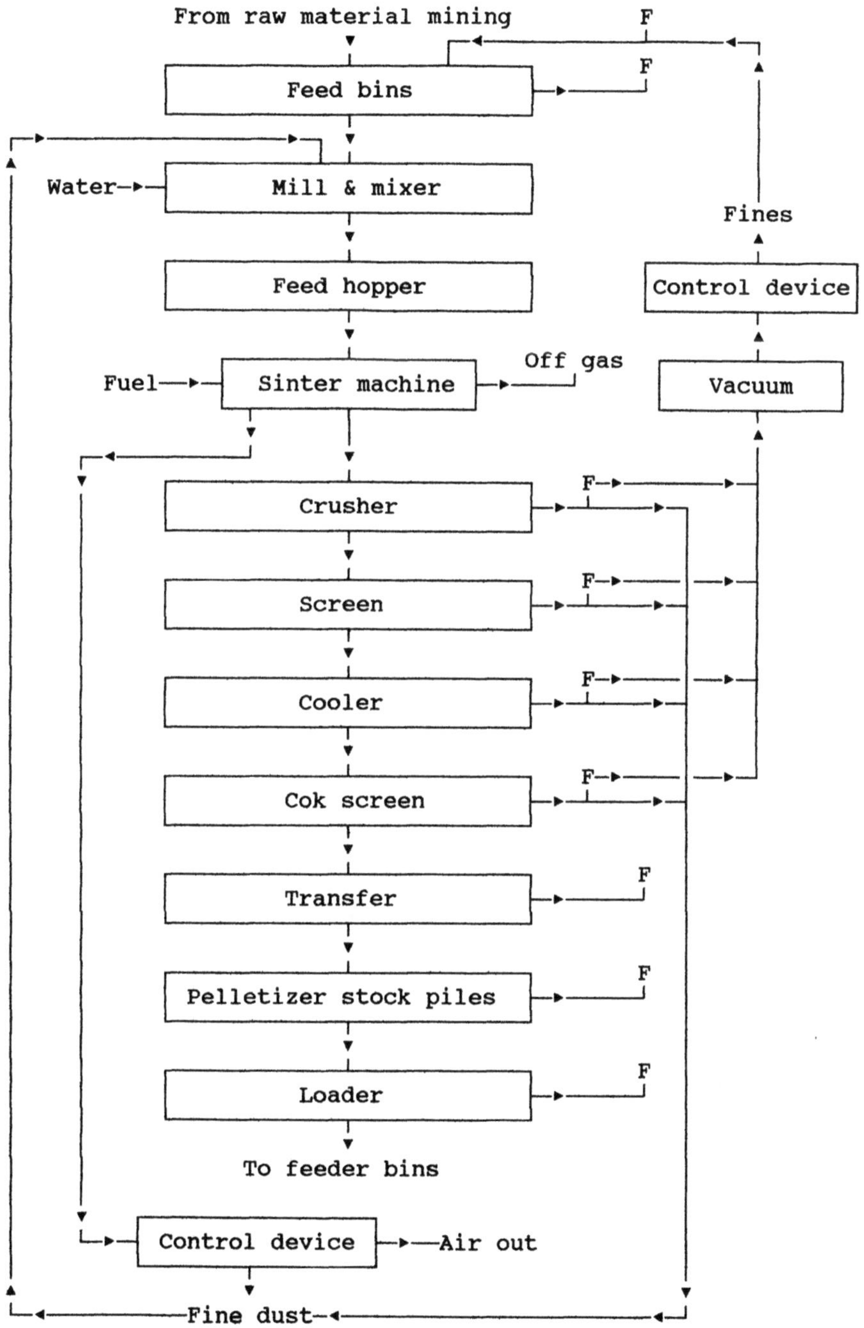


Fig.3.a. Raw material preparation for steel manufacture.

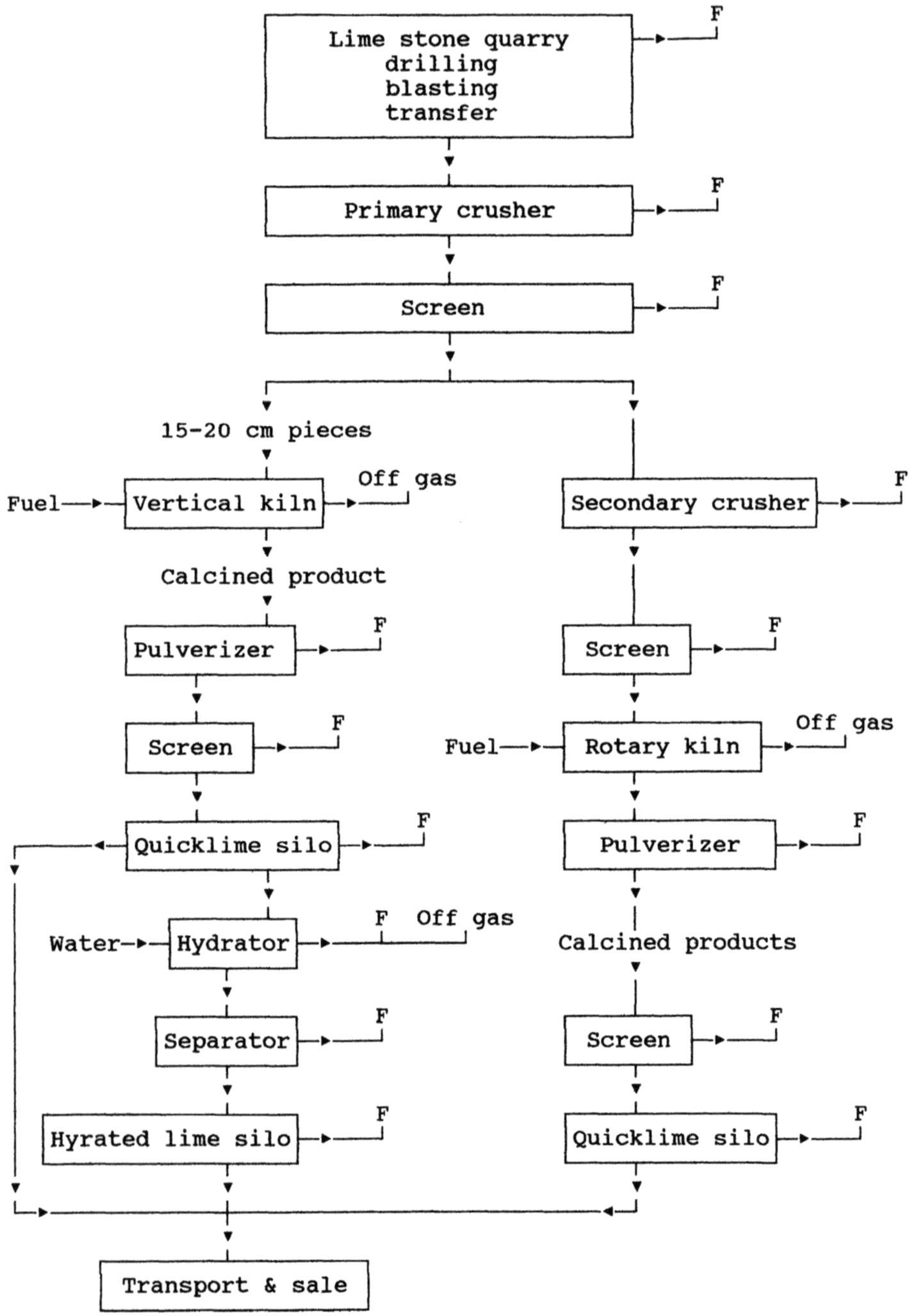


Fig.4. Flow scheme of lime manufacture.

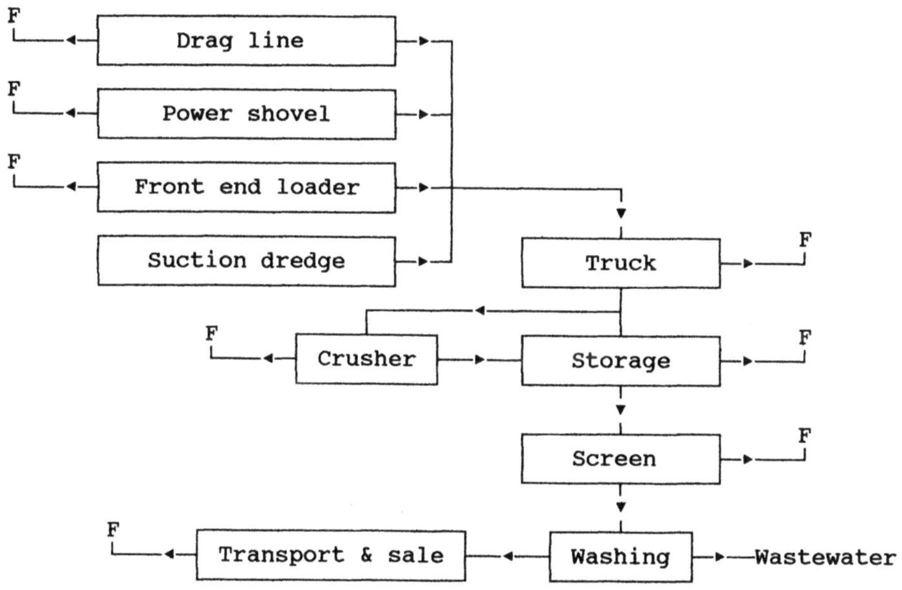


Fig.5. Flow scheme of sand and gravel production.

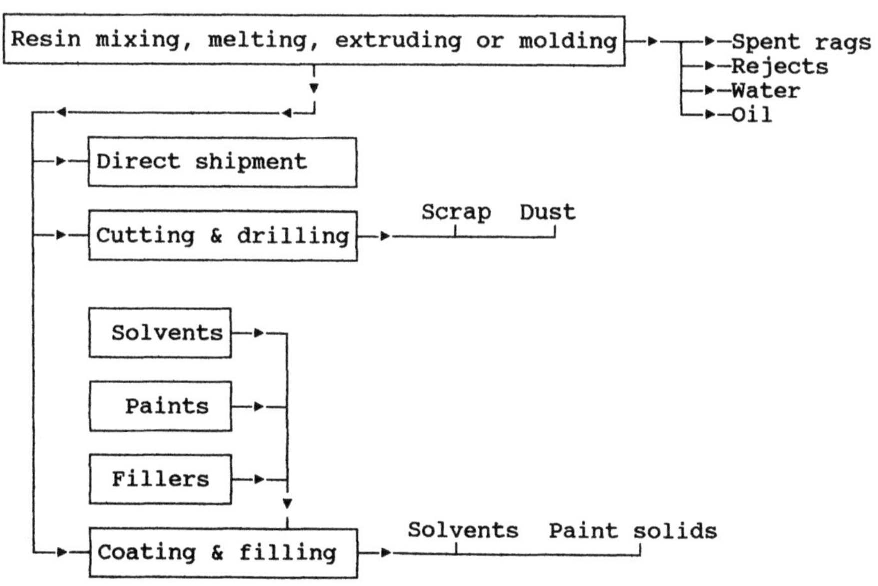


Fig.6. Flow scheme of plastic building material manufacture.

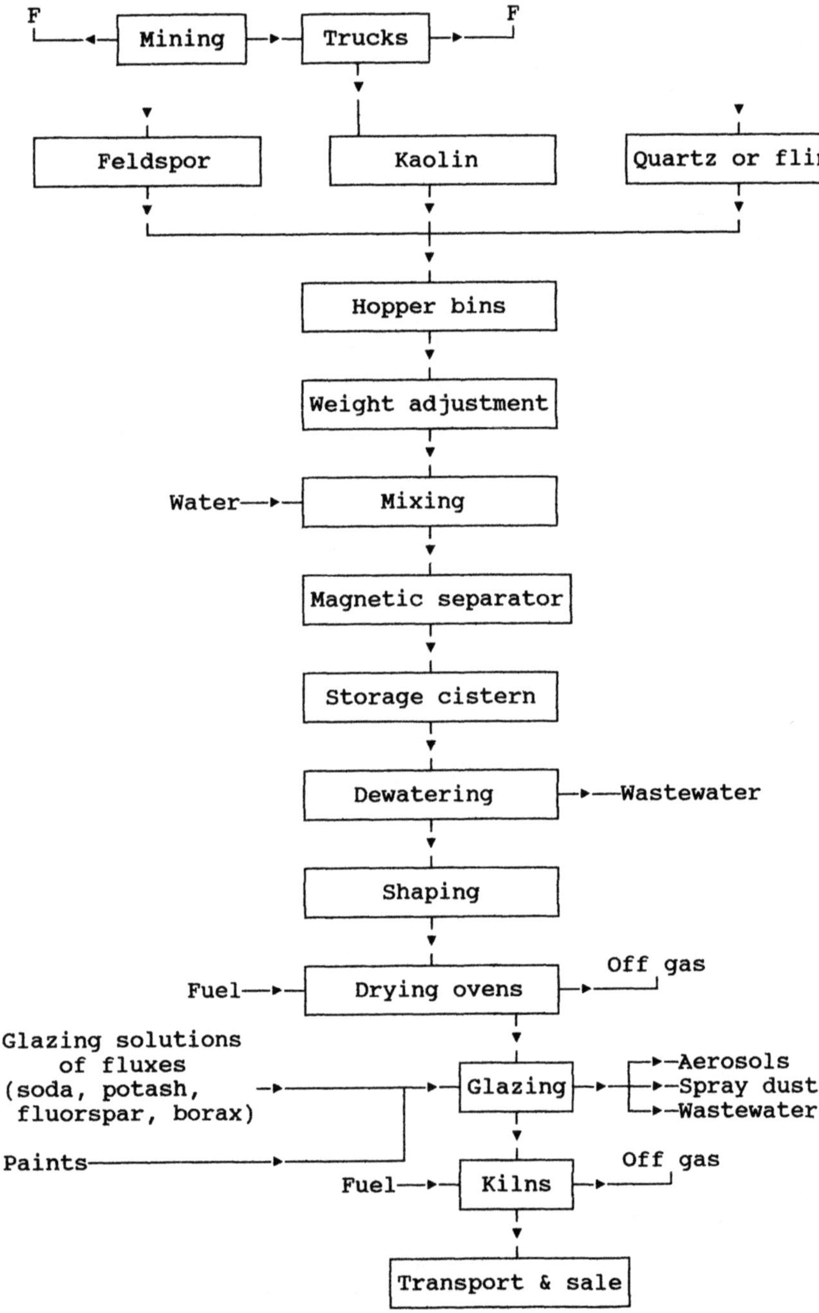


Fig.7. Flow scheme of ceramics manufacture.

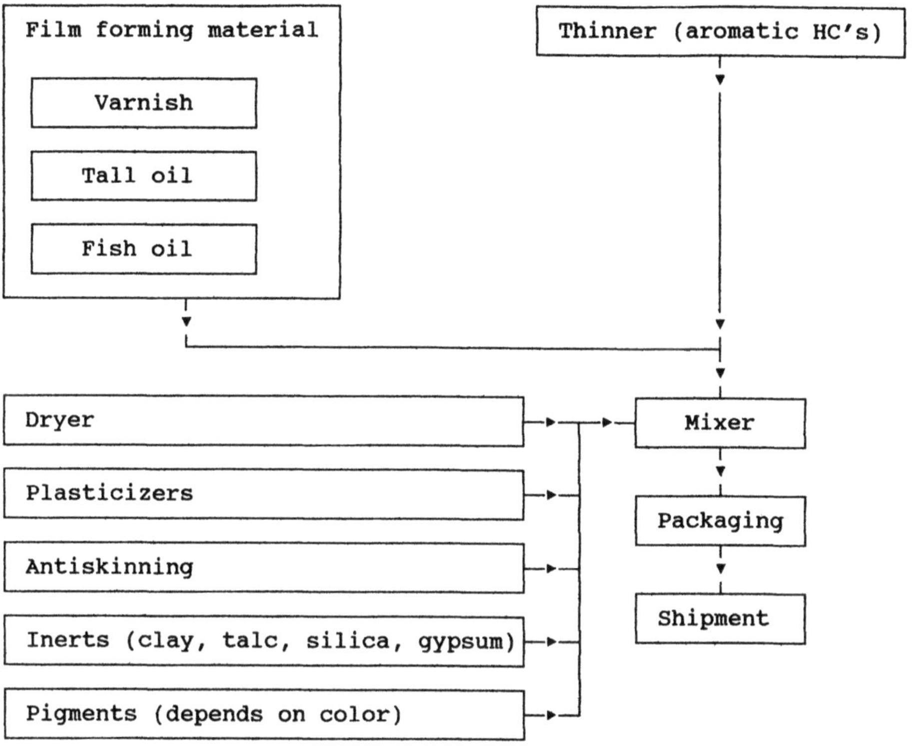


Fig.8. Flow scheme of paint manufacture.

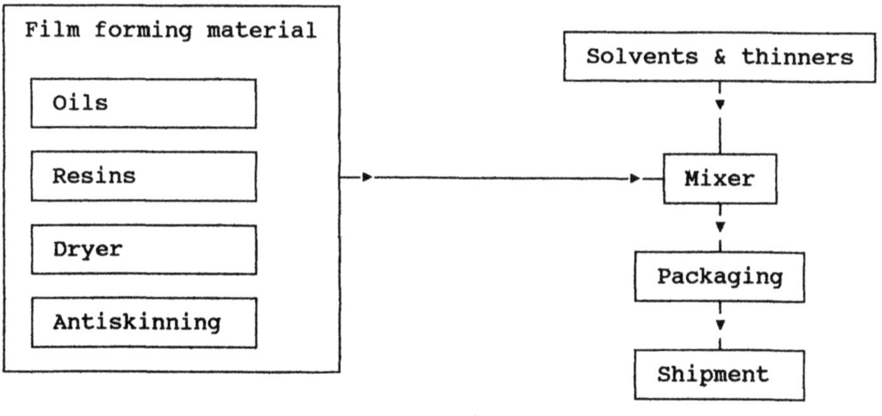


Fig.8.a. Flow scheme of varnish manufacture.

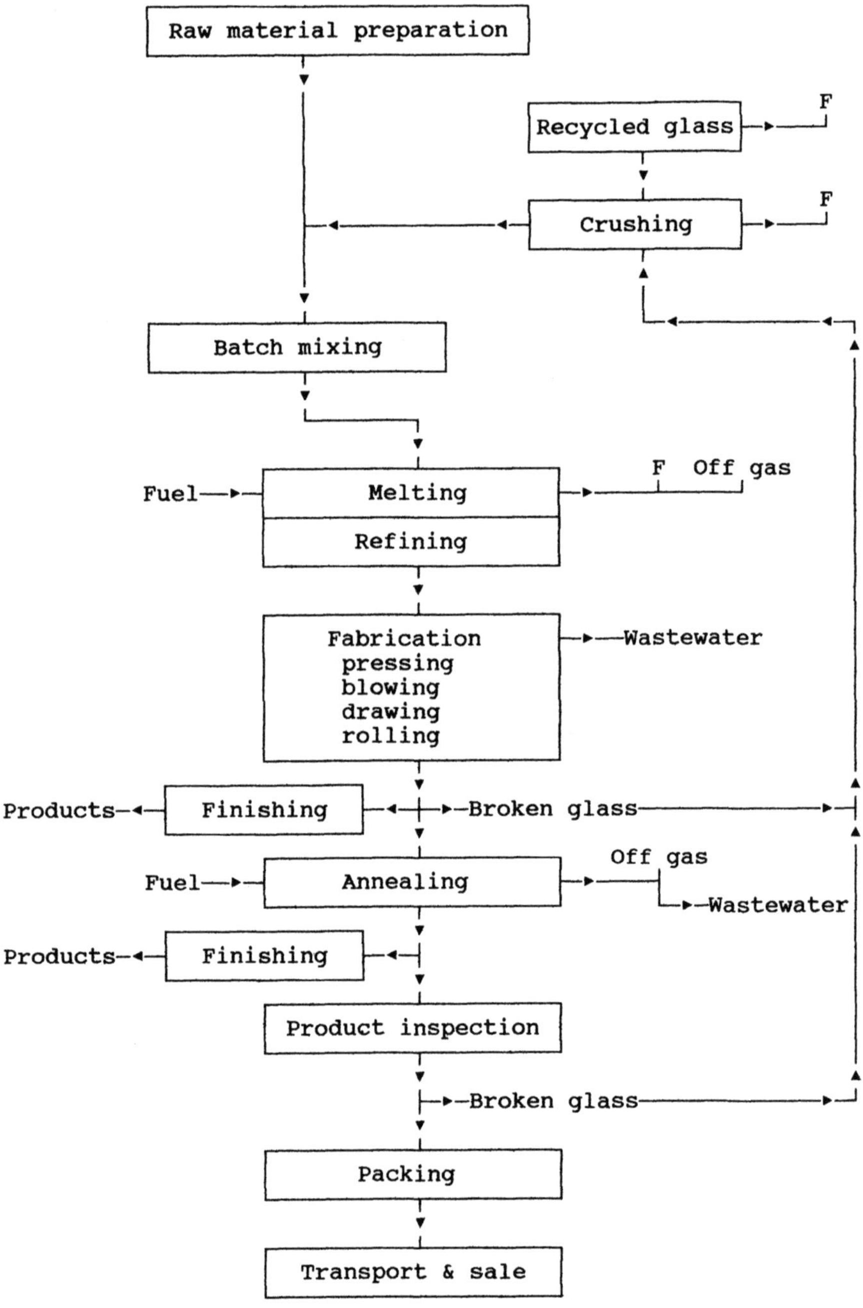


Fig.9. Flow scheme of glass manufacture.

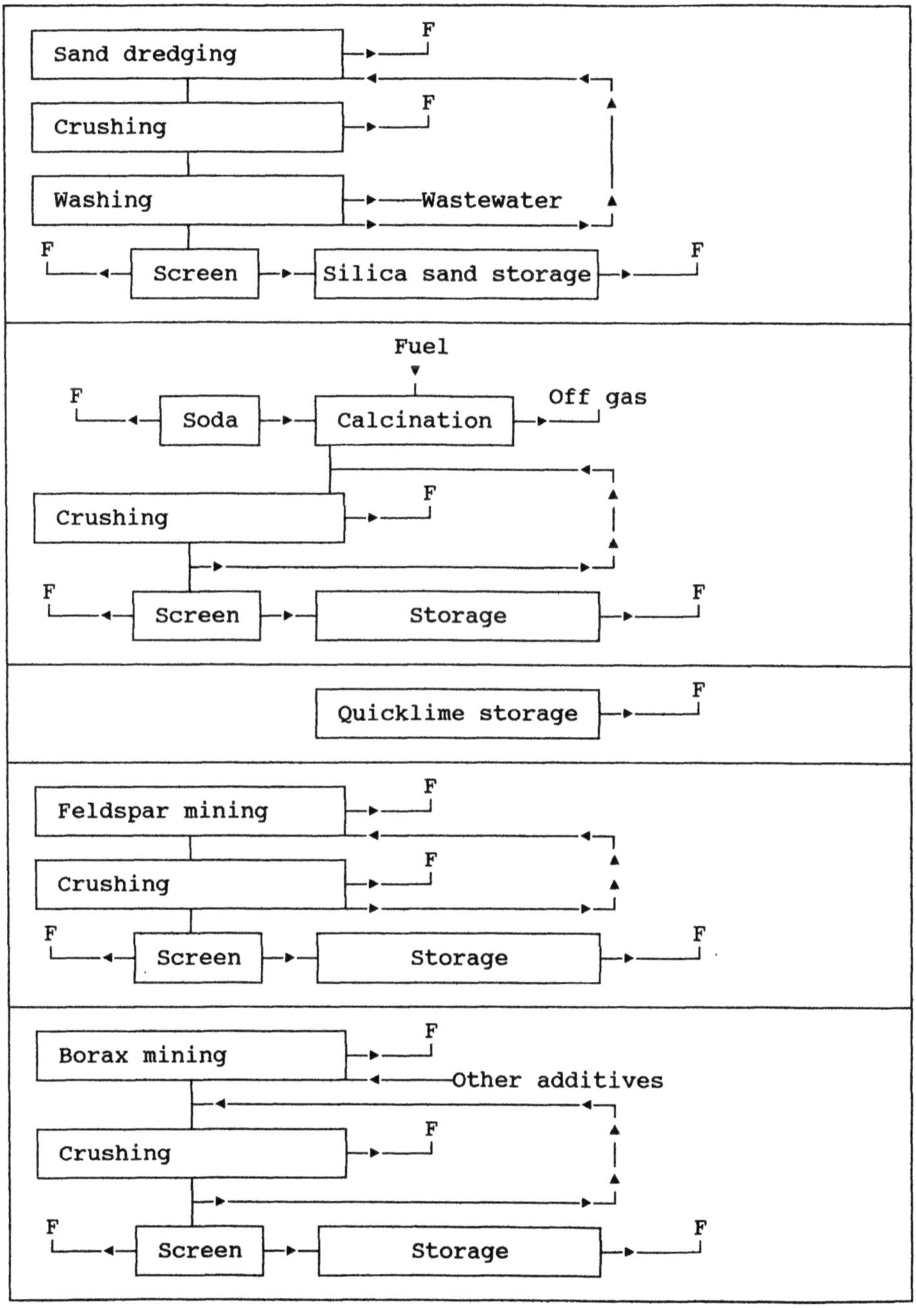


Fig.9.a. Flow scheme of raw material preparation (glass).

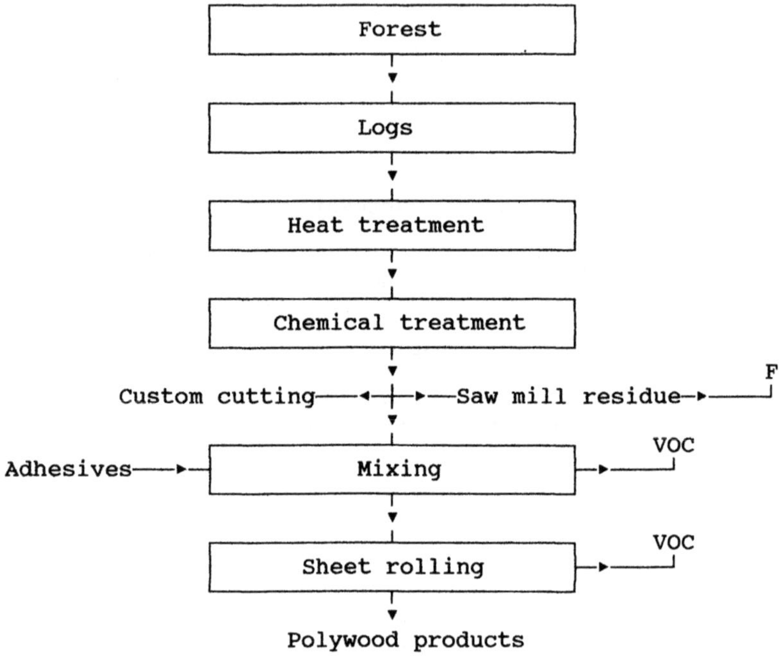


Fig.10. Flow scheme of wood and wood product manufacture.

For 119,000 tons of Portland cement consumed for building "Mavişehir" complex, energy consumption only for its production is calculated as 17,493,000 kWh. Emissions due to electrical power used in cement manufacture are given in Table 6.

Iron and steel: Basic diagrammatic flow scheme of iron and steel manufacture is shown in Figure 3. This figure outlines the basic steps in manufacturing of construction steel as applied in İskenderun Iron and Steel complex. Relevant emission factors for iron and steel production are given in Table 7.

For the construction of "Mavişehir" 27,000 tons of construction steel has been utilized. Pollutant emissions using data given in Table 7 are calculated and shown in Table 8.

Table 4. Controlled air pollutant emissions for cement production (g/g clinker) (Tekiner, 1994).

| Substance | Emission |
|---|---|
| $SO_2$ | $4.2x10^{-3}$(*) |
| $NO_x$ | $1.2x10^{-3}$ |
| Dust | $1.3x10^{-3}$(**) |
| $CO_2$ | $7.3x10^{-1}$ |
| Chromium(***) | $2.0x10^{-11}$ |
| Lead(***) | $6.0x10^{-12}$ |

(*): anything below this figure down to 0, which is our experience in the two cement factories in İzmir

(**): 81 of this comes from cement kilns, 894 comes from fugitive sources, rest from other stacks

(***): These figures are from van der Most and Veldt (1992) and are based on cement produced which is more than clinker production by a factor of about 1.005. Therefore, the figures in this table must have been smaller by this factor.

Table 5. Emissions originating due to cement production for "Mavişehir" (g).

| Substance | Emission |
|---|---|
| $SO_2$ | $4.8x10^{8}$ |
| $NO_x$ | $1.4x10^{8}$ |
| Dust | $1.5x10^{8}$ |
| $CO_2$ | $8.3x10^{10}$ |
| Chromium | $2.3x10^{0}$ |
| Lead | $6.8x10^{-1}$ |

Table 6. Electricity-originated emissions for cement production at the power plant site (g) (Müezzinoğlu, 1994).

| Substance | Emissions |
|---|---|
| $SO_2$ | $1,2x10^{9}$ |
| Carbon gases (as C) | $2,1x10^{13}$ |
| Dust | $3,9x10^{8}$ |

Table 7. Approximate cumulative emission factors for pollutants from İskenderun iron and steel plant with a combination of sinter+pellet+pig iron+open heart+continuous casting+rolling mill (g/ton-steel).

| Substance | Emission |
|---|---|
| $NO_x$ | $2,9x10^3$ |
| VOC | $7,6x10^2$ |
| $SO_x$ | $3,8x10^3$ |
| Arsenic | $4,4x10^{-1}$ |
| Cadmium | $5,5x10^{-1}$ |
| Chromium | $2,0x10^0$ |
| Copper | $3,1x10^0$ |
| Mercury | $2,8x10^{-1}$ |
| Nickel | $2,2x10^0$ |
| Lead | $3,1x10^1$ |
| Selenium | $1,3x10^{-1}$ |
| Zinc | $6,9x10^1$ |

Table 8. Emissions from steel production for "Mavişehir" (g).

| Substance | Emission |
|---|---|
| $NO_x$ | $7,8x10^7$ |
| VOC | $2,1x10^7$ |
| $SO_x$ | $1,0x10^8$ |
| Arsenic | $1,2x10^4$ |
| Cadmium | $1,5x10^4$ |
| Chromium | $5,4x10^4$ |
| Copper | $8,4x10^4$ |
| Mercury | $7,6x10^3$ |
| Nickel | $5,9x10^4$ |
| Lead | $8,4x10^5$ |
| Selenium | $3,5x10^3$ |
| Zinc | $1,9x10^6$ |

Total air pollutant emission estimates: "Mavişehir" consists of 20 blocks having 2,872 houses of which total setting area is 472,000 $m^2$. Unit emissions of air pollutant substances for 1 $m^2$ housing area are listed in Table 9.

Table 9. Emissions (grams) of air pollutants per $m^2$ floor.

| Substance | Emissions |
|---|---|
| Carbon gases (as C) | $4.4 \times 10^{7}$ |
| CO | $3.8 \times 10^{2}$ |
| Hc | $5.7 \times 10^{1}$ |
| VOC | $4.4 \times 10^{1}$ |
| $NO_x$ | $6.4 \times 10^{2}$ |
| Dust | $1.2 \times 10^{3}$ |
| $SO_2$ | $3.6 \times 10^{3}$ |
| $SO_x$ | $2.1 \times 10^{2}$ |
| $CO_2$ | $1.8 \times 10^{5}$ |
| Arsenic | $2.5 \times 10^{-2}$ |
| Cadmium | $3.2 \times 10^{-2}$ |
| Chromium | $1.1 \times 10^{-1}$ |
| Copper | $1.8 \times 10^{-1}$ |
| Mercury | $1.6 \times 10^{-2}$ |
| Nickel | $1.3 \times 10^{-1}$ |
| Lead | $1.8 \times 10^{0}$ |
| Selenium | $7.4 \times 10^{-3}$ |
| Zinc | $4.0 \times 10^{0}$ |

## RESULTS AND CONCLUSIONS

All the tables and figures that are given in this presentation prove that there is a high number of common consruction materials with serious impacts on the environmental quality "somewhere" on this world. These materials start with the manufacture and transport to the site of **cement** and its raw and supplementary materials such as **coal, other additives** and **kraft paperbags** as secondary consumption materials, **iron and steel** showing the process "Mavişehir" construction steels have been manufactured in İskenderun, **quicklime** and **hydrated lime, sand ands gravel, plastics, paints and varnishes, ceramics, glass, wood and wood products.** When all these diagrams are studied, it will be noted that manufacture of commonly used construction materials have very important impacts on the environment.

In other words in order to build even a very modest dwelling, natural resources are drastically destroyed, transformed into other forms, soils are dragged, earth is scraped, ground vegetation cover is destroyed, pollutants are discharged into the air, water and soil media. However, it is not necessary that such alterations and activities take place in the vicinity of the buildings that are being constructed, but somewhere on the nearest possible location.

When statistical data on urbanization patterns and demand for new housings due to the population growth on the earth is studied striking deviations from sustainability are found. One example is given from Turkey as with the present trends of growth of cities, all of Turkey will be a single city after about 200 years later. This shows the unsustainability of future cities with their present growth trends.

However, when this scenario is considered together with the raw material acquisition and their manufacturing technologies necessary for construction, it is seen that it is an impossible scenario for a second reason. Sustainability of natural resources for their presence and qualities is also in danger.

In order to overcome this difficulty present trends in population growth, migration into cities as well as urbanization patterns must be radically changed. Also the construction techniques must be reevaluated for sustainability and pollution prevention techniques must be applied in both the production and use of common construction materials. In order to identify the main steps and stages to apply these techniques life cycle analysis method is a helpful one. However, it must be stressed that there is a lot yet to be made in this aspect.

## REFERENCES


Buonicore J A, Daviss WT (eds.) (1992) Air Pollution Control Manual, Air and Waste Management Association, Van Nostrand Reinhold, New York.

Müezzinoğlu A (1994a) Industrial Waste Minimization Techniques: Material and Energy Balancing Method,4th.Industrial Pollution Symposium-1994, İstanbul Technical University 26-28 Sept. 1994, İTÜ publications İstanbul, Turkey (in Turkish).

Müezzinoğlu A (1994b) Gökova Symposium opening presentation, (in publication) DEU Dept. of Environmental Enginng, İzmir. (in Turkish).

State Statistics Institute (1992) State Statistics Institute Publ.No.1510, Statistical Yearbook of Turkey, Ankara, Turkey.

Tekiner G (1994) Air Pollution Potential of Cement Industry, DEU MS Thesis, İzmir, Turkey.

van der Most, Veldt (1992) TNO, Parcom Atmos. Emission Factors Manual, 92-235, the Netherlands.

# Pollution Prevention Studies in Cotton Textile Industry in Turkey

Füsun SENGÜL, Nuri AZBAR
Department of Environmental Engineering
Dokuz Eylül University
35100 Bornova-Izmir
TURKEY

## Abstract

Manufacturers are now recognizing the likely impact their activity can have on the environment and feel responsible for devising remedies to some of the existing problems. Attempts in textile sector to minimize pollution and produce "environmentally friendly textile products" have been accelerated in the last years. In several developed countries, environmentally-friendly initiatives are being taken by private and public sector industries. In this context, several eco-labelling standards which differ from each other are established.

Attempts in Turkish textile sector in this direction have been continuing for three years and some manufacturers have already started to produce and export their eco-textile products. In this study, the existing situation in Turkish cotton textile industry is presented and also environmental problems related to conventional production and pollution prevention studies needed in this sector are discussed. Minimization of textile industry's environmental effects by switching into "ECO-TEXTILE" production will be a very important step taken forward in sustainable development of Turkey.

NATO ASI Series, Partnership Sub-Series, 2. Environment – Vol. 2
Cleaner Technologies and Cleaner Products
for Sustainable Development
Edited by H. M. Freeman, Z. Puskas, and R. Olbina

## 1. Introduction

Small amounts of pollutants released by industries into the environment may cause measurable environmental impacts. In order to virtually eliminate the discharge of these pollutants to receiving media, industries should implement certain pollution prevention technologies. The concept of pollution prevention is not new but it is newly receiving recognition today.

Industrialization in developed and developing countries has proven to have a cost for public health and the environment, especially when no care is taken. Pollution prevention is the maximum feasible reduction of all wastes generated at production stages. That is, pollution prevention (sometimes referred to as source reduction) is the use of materials, processes, or practices that reduce or eliminate the creation of pollutants or wastes at the source (EPA, 1992). End-of-pipe pollution controls have not always proven to be optimal from an environmental aspect. Pollution prevention, on the other hand, is a comprehensive, preventive approach to environmental protection. It requires people to be creative and to investigate all phases of manufacturing processes and product life cycles, including product usage in offices and homes. Pollution prevention, thus, encompasses such actions as energy and raw materials conservation, eliminating toxic substances (as raw materials and as product constituents), and reducing the amount of wastes and pollutants created by processes and products, thereby lowering the amounts emitted to air, land and water. Pollution prevention should be considered the first step in a hierarchy of options for reducing the generation of pollution. The next step in the hierarchy is responsible from recycling of any wastes that cannot be reduced or eliminated at the source. Wastes that cannot be recycled should be treated in accordance with environmental standards. Finally, any wastes that remain after treatment should be disposed of safely.

## 2. Production Processes of Cotton Textile

Cotton is a fiber obtained from the cotton plant and it is cultivated and processed widely throughout the world (see Table 1). The process starts with the cotton flowering. Today there is a need that all the cotton processed should be organically grown to be able to be compatible with cleaner production. This means growing cotton without artificial fertilizers (vegetable compost and manure can supply the soil with sufficient nitrogen needs and organic materials), chemical pesticides and defoliants. General policy demands that all the cotton used in textile industry should be handpicked to avoid the use of defoliants and should not contain pesticide residues. Yarn production and cotton fabric production flowscheme is given in Figure 1-a and 1-b. Each comprises several unit operations, some of which use water while others do not. The major pollution problem presented by textile industry with wet processing units is water pollution. There are a few problems of air pollution by chemicals, lint,etc., but these are minor with respect to water pollution problems (Noyes, 1993). The major processes contributing to pollutant loads of textile industry are presented briefly below.

### 2.1. Preparation

Prior to dyeing and printing the impurities of all the textile materials must be removed. This is carried out by the following processes;

**Desizing:** The principal polymer used in sizing process before 1960 was starch. This natural polymer of glucose is easily degraded biologically and should present no problem to the conventional waste treatment plant other than BOD loading. Modified starches and alginate are also used. The development of many synthetic fibers in the 1950 s and their use in blended fabrics created the need for new sizes which were more

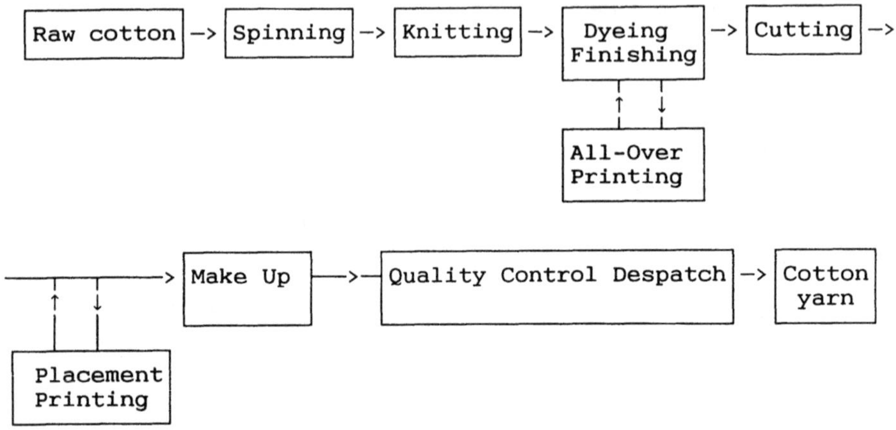


**1-a. Cotton Yarn Production Flow Scheme**

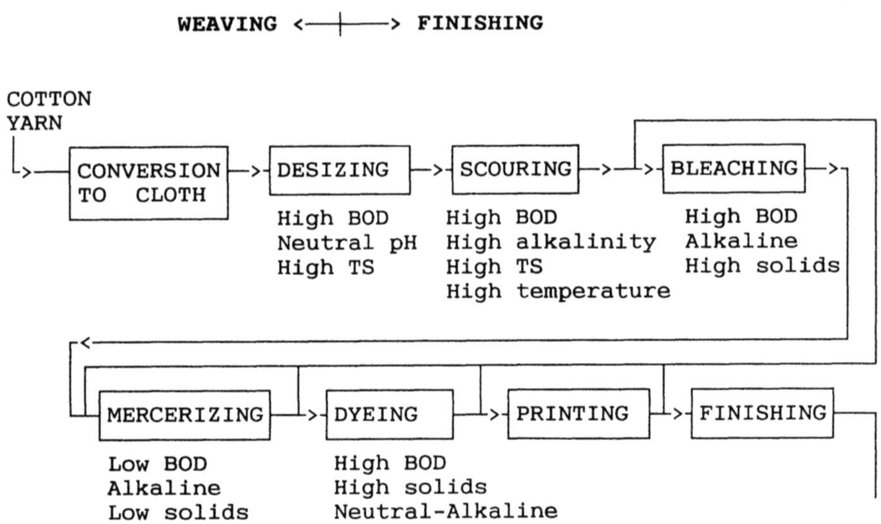


**1-b. Production Flow Scheme of Cotton Fabric**

**Figure 1. Production Flow Scheme of Cotton Yarn and Fabric (EPA, 1977)**

compatible with the hydrophobic fibers. Some of those which were developed include polyvinyl alcohol (PVA), carboxymethyl cellulose (CMC), hydroxyethyl cellulose (HEC), polacryclic acid, and polyvinyl acetate (PVAc). These are a number of auxiliary components in commercial sizing formulations that can make up 10-15 % of the total solids in the formula. These added chemicals can include humectant, lubricant, antistatic agents, biocides, glycerine, and wetting agents. A change from natural to synthetic sizes can give a substantial reduction in BOD. 70 percent of the cotton textile industry still uses starch.

**Scouring** : The change in scouring which possibly received the most attention since 1950 was the conversion to biodegradable detergents, for the removal of oil, waxes, and other impurities. Natural soaps have higher BODs than synthetic detergents. However, one must keep in mind that branched chemicals are less degradable than linear ones.

At one time solvent processing was developed but not commercialized because of its costs, and air pollution considerations. Several solvents such as mineral spirits, chlorobenzenes and perchloroethylene have been used to aid in the removal of oil born stains from synthetic fibers. Some of the solvents are inert to biological treatment and may not be removed from a waste stream by conventional waste treatment methods.

Desizing and scouring operations are the major contributors to BOD problems.

**Bleaching and Other Preparation Processes:** Bleaching contributes very little to overall plant BOD. Synthetic fibers and blends require little or no bleaching. Bleaching is an area where considerable progress has been made. Most textiles are now bleached with hydrogen peroxide which breaks down into water and oxygen instead of the chlorinated chemicals which were previously used.

### 2.2. Dyeing and Printing

There are two methods of dyeing: (1) continuous, and (2) batch. Continuous methods, including pad batch dyeing use lower quantities of water and chemicals, the waste streams can be segregated, and the chemical content of the wash water is quite low. Waste streams can contain higher chemical loadings.

The problem with most dyestuffs is that they do not give 100 % fixation on textiles, thus after the dyeing or printing process, some color is washed off which appears in the effluent. The machinery industry is working to find improved systems for the application of color, both in dyeing and printing. Good examples for these systems are jet dyeing and jet printing technologies. The advantages of this system, from an environmental point of view, is that there is less waste of color during color changes. There are also major changes in dyestuffs and auxiliaries used in dyeing process. In this context, natural dyestuffs are recommended instead of conventional dyestuffs. But one should know that the yield of natural dyestuffs from plants is exteremly low and huge areas of land would be required to grow the plants to extract the dyestuffs. According to the Indian Government statistics, 1 tonne of vegetable indigo dye is derived from 77 hectares of land (Hazel, 1994).

### 2.3. Finishing

The objective of finishing is to improve the appearance, handle and wearing properties of the textile and also to give it special properties, such as flame retardancy, water resistance or easy care. Finishing is usually a continuous process that produces little or no wastewater.

## 3. Life Cycle Analysis for Cotton Textile Industry

The life cycle of a cotton textile product consists of the phases given in Figure 2. In the life cycle analysis, internal (health and security) and external variables (energy and water conservation, waste minimization, control of emissions to the environment) have to be considered. These considerations are briefly discussed in the following section.

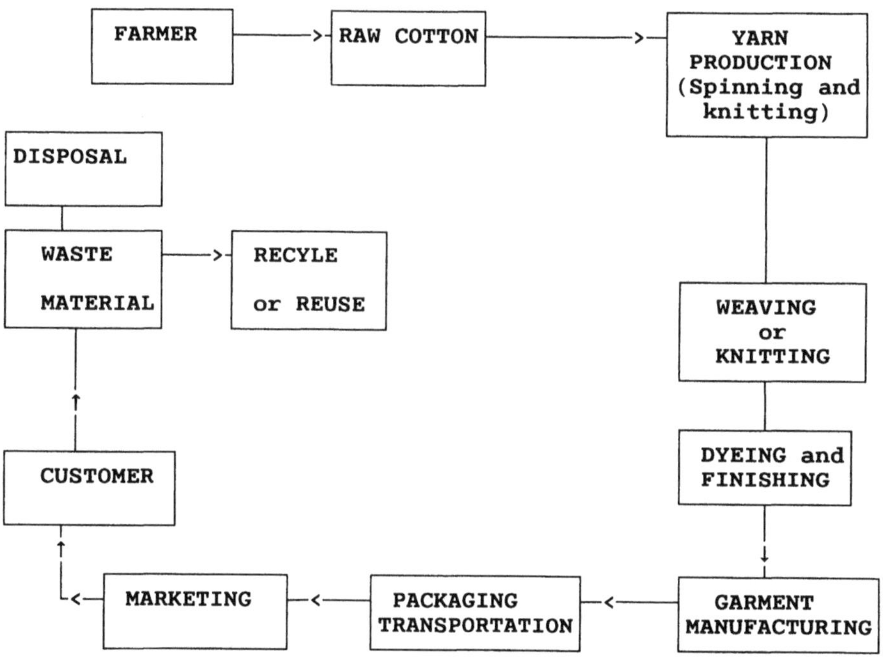


**Figure 2. Life Cycle Analysis for Cotton Textile Industry**

## 4. Pollution Prevention Strategies in Cotton Textile Industry

### 4.1. Energy and Water Conservation

There are numerous methods which have been developed to conserve energy and water at textile mills.

Some of these techniques are discussed below;

- The energy consumption can be reduced by using the computerized control system which minimizes mistakes and secures a more efficient consumption of energy.
- For the energy conservation, all knitting machines can be linked to a phase equalizing system reducing energy consumption.
- The dyeing and finishing processes are the largest energy users. The jets used should be operated at a fabric/water ratio of 1:7 which will reduce consumption of energy and water.
- Natural gas can be used for general heating purpose.
- Energy can be saved by removing water by centrifugal means rather than evaporation.
- Rapid inverse dyeing of blended fabrics to conserve water could be used
- Automatic chemical and dye-dispensing systems should be used.
- Foam processing should be used to conserve water.
- The rinse waste from the final rinse in a batch dyeing operation is fairly clean and can be used directly for further rinsing or for make-up of subsequent dyebaths (Noyes, 1993).
- The colored wastewater from the soaping operation can be reused at the backgrey washer, which does not require very high quality water. Alternatively, it can be used for cleaning floors and equipment in the print and color shop.
- Mercerizing or bleaching rinse water can be used in scouring and desizing operations as long as size recovery is not practiced.
- An automatic shutoff valve may control the flow of water into a process unit based on time, level or temperature.
- A flow or pressure reduction valve can significantly reduce the quantity of water used in a washer or clean-up step.
- A program of maintenance, inspection and evaluation of production practices should be established. Significant water reductions can be made through such areas as minimizing leaks and spills, proper maintenance of production equipment, identification of unnecessary washing of both fabric and equipment, and employee education on the importance of water conservation.
- In order to reduce the original water consumption, a second

pipeline prepared for the recycling of water can be built next to the wastewater pipe.

- Almost all of the cooling water used in process can be recycled in the plant.

## 4.2. Chemical Substitution and Conservation for Waste Minimization

Many textile mills use excessive and even unnecessary chemicals. Chemicals are often added to a dye recipe to counteract a negative side effect of another ingredient instead of finding a substitute for the offending chemical. Excess chemical usage also results from lack of precise measurement techniques. By providing specific instructions, metering/measurement equipment, both chemical and water requirement can be reduced.

Substitution should ideally assume the direction of easily treatable or reusable materials in terms of waste-treatment technology, recoverability, or reuse. Cutting down excessive use of chemicals will also help in overcoming the pollution problems and lowering the costs.

Some examples for waste minimization application are given below;

- Recovery of carboxymethyl (CMC) size material has been demonstrated over starch and polyvinyl alcohol.
- Eliminate phosphate wherever possible.
- Evaluate all maintenance chemicals for metals and toxicity.
- Recover caustic from mercerizing step.
- Use liquid ammonia mercerization instead of caustic.
- Replace all chromates used for oxidation of vat dyes.
- Use pad patch dyeing to eliminate salt and other chemical specialties. Pad batch dyeing can also conserve energy, water, dyes, chemicals, labor, and floor space.
- Reuse the residual portions of finish mixes, if possible, by adding the required components to make up the next mix.

- Consider substituting supercritical carbon dioxide for water as a dye solvent.
- Recovering CMC by precipitation with aluminum sulfate allows reuse of the CMC
- Work on recovery of starch and polyvinyl alcohol has not demonstrated an attractive procedure.
- Carding oils and antistat lubricants should be replaced by mineral oils with nonionic emulsifiers and other low-BOD substitutes (Masselli, 1956).
- The primary source of waste from a knitting mill is the production imperfections. All waste can be recycled as rags.
- All cotton fabrics can be dyed on jet-dyeing machines, which are considered to be the most efficient under the conditions given
- Substitution of formic acid for acetic acid in dyeing gives a substantial reduction in BOD and, because of its lower equivalent weight, can give a cut in costs (Little, 1967).

| Chemical | BOD equivalent (kg/kg) |
|---|---|
| Acetic Acid | 0.64 |
| Formic Acid | 0.12 |

## 4.3. Control of Emissions to the Environment

Emissions of some pollutants can be controlled with following measures;

- Spinning and knitting have always been considered a hazard to health because huge quantities of dust are produced. Air filtered system can be installed on all-knitting machines in order to catch the dust immediately upon formation.
- There are minor emissions to the air of acetic acid, formaldehyde and other chemicals primarily from high drying and curing ovens. They are usually captured by the appropriate abatement equipment. Proper selection of preparation, dyeing, and finishing agents can help reduce these emissions. Air filtering

systems are installed everywhere dyestuffs are mixed, thus lowering the risk of health problem for the employees.

- All cotton fabrics are given a mechanical finish. Formaldehyde containing finishing agents should not be used in the finishing process.
- All wastewater which is unavoidable after proper pollution prevention strategies should be carefully monitored and discharged into a pipeline leading to a treatment plant. All dyeing machines, scouring machines etc. should be connected to this pipeline according to the characteristics of wastewater.
- The production imperfections from a knitting mill which cause wastes can be minimized by
  - -using combed yarn only which is better quality material
  - -avoiding dust in the machinery by air filtering system
  - -softening the yarn by vaporizing system.
- Wastes produced by garment manufacturing like paper, plastic, fabric can be recycled. Domestic solid wastes can be collected by the municipal refuse collection truck and disposed to a sanitary landfill.

## 5. Existing Situation of Turkish Cotton Textile Industry from The Cleaner Production Point of View

In the early years of the Turkish Republic, there was only a few textile mills. These factories were established by the government to meet military textile demands. These first textile mills built in ijstanbul in 1922-1923 were then joined to SUMERBANK in 1933. The organized textile industry in Turkey progressed by the support of The Government as a civil sector within Sümerbank until 1950. After 1950 private sector took its place in Turkish textile industry. Cotton yarn production which was around 83,000 tones in 1960 increased to 479,000 tones in 1988. On the other hand, cotton fabric production has increased from 63,200 tones to 300,000 tones. After 1980's an important increase is seen in the export of textile products. Today textile industry has an important place in overall Turkish industry.

Turkey is taking an important place among the biggest textile manufacturer countries. The reason is that Turkey has a very high cotton production capacity of which 90 percent is cotton with long fibers and the rest having moderately long fibers. Cotton production efficiency was about 885 kg per hectare in 1987. Turkey meets more than 3 percent of the world cotton demand and is the seventh biggest cotton producing country after China, USA, Soviet Unions, India, Pakistan and Brazil. Cotton production of Turkey by the year of 2000, following the completion of the GAP project (South East Anatolian Project) is estimated to increase 128 percent. As a result of this project Turkey will be the third biggest country all over the world in cotton production. Yearly world wide cotton production and consumption values are given in Table 1.

**Table 1. Yearly World Wide Cotton Production and Consumption Values (1000 tons) (DPT ; Simpson & Anson, 1990)**

| | 1985/86 | | 1990/91 | |
|---|---|---|---|---|
| | Production | Consumption | Production | Consumption |
| Republic of China | 4200 | 4110 | 4470 | 4355 |
| USA | 2925 | 1393 | 3375 | 1829 |
| Soviet Unions(*) | 2792 | 2772 | 2635 | 2403 |
| India | 1955 | 1546 | 1955 | 1988 |
| Pakistan | 1216 | 533 | 1414 | 1257 |
| Brazil | 793 | 692 | 675 | 729 |
| Turkey | 518 | 456 | 670 | 505 |
| Total | 17443 | 16603 | 18848 | 18714 |

(*) Including East European Countries

Almost all yarn production of Turkey is accomplished by 100 organized cotton yarn factories. On the other hand, 40 percent of the cotton textile manufacturing is produced by organized civil and private sector textile mills and the rest is produced by unorganized and small enterprises. While the capacity of cotton weaving sector was 225,000 tons in 1980, it reached to 299,000 tons in 1987 and 363,000 tons in 1991. The use of production capacity according to the ratios given above is summarized in Table 2.

As mentioned above, the bigger percentage of production capacity of cotton textile belongs to small enterprises. This industry had a chance to progress in the regions of Mediterranean, South East Anatolian, Marmara and Aegean due to the climate suitable for cotton growing.

**Table 2. The Use of Cotton Textile Weaving Capacity among The Organized, Unorganized and Sümerbank Factories in 1987 (Parilti, 1994).**

| | The Number of Loom | Contribution (%) |
|---|---|---|
| Organized Private Sector | 12,438 | 23,0 |
| Sümerbank | 8,751 | 16,1 |
| Unorganized | | |
| Small enterprises | 33,035 | 60,9 |
| Total | 54,224 | 100,0 |

At present some agricultural areas (mainly South East Anatolian) in Turkey are used only for growing organic cotton but the amount is small (80 tons per year and at the moment it is all exported). It seems that Turkish cotton is very suitable for eco-textile production from point of view raw material, since it is not needed to use large amount of pesticides because of suitable climatic conditions and also they are handpicked. On the other hand, Some Turkish textile manufacturers producing eco-textile garments have very close contact with some European companies which are experts on eco-textile production. These companies choose the right raw material (cotton growth under the conditions that satisfy the ecological aspects) and also they care for textile processing steps and try to minimize likely environmental impacts due to their activities.

The textile industry is formed of a series of inextricably interlinked processes which are usually carried out in separate factories and the process in one factory may affect the environmentally friendliness of a process further down the chain. This situation is also valid for the Turkish cotton textile industry. In general, production processes are accomplished in mainly four subcategories as follow;
1) Cotton yarn production, 2) Dyehouse having dyeing and finishing processes 3) Printing works and 4) Garment manufacturing. Cotton yarn production is usually done in separate factories while the other processes can be carried out in same factory.

In Turkey, usually cotton garment is produced with the help of some commission textile works such as commission dyer and printer company, commission bleacher, washer. These companies are usually small enterprises and they prefer to use cheap chemicals, dyestuffs and auxiliaries in their nonecological production processes. But on the other hand, these commissioner companies are able to shift their production to ecological textile production according to the demand of the market.

Cotton textile production has mainly three processes such as preparation, dyeing and finishing of textile goods. In the yarn and fabric manufacture there is no important environmental problem except dust formation. Prior to dyeing, impurities are removed by washing and bleaching. During washing textile factories use detergents, soaps and washing agents which are mostly biodegradable. Some of these agents are imported from abroad and some are produced in Turkey. At present, for bleaching chlorinated chemicals are generally used. Almost 80% of cotton textile industry started to use hydrogen peroxide instead of chlorinated chemicals in order to meet the eco-textile standards foreseen by EC countries. Other pollutants of textile industry are discharged from dyeing and printing processes. Some of the dyestuff used can reach the wastewater because they do not give 100 % fixation. As a result of dyestuff discharged, the effluent becomes colored which is difficult to treat. Nowadays, use of natural dyestuffs as an alternative to synthetic dyestuffs is still under research. Since natural dyestuffs have relatively poor fastness to washing, Turkish cotton textile industry prefers using synthetic dyestuffs and use of natural dyestuffs is not common at present. Synthetic dyestuffs are imported from both European and East Countries (India, Taiwan, China, Korea, Japan, etc.). Due to different quality of the dyestuff imported from various countries, the fixation strength of them are different. Depending on the fixation strength of the dyestuff, the amount of the dyestuff washed off with water will be different. Consequently their waste problems and treatability differ. For the application of the color, new technologies like jet-dyeing and jet-printing are available and a few textile factories in Turkey use these technologies. But mostly used dyeing and printing methods especially in small enterprises are technologies having a higher pollution potentials.

The principal sizing agent used in sizing process is starch in 70 percent of Turkish Cotton Textile Industry.

Most textile goods are now bleached with hydrogen peroxide instead of chlorinated chemicals which were previously used in Turkey. But some textiles still require use of hypochlorite and other chlorine derivatives for bleaching process. For instance in the processing of blue jeans, the chlorinated chemicals are still used in Turkey.

The objective of finishing is to improve the appearance, handling and wearing properties of the textile and also to give it special properties, such as flame retardancy, water resistance or easy care. Finishing involves the use of chemical and/or mechanical processes. Manufacturers producing cotton textile goods complying with Eco-Labelling standards use finishing chemicals given in Table 3. Nowadays, there is an intention in Turkey to diminish the use of formaldehyde containing finishing chemicals as it is foreseen in Table 3.

There are two methods to look at the pollution prevention process in textile industry, firstly "from the cradle to grave" of a product, that is from fibre production to disposal of finished product. This is the Danish approach and some of the EC countries also adapt this approach for Eco-production. The second method is to consider the final product in order to give it an eco-label. In this approach, some chemical and physical tests are carried out on the product which are required in Eco-Tex Standards such as Eco-Tex 100, Eco-Tex 1000, ISO 9000, etc. Nowadays in Turkey the second approach is preferred by textile manufacturers.

Although the ideal situation is to treat the wastes remaining after proper pollution prevention measures, this is only valid for the industries which produce ecological products. The general application in Turkish textile industry except for eco-textile manufacturers is to pretreat wastewater generated during production processes and to discharge it into local sewerage system to be treated together with domestic wastewater. But central treatment plants especially in big cities are still under construction. On the other hand, some organized industrial districts have their own collective treatment plant in order to treat their wastewater. The textile mills which are located in the organized industrial districts use the collective sewerage system and treatment plant in order to treat and dispose their wastewater.

**Table 3. Requirements for Cotton Production under Cleaner Production Rules (H&M Standards,1993)**

| **PHASE** | **Minimal Requirement** | **Preferable Requirement** |
|---|---|---|
| <u>Cotton growing</u> | -Integrated pest management (x) | no pesticides (biological grown)(x) |
| | - no artificial fertilizers (x) | |
| | - no defoliants (x) | |
| <u>Spinning and knitting</u> | -exhaust hoods on machines (x) | |
| <u>Processing</u> | | |
| *strengthening* | -starch instead of polyalcohol | |
| *bleaching* | -hydrogen peroxide instead of sodium hypochlorite | - unbleached |
| **dyeing** | | |
| **dyestuff** | - no heavy metals | |
| | - no azo dye derived from an aromatic amine | |
| | - no sulphur dye | |
| | - no copper-complex dye | |
| | - no naphthol dye | |
| | - no benzidine dye | |
| | - no dye containing traces of PCB's | |
| | | - C type reactive dyestuffs |
| | | -membrane recovery of indigo dyestuffs |
| | - dyebath exhaustion= 70% | -dyebath exhaustion >90% |
| | - high washability | |
| | - no organochlorine carriers | -no carriers |
| **equipment** | - liquor ratio 1:8 | -low liquor ratio,1:5 |
| | - counterflow technologies | -dosefying units |
| | | - automatic color preparation |
| additives | - critical analysis of the use of additives | - minimization of BOD and COD |
| | | -minimization of salt |

**Table 3. (Continued)**

| | | |
|---|---|---|
| *finishing* | | |
| **waterproof** | - no waterproof treatment | |
| **flameproof** | - no organohalogen or organophosphorus flame retardants | -no flameproof |
| **no-iron** | - < 300 ppm formaldehyde in clothes | - no formaldehyde(x) |
| **shrink-proof** | -mechanical treatment | |
| **stonewashing** | | - mechanical treatment<br>- reuse of materials |
| *cleaning* | - no fabric softeners<br>- LAS, semi synthetics are acceptable | |
| *General* | - on site wastewater treatment<br>- water recycling(x)<br>- minimization of number of processing steps | -reuse of dyebath(x) |
| **Manufacture** | | |
| | - reuse of cotton waste | -minimization of cotton waste (for example: knitting in a circle (t-shirts) (x) |
| **Transport** | | |
| **Packaging** | - no PVC<br>- on site reuse of packaging (x) | - returnable packaging (x) |
| **Maintenance** | - no chemical cleaning required | |
| Financial requirement | price increase=< 10% | |

**Results**

The cotton growing is being focused on organic cotton thus avoiding the use of defoliants, the chemical pollutants such as pesticides and fertilizers. Although organic cotton production is possible in Turkey at present, the amount of production is not satisfactory for the actual organic cotton demand of Turkish cotton textile industry. But by the completion of GAP project, the production rate of organic cotton will be able to raise to a satisfactory amount complying the demand of the Turkish textile market. At present time, Turkey has already an advantage that almost all cotton are handpicked which is very suitable to ecological textile production. Handpicking is also an energy preserving method and it might have a future social impact since an increase in demand will create a greater need for workers.

Natural dyestuffs can be recommended as an alternative to synthetic ones, but the yield of natural dyestuffs from plants is extremely low and huge areas of land would be required to grow the plants to abstract the dyestuffs from. Reactive dyes should be preferred and dyes should not contain benzidine. Waterbase dyes should not give the AOX to the water. All dyestuffs used in dyeing should be well-known trade marks. The excessive use of dyestuff containing heavy metals can be reduced by computerized dyeing processes which also minimize the heavy metal content of actual fabric as well as the wastewater and sludge.

Formaldehyde containing finishers should be reduced or not be used in the finishing process in Turkey. To be able to make a cleaner textile production, criteria given in Table 3 should be considered in production.

**References**

Audit and Reduction Manual for Industrial Emissions and Wastes (1991) Unep & UNIDO Technical Report Series No 7, U.S.A.

DPT (1989) VI. Developments before the Five Years Development Plan, s.189.

EPA, (1977) In-plan Control of Pollution, Upgrading Textile Operations to reduce Pollution, EPA Technology Transfer Seminar Publication, 1, USA .

EPA, (1977) Wastewater Treatment Systems, Upgrading Textile Operations to reduce Pollution, EPA Technology Transfer Seminar Publication, 2, USA .

Facility Pollution Prevention Guide (1992) EPA/600/R-92/088, Cincinnati,OH.

Hazel, B.G., (1994) When will the Textile become Environmentally Friendly, Econgress 1994.

Little, A.H., (1967) Treatment of Textile Waste Liquors, J.Soc. Dyers Colorists, 268-273, July.

Masselli, J.W., Burfird, M.G., (1956) Pollution Sources From Finishing of Synthetic Fibers, New England Interstate Water Pollution Control Commission, Boston, Mass., June 1956.

Noyes, R., (1993) Pollution Prevention Technology Handbook, Noyes Publications, U.S.A.

NOVOTEX (1992) Environmental Audit, Dafolo Development, 13p, Denmark.

Parilti, H. (1994) Ihracati Tesvik Politikalari ve Tekstil ve Hazirgiyim Sektörünün Tesvigi, T.C. Basbakanlik Hazine ve Dis Ticaret Müstesarligi, Istanbul.

Simpson, P. , Anson, R., (1990) World Textile and Production Trends, EIU, Textile Outlook International, January,1990.

# Utilisation of Copper Slags for Chemicals Production

H. Sesigür, E. Açma , O. Addemir , A. Tekin , E. Ekinci
Faculty of Chemical and Metallurgial Engineering,
Maslak, Istanbul 80626, Turkey

## Abstract

In this study production of Cu , and Cu as mainstream wastes and magnetic iron oxide has been contemplated. For the purpose of recovering Cu and Co an arc melting furnace has been designed and developed. The alloy obtained from the arc furnace is leached by sulphuric acid and precipitated by $H_2S$. Magnetic iron oxide is produced from the leachate using $FeSO_4$ , geothite precipitation, dehydration, reduction and reoxidation. The resultant end product was a market value magnetic iron oxide. At present studies an alternative processing routes are under consideration for optimisation purposes.

## Introduction

One of the definitions of waste is substance that has got no use at a given time and point of its presence. In Kure region of Turkey there remains approximately 2 million tonnes of copper slags from Genevisian times as waste from ancient copper

NATO ASI Series, Partnership Sub-Series, 2. Environment – Vol. 2
Cleaner Technologies and Cleaner Products
for Sustainable Development
Edited by H. M. Freeman, Z. Puskas, and R. Olbina

works.For centuries these remains were considered as waste since no use for these slags has been found. Meantime these wastes affected the quality of ground and surface water at its surroundings due to natural rain wash.

The chemical composition of the slags at Kure region was found to be about 0.4 - 0.5 % Co , 1% Cu , 0.35 % Zn , 70 % FeO and 26 % $SiO_2$ . Even though ancient copper slag is a unique case there are various studies on recovery of valuable metals from copper slags from recent operations.The target metals that are recovered are mainly Co, Ni, Cu. Carbothermic reduction of the slags cause partial reduction of iron and concentration of non-ferrous metals is a main route for Co,Cu, and Nickel recovery ( Das et. al. 1987;Imris and Komorava, 1982;Reddy, 1982;Snelgrove and Taylor, 1981;Fugleberg, 1985). The high market value of Co makes the Co recovery process an economic possibility , (Açma et.al., 1993). In order to improve the feasibility of ancient slag utilisation process integrated processing leading to multitude of products such as cobalt salts, copper and Zn should be considered.Also Kure slags contain minor amount of impurities in Co processing.Resultant cobalt containing iron sulphide solution may partially be utilised for production of magnetic iron oxide , (Fugleberg, 1985;Açma et.al., 1992). Magnetic iron oxide ($\gamma$, $Fe_2O_3$) is used for production of video and audio tape bands.The price, magnetic properties, aspect ratio (1/8), surface area (26 $m^2/g$) and coercivity (330-380 $O_e$) classify the iron oxide crystals as a favourable material for tape band industry .

## EXPERIMENTAL

The ancient copper slags are obtained from the site by collecting samples from predetermined 37 locations covering 10 hectares.The slags are characterised in our laboratories and the chemical composition of the samples are found to be in between the limits reported table 1.

The average values indicate that recovery of Co may be potential economical process and the level of $SiO_2$ is with in the suitable range for an arc- resistance furnace - smelting ,Whyte et. al., 1977.

Table 1. Chemical composition analysis of Kure slag (wt%)

| | % | | |
|---|---|---|---|
| Component | min | max | ~Av. |
| Cu | 0.53 | 1.11 | 1.00 |
| Co | 0.35 | 0.51 | 0.43 |
| Ni | 0.02 | 0.05 | 0.04 |
| Zn | 0.23 | 0.45 | 0.35 |
| Pb | 0.01 | 0.13 | 0.01 |
| S | 1.36 | 1.72 | 1.54 |
| $SiO_2$ | 21.36 | 26.04 | 26.00 |

In order to reduce the slag a D.C. 100 kVA single phase arc furface has been designed and constructed in our laboratories.This furnace has been designed for a charge capacity of 50 kg keeping in mind scale up considerations for industrial applications.

## RESULTS and DISCUSSIONS

### Co Production

The Kure slags are crushed below 4 cm size prior to charging to arc- furnace. After 50 kg of charge is melled 8-10 kg of coke is added to reduce the iron partially to concentrate the cobalt and copper. In the phase partioning the cobalt concentration in the alloy increased to about 3 % after 30-40 minutes of reduction time.

During the experiments it has been found that there is an optimum power input. A power input of 40-43 kVA is sufficient to have about 85 % copper and cobalt recovery in 40 minutes.The decrease in concentrations of Co and Cu in the alloy is distributed to entraintment of Co in the slag at higher temperatures , Yücel et.al., 1992.

The produced alloy contained iron which is another environmental concern for waste disposal. Therefore processing the iron to valuable products is desired.Therefore ,a scheme has been developed in which some of the iron is converted into a magnetic oxide.

In order to extract Co and Cu from the alloy , a granulator has been designed and constructed which operates by spraying the molten alloy through an air jet. For this case nozzle size used was 1mm and the air back up pressure was $1.2\times10^6$ Pa. More than 90% of the resultant particle sizes were between 0.295-0.208 mm.

The granules were leached using a (350 g/l) sulphuric acid at 70 C° in a stirred vessel. The reactions that are operative at this stage are given below.

$$Co + H_2SO_4 \rightarrow CoSO_4 + H_2$$

$$Fe + H_2SO_4 \rightarrow FeSO_4 + H_2$$

$$Cu_2S + 2H_2SO_4 + 1/2\ O_2 \rightarrow 2CuSO_4 + H_2S + H_2O$$

$$CuSO_4 + H_2S \rightarrow CuS + H_2SO_4$$

As a result of these reactions all of the copper was precipitated as copper sulphide. Cobalt was precipated a CoS by controlling the pH at 4 by addition of $H_2S$. A small amount of cobalt was left in the solution for the purpose of obtaining cobalt dotted magnetic iron oxide in the latter stages of the process.

## Magnetic Iron Oxide Production

A small part remaining solution was treated with NaOH for 24 hours to prepare colloidal seed material for geothite cyristallisation.The reaction involved in seed material production is:

$FeSO_4 + 2NaOH \rightarrow Fe(OH)_2 + Na_2SO_4$

Oxidation of $Fe^{+2}$ to $Fe^{+3}$ takes places on $Fe(OH)_2$. Thus, surface area of this material is determined by BET method as 35-36 $m^2/g$ ( Açma, Sesigür).

In order to oxidise CoS the stripped mother liquor, it is mixed with scrap iron for neutralisation of evolved sulphuric acid and seed material at 70C° at pH 2.5 and for a reaction time of 6 hours. In order to assume high oxidation rates $O_2$ was bubbled through the reactor. On the surface of seed material acicular form geothite is formed. The reactions involved in this process are

$4 Fe(OH)_2 + O_2 \rightarrow 4 FeOOH + 2H_2O$

$2 FeSO_4 + 1/2\ O_2 + H_2O \rightarrow 2 FeOOH + 2H_2SO_4$

Geothite product was subjected to heat treatment at 250 C° to obtain $\alpha$-$Fe_2O_3$ for about 30 minutes.The dehydration behaviour of $\alpha$-FeOOH is shown in Figure ( 1 ).

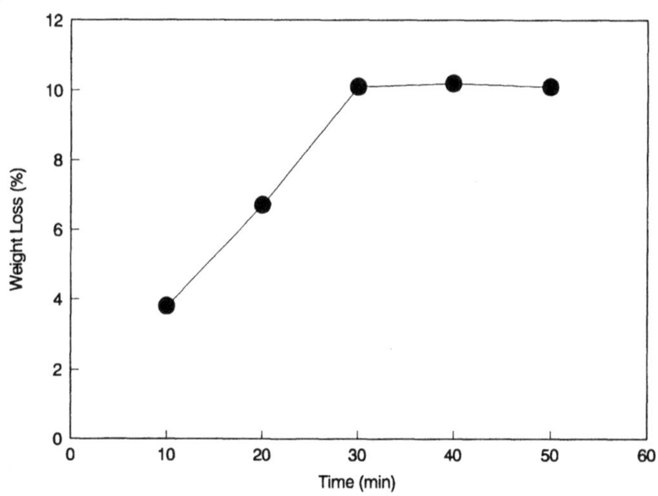


Figure 1. Weight loss to changing dehydration time

$\alpha$-$Fe_2O_3$ was then subjected to reduction procedure under hydrogen athmosphere between temperatures 360 and 390 °C for 15 minutes.During the experiments it was found that with increasing temperature coercivity increased from 240 to 290 for temperatures 360 and 390 °C respectively. In this stage in order to avoid deteriation of magnetic properties reaction time ( 15 minutes ) and temperature ( 390°C) were not superseeded.

As the last step magnetite is annealed in a tube furnace at different temperatures and times to generate yellowish acicular particles of $\gamma$-$Fe_2O_3$. In order to meet the limiting value of 1% residual FeO content , it was necessary to use air draft at the tube furnace. 225C° and 15 hours were choosen as optimum condition for FeO content and coercivity considerations.

The $\gamma$-$Fe_2O_3$ obtained in this study is characterised as a product had 20 $m^2/g$ surface area, 1:8.3 aspect ratio (L= 0.5 $\mu$m , w= 0.06 $\mu$m ) and a coercivity of 330-380 Oersteed. A typical chemical analysis of the product was shown to be : 0.65 % FeO, 0.2 % Co , 0.02 % Cu , 0.02 % Ni , 0.2 % S, 0.02 % M and 0.18 % $SiO_2$.

## CONCLUSIONS

An integrated process for the utilisation of ancient Küre slags of Turkey has been developed. The arc furnace reduction of the slags gave way to concentrated Co, and Cu as an alloy Co and Cu are seperated by $H_2SO_4$ leaching. The leachate is used for the production of magnetic iron oxide which showed fafourable technical characteristics for the manufacture audio, video, and data recording industries..

REFERENCES

Das, R.P., Anand, S., Rao, K.S., and Jena, P.K., 1987, Leaching Behaviour of Copper Converter Slag Obtained Under Different Cooling Conditions, Trans. Inst. Min. Metall., Section C, 96, pp C156 - C162.

Imris, I. and Komorova, L.,1982, Cobalt Extraction from Copper Converter Slags, Metalurgie, XXII, 3,pp 143 - 156.

Reddy, R.G., 1982, The Recovery of Non-Ferrous Metals from Primary Copper Smelter Discard Slags, Resources and Conservation, 9, pp 33 - 342.

Snelgrove, W.R.N, and Taylor, J.C., 1981, The Recovery of Values from Non-Ferrous Smelter Slags, Canadian Metallurgical Quarterly, vol. 20, no 2, pp 231 - 240.

Fugleberg, S., Processing of a Copper Shaft Furnace Residue, 1985, Presented in 17th Metalurgisches Seminar GDMB, 21 - 23 Nov., Nordkirschen, Germany.

Açma, E., Sesigür, H., Arslan, C., Addemir, O., Tekin, A., 1994, Production of Copper, Cobalt, and Magnetic Iron Oxide from the Ancient Copper Slags, Metallurgical Transactions (Submitted).

Açma, E., Toy, C., Addemir, O.,Tekin, A., 1992, Beneficiation of Copper Slag for the Production of Magnetic Oxide, Proc. of the 1992 Conference on Recycling of Metals, ASM, Düsseldorf, Germany.

Whyte, R.M., Orjans, J.R., Harris, G.B., and Thomas, J.A., 1977, Development of a Process for the Recovery of Electrolytic Copper and Cobalt from Rokana Converter Slag, in Advences in Extractive Metallrugy, Ed. by Sones , M.S., IMM , pp 57 - 68.

Yücel, O., Addemir, O., Tekin, A., Nizamoğlu, S., 1992, Recovery of Cobalt from Copper Slags Mineral Processing and Extractive Metallurgy Review, vol.10, pp 99-107.

# AN APPROACH FOR REDUCING AOX IN THE BLEACHERY EFFLUENTS OF A TURKISH PULP AND PAPER PLANT

Ülkü Yetiş, Aylin Selçuk, Celal F. Gökçay

Middle East Technical University
Environmental Engineering Department
06531 Ankara, TURKEY

## INTRODUCTION

SEKA Dalaman Pulp and Paper Plant which is located on the Aegean Sea coast, is the largest mill in Turkey with a pulping capacity of 100.000 tons/year. The mill mainly produces kraft pulp, and also linter cellulose. One of the most urgent problems faced in this plant, as the other pulp and paper plants in Turkey is that of decreasing the pollutionary load of the bleaching effluents. In all the countries producing bleached pulp, the environmental authorities are focusing on chlorinated organic compounds in the bleacheary effluents (Heimburger et al., 1988; Reeve et al., 1989). Many governments have already set regulatory limits to the amount of chlorinated organics, measured as adsorbable organic halogens (AOX) test (Hart et al., 1991). There is no limit set for AOX, in the Turkish Water Pollution Control Regulations published by the Ministry of Environment, yet. However; State Enterprise SEKA, which is the sole pulp producer in Turkey has initiated a research towards reducing AOX discharge levels in bleaching effluents.

Many developed countries including Canada, Sweden, Finland and Germany are aiming at reduction and ultimately complete elimination of chlorinated organic compounds. Canadian MISA Expert Committee, has set the limit for 1993 as <1.5 kg/a.d.t.(air dried) pulp; and has also suggested the minimization of AOX rather than end-of-pipe treatment. The National Environmental Protection Board of Sweden has established a limit for chlorinated organic compounds as 1.5-2.0 kg TOCl (total organic chlorine)/ ton of pulp produced starting

NATO ASI Series, Partnership Sub-Series, 2. Environment – Vol. 2
Cleaner Technologies and Cleaner Products
for Sustainable Development
Edited by H. M. Freeman, Z. Puskas, and R. Olbina

from 1989. Germany has regulated the discharge of organically-bound chlorine as 1.0 kg AOX/ton of pulp produced since 1990. Similarly, European Community sets an AOX limit of 1.5 kg/ton of pulp produced for bleaching wastewaters (Reeve et al., 1989).

## PULP BLEACHING AND PULP BLEACHING EFFLUENTS

The bleaching of kraft pulp is the process of removing the lignin still remaining in the pulp after cooking. Chlorination, which has historically been used in kraft pulp bleaching, selectively and economically completes the process of lignin removal and thus renders the pulp fibres responsive to final brightening treatments. Alkaline extraction following chlorination completes the solubilization of chlorine-treated lignin fragments. Thus, the effluents emanating from kraft mills that employ chlorination as bleaching process are characterized by their high AOX contents. The quality of these effluents are greatly affected by operating conditions in chlorination and alkaline extraction stages (Allison et al., 1990).

There are a number of process options for reducing the generation of AOX in kraft mill bleach plants (Heimburger et al., 1988; Du Manoir et al., 1989). They all rely on reducing the amount of elemental chlorine used in the initial chlorination stage since the vast majority of organically bound chlorine is produced through the electrophilic addition of chlorine to the aromatic ring of residual lignin (Singh, 1979). Chlorine use in the bleach plant can be reduced by replacing the bleaching power of chlorine in the first stage with that of chlorine dioxide. Chlorine dioxide also produces organically bound chlorine, but only one-fifth that of chlorine when compared on an active chlorine basis (Germgard et al., 1982; Axegard, 1989). Therefore, substitution of chlorine with chlorine dioxide (DC) appears as the most promising option for reducing AOX emissions. But, final brightness reached in the chlorine dioxide bleach is unacceptable, if this is not accompanied by reinforced extraction following the chlorine dioxide treatment (Allison et al., 1990). Hence, a lower the chlorine charge may be employed while still maintaining the target permanganate number.

Table 1. Bleaching Conditions for the CEHDED Sequence; Softwood Kraft Pulp

| Operational variable | C | E | H | D | E | D |
|---|---|---|---|---|---|---|
| Pulp consistency | 3% | 10% | 10% | 10% | 10% | 10% |
| Reaction time,(h) | 1 | 1.5 | 4 | 4 | 1.5 | 4 |
| Temperature, $^{0}C$ | 20 | 70 | 40 | 75 | 70 | 75 |
| Chemical dosage, kg/ton | 60-70 ($Cl_2$) | 50-60 (NaOH) | 30-40 (NaOCl) | 12 ($ClO_2$) | 50-60 (NaOH) | 6 ($ClO_2$) |
| Effluent AOX, mg/l | 42.90 | 92.00 | 66.20 | 17.22 | 4.28 | 2.92 |

In this way, it becomes possible to shift delignification from the chlorination stage to the first extraction stage and thereby decrease the total active chlorine charge (Basta et al., 1990). In SEKA Dalaman Pulp and Paper Mill, softwood kraft pulp is normally bleached by a CEHDED (chlorination, extraction, hypochlorite, chlorine dioxide, extraction, chlorine dioxide) bleaching sequence. Bleaching conditions and the AOX discharge levels for all stages of this sequence are given in Table 1.

To reduce the discharge levels of AOX in the bleach plant effluents, the mill is now considering the use of chlorine dioxide together with chlorine in the chlorination stage and oxidative extraction in the alkali extraction stage denoted by "DC(EOP)DED" (DC:sequential chlorine dioxide and chlorine application; EOP: oxygen and hydrogen peroxide reinforced extraction) bleaching sequence. The ultimate purpose of the research initiated and supported by SEKA is to adapt these options to SEKA Dalaman Pulp and Paper Mill bleach plant in the near future to reduce the discharge AOX levels. This paper which gives the results obtained from the initial stages of the study is regarding the determination of the effects of operational parameters of chlorination and extraction stages on pulp quality and AOX emission.

## EXPERIMENTAL

### Source pulp

Kraft pulp at 17.6 permanganate number, 7.8 cp. viscosity and with dry matter content of 19.52 % was supplied by SEKA Dalaman Pulp and Paper Mill.

### Chlorination

The DC treatments were performed on 75 a.d. g pulps in sealed plastic bags. Desired quantities of bleaching chemicals were added sequentially under the following standard conditions: 3% pulp consistency, $30^0$C reaction temperature and 60 min reaction time. Constant temperature was provided by placing the bags in a water bath. Pulp samples were preacidified to pH 1.6 with 4N sulfuric acid before adding the bleach chemicals.

After treatment, pulps were drained and washed with 1.0 l of tap water. Collected wash waters were analyzed for residual available chlorine, final pH, and AOX. Half of the washed pulps were analyzed for permanganate number and pulp viscosity, while the other halves were subjected to oxidative extraction treatment.

### Oxidative extraction

Oxidative extraction was performed in a specially designed pressurized laboratory reactor equipped with a stirring mechanism. The experimental conditions were: 6% NaOH; 3 atm oxygen pressure; 10% pulp consistency; 0.05% $MgSO_4$; $70^0$C; 60 min. Following the treatment, pulps were drained and washed with 1.50 l of tap water. Effluent filtrates for each pulp were analyzed for residual alkaline, final pH, and AOX. Washed pulps were analyzed for extracted permanganate number and pulp viscosity.

### Hydrogen-Peroxide Reinforced Oxidative Extraction

To see the effects of adding hydrogen peroxide to the EO-stage on the extracted pulp quality and AOX discharge levels in the bleaching effluents, $H_2O_2$ charges of 0.4 and 0.5 % on dry basis were tested in the EO-stage. In this test, the pulp chlorinated under 90% chlorine dioxide substitution and total available chlorine charge of 40 kg/ton was used.

### Analytical Methods

Effluent filtrates from chlorinations and extractions were analyzed for AOX using an Euroglass Analytical Instrument according to DIN 34809. Pulp consistency was measured in terms of permanganate number according to the method "TAPPI 214m-50".

## RESULTS AND DISCUSSION

### Base-Line Study

In this base-line study, the experimental conditions that are currently being employed in the plant were tried to be maintained (Table 2). Pure $Cl_2$ application (C-stage) (100 % $Cl_2$) with a total available chlorine charge of 60 kg/ton was used. The pulp sample was initially chlorinated at a constant temperature, and then treated with NaOH at atmospheric pressure for extraction (E-stage). The average AOX concentrations, measured in the base-line tests, are also given in Table 2.

Table 2. The experimental conditions and the Effluent Quality in the Base-Line Test

| Operational variables | CE |
|---|---|
| % $Cl_2$ | 100 |
| Total available chlorine, kg/ton | 60 |
| Temperature, °C | 30 |
| % NaOH | 4 |
| C-AOX, mg/l | 49.40 |
| E-AOX, mg/l | 47.49 |

**Chlorine Dioxide Substitution and Oxidative Extraction**

In the second part of the experimental work, three different chlorine dioxide substitutions and three different total chlorine application rates were examined and nine tests (total of 18 tests) were performed in parallel. Pulp samples were initially bleached under the conditions shown in Table 3. Then, the bleached pulps were sent to extraction stage where oxidative extraction is employed. The AOX concentrations measured in the chlorination stage (DC-AOX) and extraction stage (EO-AOX) effluents at different $ClO_2$ substitution levels and at different total available chlorine charges are also given in Table 3. As can be seen from this table that, the lowest AOX concentrations were achieved when the total available chlorine dosage was the lowest (10 kg/ton) and the $ClO_2$ substitution level was the highest. As the applied chlorine dose increases, DC-AOX and EO-AOX increase, despite the high proportion of chlorine dioxide used. This is because of the fact that the excessive chlorine charge required, when bleaching to low permanganate number at high substitution, may promote more chlorine attachment to dissolved lignin fragments.

Table 3. The Conditions and the Effluent Quality in the (DC)(EPO) Tests

| Tot.Av.chlorine kg/ton | 10 | 10 | 10 | 25 | 25 | 25 | 40 | 40 | 40 |
|---|---|---|---|---|---|---|---|---|---|
| % $ClO_2$ | 90 | 50 | 10 | 90 | 50 | 10 | 90 | 50 | 10 |
| % $Cl_2$ | 10 | 50 | 90 | 10 | 50 | 90 | 10 | 50 | 90 |
| DC-AOX, mg/l | 2.7 | 2.2 | 5.2 | 4.76 | 8.76 | 11.2 | 11.5 | 10.27 | 24.3 |
| EO-AOX, mg/l | 6.7 | 5.5 | 8.3 | 2.40 | 2.48 | 20.9 | 8.61 | 22.04 | 27.9 |

Figures 1 and 2 show schematically how DC-AOX and EO-AOX decreases respectively as the relative amount of chlorine dioxide increases. As shown in both of these figures, an increase in the percentage of chlorine dioxide leads to a substantial reduction in the AOX concentration in the bleaching effluents. A chlorine dioxide substitution of 90 %, when the chlorine charge is 40 kg/ton of pulp, causes a 77 % decrease in the DC-AOX, and a 82 % decrease in the EO-AOX. However, as the chlorine dioxide substitution level reduces to 10 %, the decrese in the the AOX concentration becomes less and only 51 % and 41 % reductions in DC-AOX and EO-AOX are achieved respectively.

Conditions which produce minimal AOX are known to achieve limited bleaching; and produce pulps with high extracted permanganate numbers. Therefore, when assessing the effects of chlorination conditions on effluent properties, the concomitant effects on

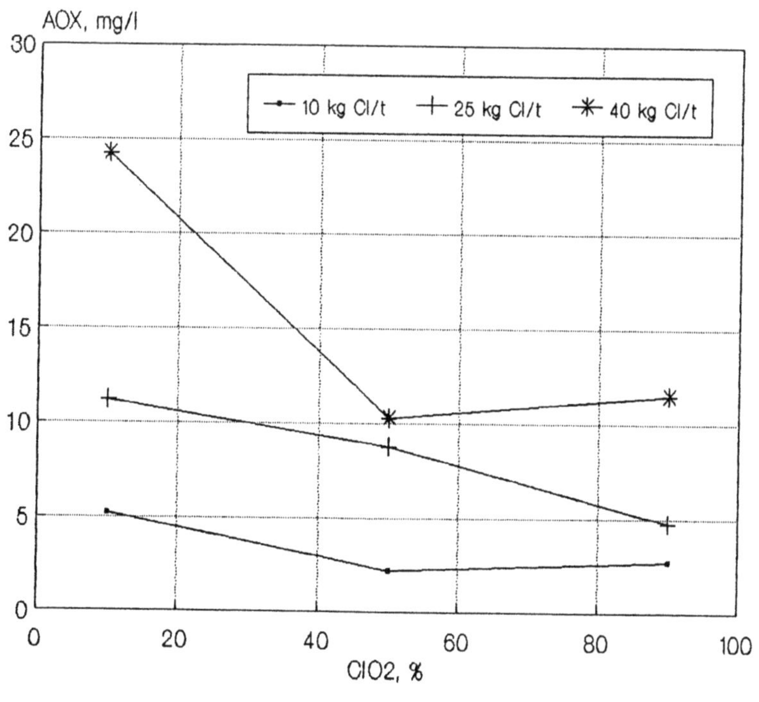


Figure 1. DC-stage AOX vs. $ClO_2$ substitution level.

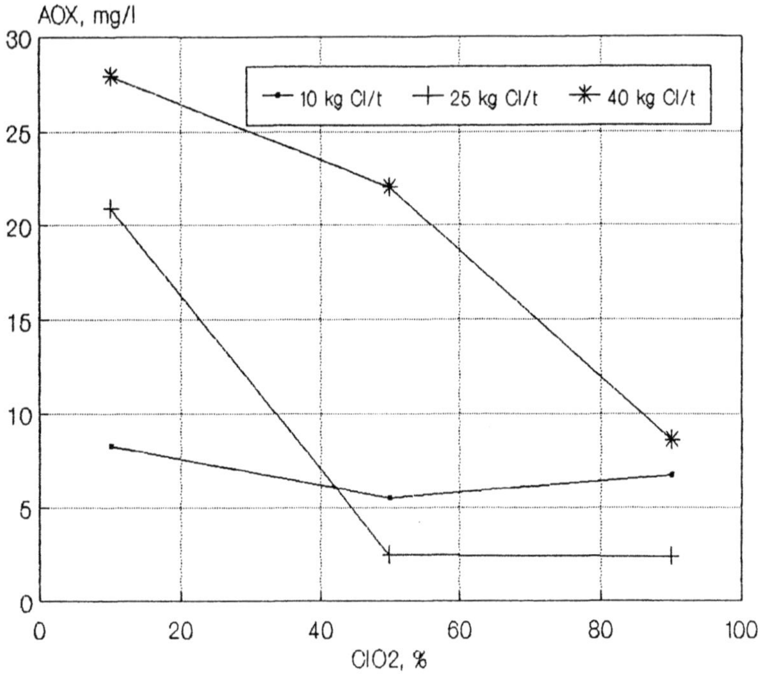


Figure 2. EO-stage AOX vs. $ClO_2$ substitution level.

pulp bleaching must also be considered. Figure 3 shows EO-AOX predictions when bleaching from 5 to 13 extracted permanganate number with 10 to 90% chlorine dioxide substitution. As can be seen from Figure 3, EO-AOX was clearly reduced by a decrease in the extent of bleaching (i.e., to higher extracted permanganate number) and by an increase in the proportion of chlorine dioxide in the applied chemical. However, at 90% chlorine dioxide substitution, bleaching to low extracted permanganate number was predicted to produce as much AOX as bleaching with 50% substitution. This effect was due to a sharp decrease in bleaching effectiveness at high substitution. Therefore, with high degrees of chlorine dioxide substitution, the total available chlorine charge in the chlorination stage must be substantially increased, or the subsequent first extraction stage must be reinforced in order produce a pulp with an acceptable extracted permanganate number (Allison et al.,1990).

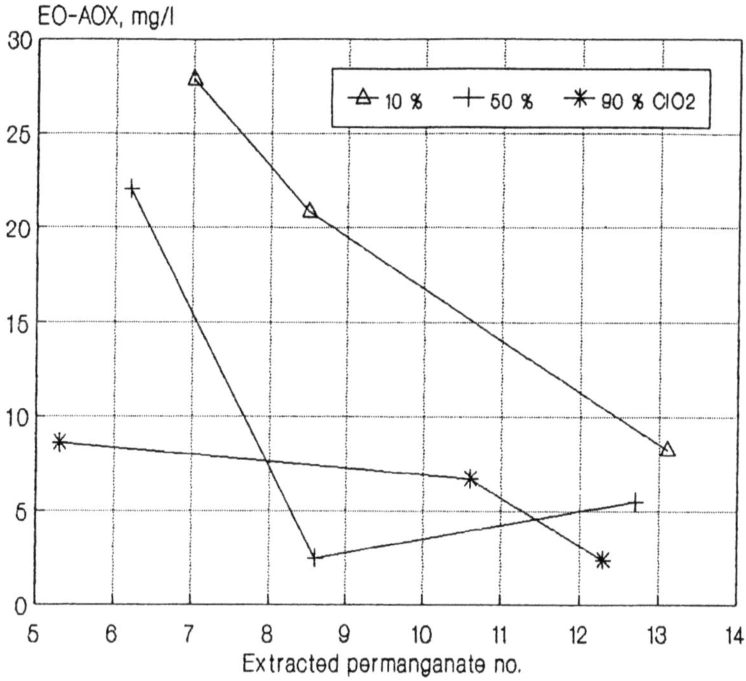


Figure 3. The dependency of EO stage AOX on extracted permanganate number at different chlorine dioxide substitution levels.

In Figure 4, the influence of total available chlorine charge on extracted permanganate number was shown for varying degrees of chlorine dioxide substitution. This plot confirms that 90 % substitution requires high chemical charges especially when bleaching to low extracted permanganate number.

**Hydrogen-Peroxide Reinforced Oxidative Extraction**

In the second phase of the experimental work, hydrogen-peroxide reinforced oxidative extraction sequence was tried to be adopted to the pulp treated in the DC bleaching sequence. The pulp used in this phase of the study was the one bleached with a total chlorine charge of 40 kg/ton and a chlorine dioxide substitution of 90 %. In these tests, two different hydrogen-peroxide doses which are 0.4 and 0.5 % (on dry pulp basis) were employed. The results obtained are presented in Table 5. As can be seen from this table, the

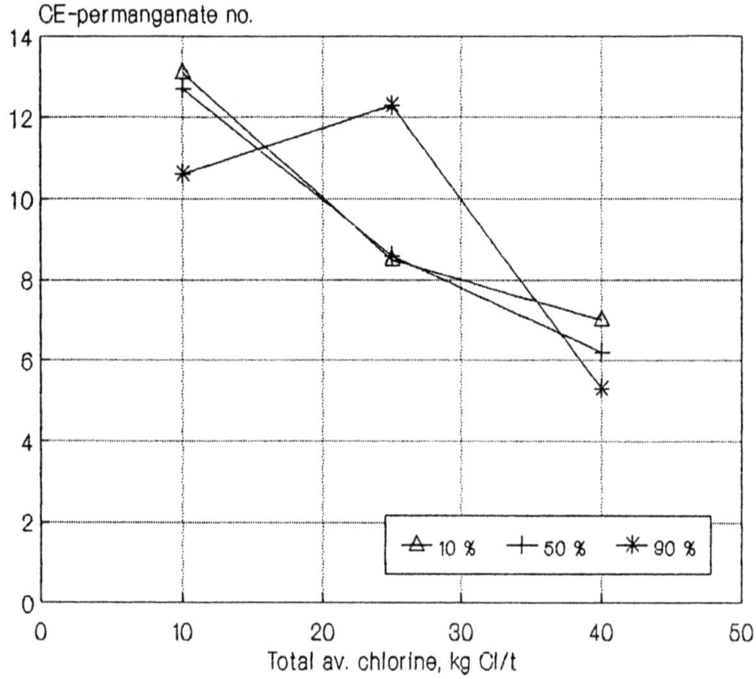


Figure 4. Extracted Permanganate number versus total available chlorine.

addition of hydrogen-peroxide has provided an acceptable pulp quality with low AOX concentrations in both chlorination and extraction effluents, as compared to the reference CE pre-bleaching sequence. The C-stage AOX concentration has decreased from its earlier value of 49.40 mg/l to 10.40 mg/l while the pulp quality - in terms of extracted permanganate number- has remained almost unchanged. Similarly, E-stage AOX concentration has also decreased from 47.49 mg/l down to 13.82 mg/l.

A comparison of the permanganate numbers measured after (DC)(EO) and (DC)(EPO) sequences reveals that $H_2O_2$ addition to oxygen-reinforced extraction improves delignification work. However, this effect has not appeared for the $H_2O_2$ dose of 0.4 %; which has indicated that 0.4 % $H_2O_2$ was not sufficient for extending delignification.

Table 5. Comparison between (DC)(EPO) Pre-bleaching and CE Pre-bleaching

| Operational variables | CE | (DC)(EO) | (DC)(EPO) | |
|---|---|---|---|---|
| | | | % 0.4 $H_2O_2$ | % 0.5 $H_2O_2$ |
| $\%Cl_2$ | 100 | 10 | 10 | 10 |
| $\%ClO_2$ | - | 90 | 90 | 90 |
| Tot. Av. chlorine, kg/ton | 60 | 40 | 40 | 40 |
| % NaOH | 4 | 6 | 6 | 6 |
| Extracted permanganate No. | 4.3 | 5.5 | 5.6 | 4.2 |
| C stage AOX, mg/l | 49.4 | 12.13 | 11.48 | 10.40 |
| E stage AOX, mg/l | 47.49 | 10.41 | 13.37 | 13.81 |

CONCLUSIONS

The following conclusions can be drawn from this study:

1. The AOX formation has been shown to be proportional to the amount of elemental chlorine in oxidants consumed during delignification, regardless of whether it is from chlorine or chlorine dioxide. While both chlorine or chlorine dioxide have contributed to the AOX formation, the molecular chlorine has so to a much greater extent.

2. Substitution of chlorine dioxide for chlorine, combined with reinforcement of the extraction stage with oxygen can reduce AOX emissions substantially from a fully bleached, softwood kraft pulp. The lowest AOX concentrations are obtained with the highest chlorine dioxide substitution. However, the target pulp quality can not be reached with high substitution levels of chlorine dioxide for chlorine, if the oxidative extraction is not reinforced with hydrogen peroxide.

3. (DC)(EPO) technology seems to be a convenient process alternative for SEKA-Dalaman pulp and paper plant.

4. To implement the proposed process modification, further efforts has to be devoted.

## ACKNOWLEDGEMENTS

The financial support from the Turkish National Science Foundation (TÜBITAK) and The SEKA is gratefully acknowledged. Thanks are expressed to Kemal Demirtaş and Gülfem Köse for their analytical contributions.

## REFERENCES

Allison, R.W., McFarlane, P.N. and Clark, T.A. "Predictive Models for Effluent and Pulp Properties After Kraft Chlorination." Preprints 44th APPITA Annual General Conference, Rotonia, New Zealand, A1.1-A1.23, 1990.

Axegard, P., Pulp Paper Canada, 90(5):T183, 1989.

Basta, J., Holtinger, L., Hook, J. and Lundgren, P., Tappi J. 73(4):155 (1990).

Du Manoir, J. and Dubelsten, P., "Chlorination Strategies for the 1990s: A Review of Recent in Chemical Pulp Bleaching and their Environmental Impact", Proceedings of the 43rd Appita Annual General Meeting, Brisbane, 1989.

Germgard, U. and Larsson, S., Paperi Puu, 65(4):287, 1982.

Hart, P.W. and Hsieh, J.S., Tappi J. 73(11):117, 1990.

Heimburger, S.A., Blevins, D.S., Bostwick, J.H. and Donini, G.P., Pulping Conference Proceedings, TAPPI PRESS, Atlanta, 1988.

Reeve, D.W. and Earl, P.F. Pulp and Paper Canada 90(4):T128, 1989.

Singh, R. (ed) "The Bleaching of Pulp", TAPPI Press, Atlanta, 1979.

# Clean Synthesis as Environmentally Friendly Chemistry

Colin J. Suckling
Department of Pure and Applied Chemistry
University of Strathclyde
295 Cathedral Street
Glasgow G1 1XL
Scotland

## 1. Setting the Scene

I grew up in the North of England not far from major centres of chemical manufacture on Merseyside. On sites adjacent to these factories it was common to see brightly coloured sediments: greens, presumably copper, yellows, perhaps chromates, and reds, probably from iron(III) salts were all visible. These deposits together with characteristic smells have contributed to giving the chemical industry its low public esteem and today, more sophisticated objections couched in molecular scientific terms are common. Much of what causes concern comes from an era of the chemical industry now passed. Chemistry, although a mature science, is far from being exhausted in providing solutions to environmental problems and moreover to developing new reactions and processes with today's environmental concerns in mind from the beginning of research. This is not just a matter for the attention of the industrialist; it is quintessentially what academic research is about: the development of ideas and the creation of opportunities. Therefore I shall not be concerned with whether a particular product or material in a given application is itself environmentally friendly or acceptable nor shall I consider so-called 'end of the pipe' solutions to current pollution problems. Others in this book deal strongly with these aspects. The discussion in this paper derives from a report prepared for two of the then Research Councils in the UK by myself and colleagues from the University of Strathclyde defining themes for research in clean synthesis as part of a Clean Technology Initiative (Suckling et. al., 1992). I have prepared the text in such a way as to make as much of the discussion as possible accessible to non-expert chemists.

NATO ASI Series, Partnership Sub-Series, 2. Environment – Vol. 2
Cleaner Technologies and Cleaner Products
for Sustainable Development
Edited by H. M. Freeman, Z. Puskas, and R. Olbina

*Clean up* can be clearly contrasted with *clean synthesis*. The essence of clean synthesis is that the reactions are designed so that clean up is simply not necessary. In order to look at the underlying concepts and examples of successes in clean it is a good idea to spend a short time understanding the properties of traditional organic chemical synthesis and industrial manufacture.

The scene for industrial synthesis is largely set by the feedstocks available. In this century there has been a shift from coal, which afforded tars and acetylene and a lot of smoke and smells, to oil whose corresponding products are naphtha and ethylene in general without obviously nasty gaseous emissions. Through various hydrogenation, cracking and isomerisation reactions, large quantities of intermediates have become available at low prices. Modern high-tonnage petrochemical operations are clean in many of the respects that we shall consider below. However all of these operations comsume fossil resources. One of the background thoughts in clean synthesis must therefore be the use of renewable resources, not only with the carbon source in mind, but also conscious of the energy costs associated with chemical processes. Such considerations, which I shall not explore much further, are set in the economic and fiscal environment imposed by national and supranational governments. Renewable resources come strongly into the picture when we come to examine biological catalysis.

The problem of feedstocks has been expressed another way by Jentzsch (Jentzsch, 1990). He points out that feedstocks also determine the technology to be used. Moreover, although one might imagine other primary carbon feedstocks the chemistry based upon steam cracking that converts feedstock compounds into useful synthetic building blocks for thermodynamic reasons tends to afford a similar mixture of compounds whatever the primary source. This is another background problem for clean synthesis.

I have used the term *synthetic building block*. For clarity, let us call to mind schematically what typical organic synthesis does (figure 1). Its chief aim is to construct new skeletons of carbon bearing the right substituents to have the desired properties for the study itself or for an application. In the majority of carbon compounds, however, you cannot simply join one selected atom in one molecule to another without some cost. Some sort of *activation* is usually necessary. As a bond forms, another often breaks (so that the valency of carbon does not exceed 4), and the atoms that come away are referred to as the *leaving group*. Thus there are two costs associated with making new bonds - energy of some sort to get the reaction going and waste in terms of the atoms thrown away. The examples selected, Friedel Crafts compared with Diels Alder reactions (figure 1) show that well established chemistry can be clean in the context of this discussion as well as in need of improvement. It is therefore not at all unreasonable to expect chemists to develop more intrinsically clean processes.

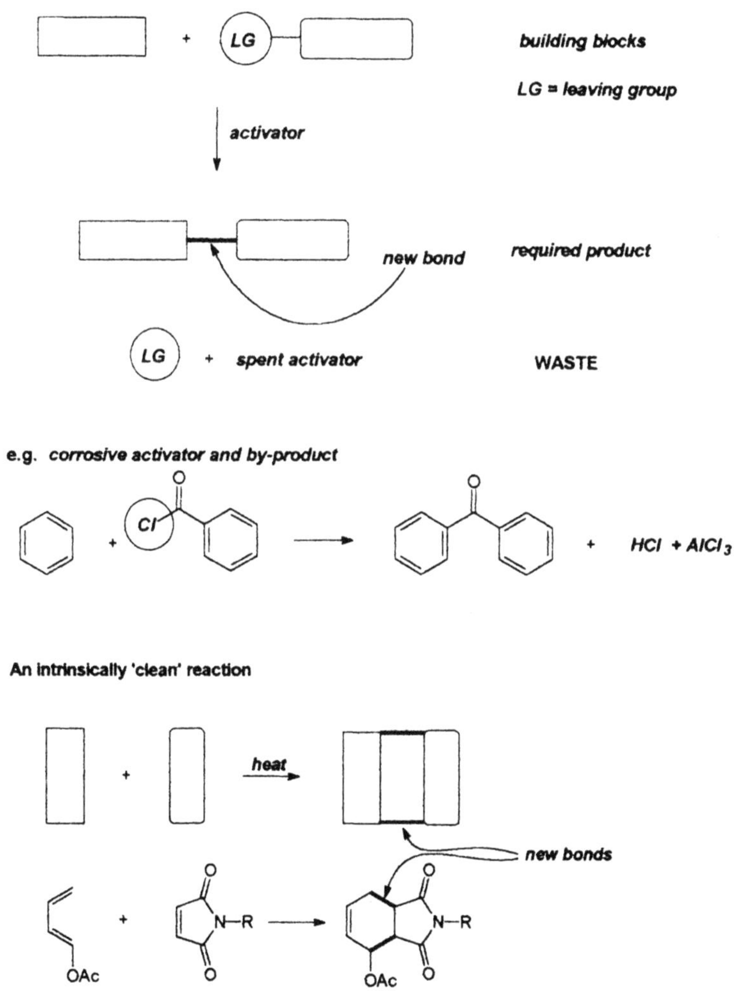


**Figure 1.** Principles of clean synthesis

## 2. Clean Synthesis and the Leaving Group and Activating Group Principles

In the preparation of our report (Suckling, et.al., 1992), one company told us that they imported tens of tons of bromine into one of their sites each year, none of which ended up in the product sold. Bromine and chlorine are typical highly reactive elements that provide the activation referred to above. Once in the synthetic sequence, they also provide leaving groups because of their electronic characteristics; hence they end up in effluent and back in the sea. Clean synthesis, in one manifestation, is about avoiding such cycles (figure 2).

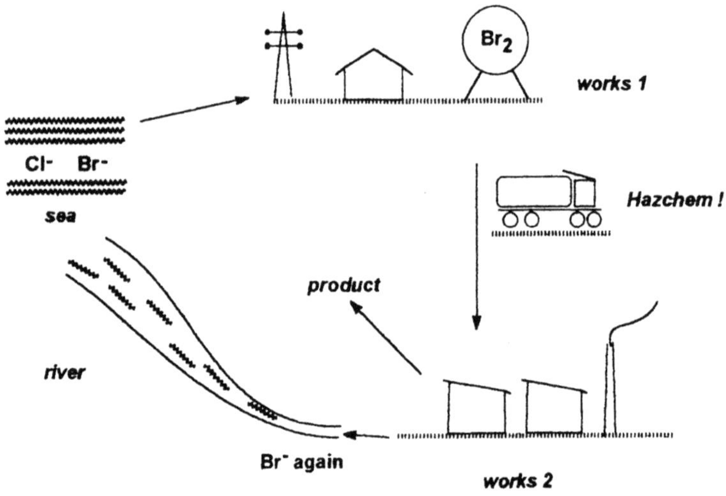


**Figure 2.** A cycle of industrial activity to avoid.

In order to progress the argument further, I shall need to use some technical organic chemical terms. To take an extreme view based upon the above, nucleophilic substitution and organometallic reactions are intrinsically unclean for the formation of new carbon-carbon bonds. On the other hand, catalysed aldol and many other reactions would be intrinsically cleaner because they do not necessarily require the intervention of activating atoms such as halogens or metals (figure 3).

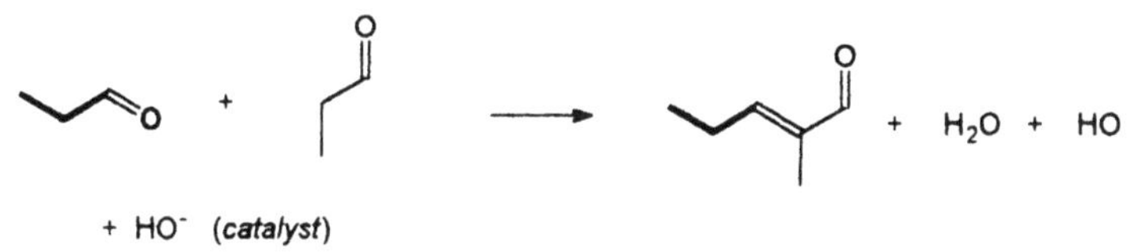


***Aldol reaction: base catalysed***

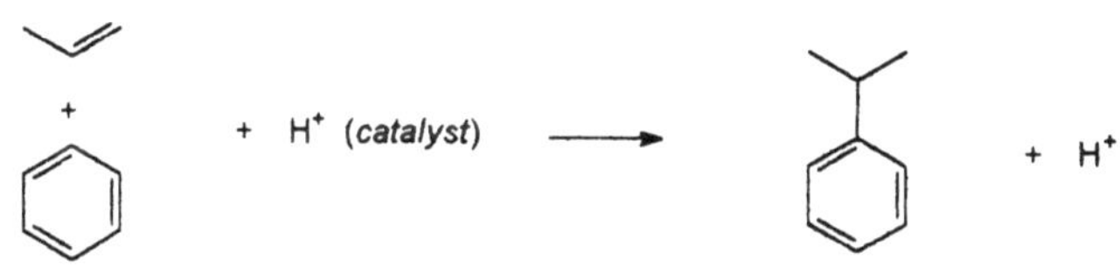


***Benzene alkylation: acid catalysed***

**Figure 3.** Clean synthetic reactions: two catalysed prototypes.

Let us take an example using these reaction types of an industrial process that violates what we could call the *leaving group principle* and how it can be avoided (figure 4). α,β-Unsaturated compounds are important monomers in the plastics industry; in particular, methacrylic acid and its derivatives are widely used in the manufacture of transparent plastic sheeting. A route based upon traditional industrial chemistry made use of acetone available from the cumene process for phenol manufacture. This route extends the carbon chain of acetone using cyanide (toxic) which is then hydrolysed using sulphuric acid (corrosive) to give acrylamide (a danger listed compound) and ammonium sulphate as a by-product. So apart from the toxic or corrosive reagents and hazardous intermediate, nitrogen introduced in the cyanide is thrown away with the sulphate of sulphuric acid. An alternative has been developed by BASF using technology associated with the oxo-process for hydroformylation (Malle, 1993). Here, the organometallic catalyst provides the necessary activation to prepare the one and three carbon intermediates. Aldol condensation then sets up the necessary carbon skeleton. The rest is routine oxidation and esterification. Notice that all of the atoms in the starting materials end up in the products, except water. If anything is acceptable as a leaving group in clean synthesis it is (clean) water.

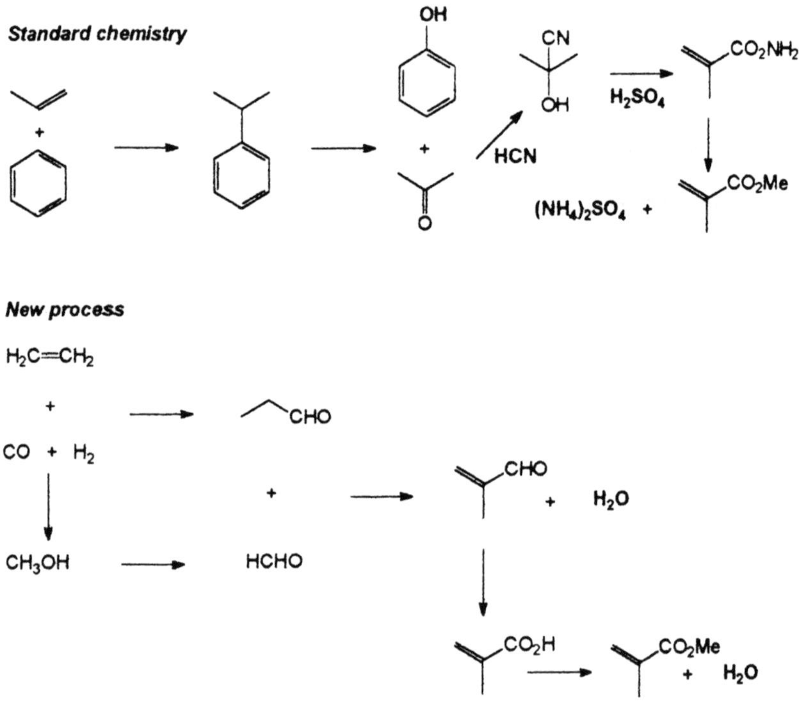


**Figure 4.** Contrasting routes to methyl methacrylate

This is being a little hard on the traditional process because it can be carried out cleanly and economically. A significant investment has been made by one manufacturer in the recycling of the sulphate to sulphuric acid thereby removing the the waste (Wright, 1993). Nevertheless, that is end of the pipe and not clean synthesis. Conversely, BASF was not in the methacrylates business until the new process was discovered. Thus clean synthesis can lead to new business and is not simply an add-on cost.

The above discussion leads to what can be called the *Leaving Group Principle of Clean Synthesis:*

For clean synthesis, all leaving groups should be either environmentally neutral compounds such as water or nitrogen, or not released into the environment at all.

The complementary principle of clean synthesis is the *Activating Group Principle* namely:

For clean synthesis, all activating groups should be carbonyl derivatives (or equivalents such as epoxides) within the substrate, or should be catalysts that are not released into the environment at all.

To satisfy both of these principles is difficult but they are a good stimulus for creative chemical thought.

An example from a different field of chemistry concerns research carried out at the University of York into catalysis of aromatic substitution reactions. They had recognised that traditional Friedel Crafts chemistry was intrinsically unclean. Acylation reactions are particularly poor because they require stoichiometric quantities of aluminium chloride (or a similar Lewis acid), a suitable inert solvent, and an activated acid in the form of the acid chloride (figure 5). During the reaction, hydrogen chloride gas is produced and at the end of the reaction during work up, substantial residues of aluminium salts are produced. All of these problems can be readily associated with the use of anhydrous aluminium chloride. The preparation of aryl ketones thus becomes a prime candidate for research in clean chemistry (Clark et.al., 1989, 1992).

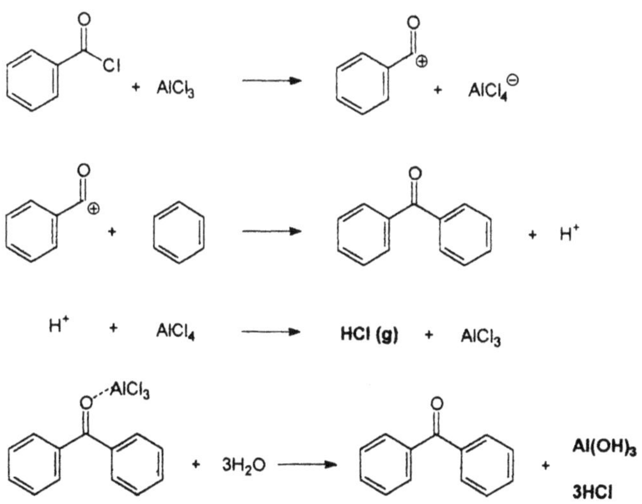


**Figure 5.** Intrinsic waste problems associated with aromatic acylation.

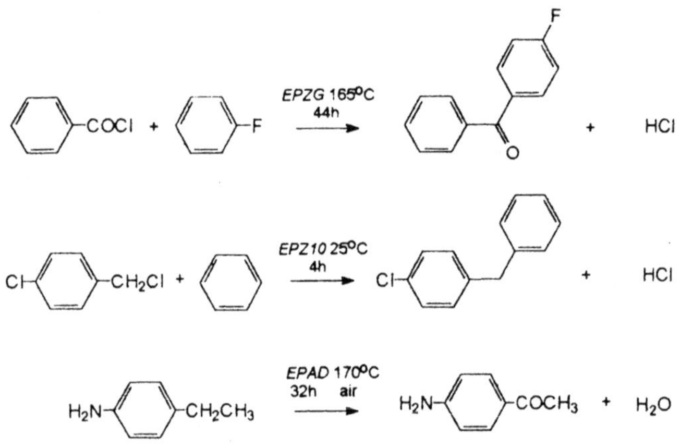


**Figure 6.** Aromatic substrates transformed with the aid of *Envirocats*

The solution to the problem comes from suitably restructured montmorillonite clays, the chemical properties of which are essentially those of strong acids (figure 6). However there is no labile chlorine in the catalyst. The University of York and their partner industrialists have felicitously called their products *Envirocats*. Several important reactions can be catalysed by the appropriate *Envirocat* including benzylation, benzoylation, and even benzylic oxidation. Oxidation is, as we shall see, a most important reaction in the context of the activating and leaving group principles because oxidised compounds can contain activated groups (carbonyl groups) or leaving groups (alcohols). The oxidation illustrated in figure 6 could not be carried out directly using a halogen based method because of competing substitution in the benzene ring. *Envirocats* therefore score on grounds of chemical selectivity. Whilst *Envirocats* have saved three equivalents of hydrogen chloride and one of aluminium salts being discharged, there is still a need for an activating chlorine in benzylation and benzoylation reactions. Other related reactions including sulphonation suffer from the same limitation but alkylation with alkenes or aldehydes, which are activated by protonation by the *Envirocat* to give the electrophile, can also be carried out. Of course, the *Envirocat* can be recovered and reused. At the end of its life, being a clay, disposal in landfill sites is more acceptable than with many other chemical wastes.

### 3. Avoid overkill

In considering the energetics of a process, chemical engineers are familiar with problems of heat transfer and the need to minimise the energy input into a process. It is possible to draw a parallel to this in clean chemistry, the logic of which is suggested in part by the previous example of the *Envirocats*. One of the problems of the industrial development of nineteenth century aromatic substitution chemistry is that you always get mixtures of products (figure 7). This is partly due to the extremely high reactivity of many of the reagents used such as strong Lewis acids (aluminium trichloride) or the mixtures of acids used for nitration and sulphonation. These reagents, with their high chemical reactivity, which is another way of saying high energy cost, have to be manufactured. It would clearly, therefore, be a good idea if the reactivity of the reagent is made to match that of the substrate as closely as possible to give a selective reaction. There is nothing new in this idea because the natural product synthetic chemists would frequently put a lot of effort into selecting, or even developing, a reagent of suitable reactivity for a key step in a synthesis. What is new is the emphasis in the context of clean synthesis.

Several groups have demonstrated the modulation of reactivity of highly active reagents in aromatic substitution (Guy et.al., 1982; Breslow and Campbell, 1969) using chlorination as an example. In our own work, we discovered a selective nitration reaction that has many of the properties appropriate to clean synthesis (figure 8) (Pervez et.al., 1988). This chemistry was developed from a programme of biomimetic chemistry (see below). Our primary interest was in controlling selectivity and we had found that it was possible to control the orientation of chlorination of aromatic compounds, especially phenols which are highly reactive, using functionalised surfactants. These materials themselves were not suitable for preparative application but we were able to develop derivatives that were. The potentially useful materials illustrate two principles of clean chemistry namely *localised reagents* and *immobilised reagents* together with the main point of this section *avoid overkill.* Localised and immobilised (or supported) reagents often go together because the inert support of a reagent can supply a controlled environment for reaction equivalent to the binding site of an enzyme. Conceptually at least, these reagents together with their associated environments are the artefactual analogues of enzymes. Much of the development of bioorganic chemistry has centred on this parallel.

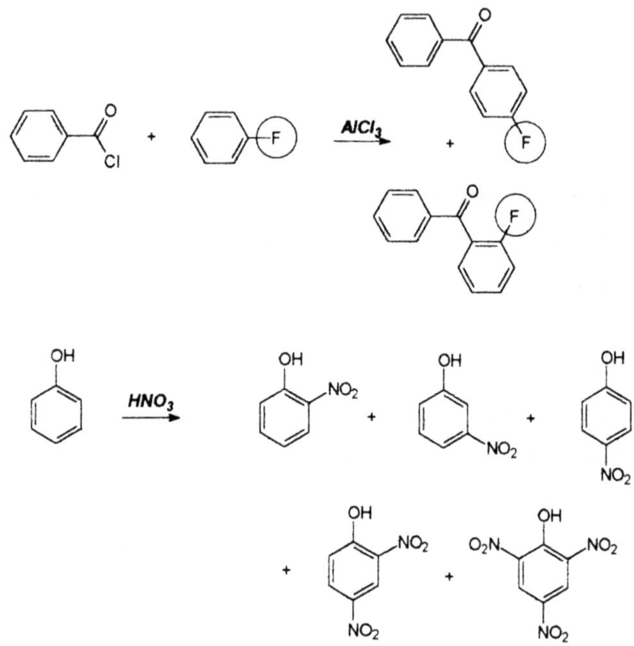


**Figure 7.** Mixture formation in typical aromatic substitution reactions.

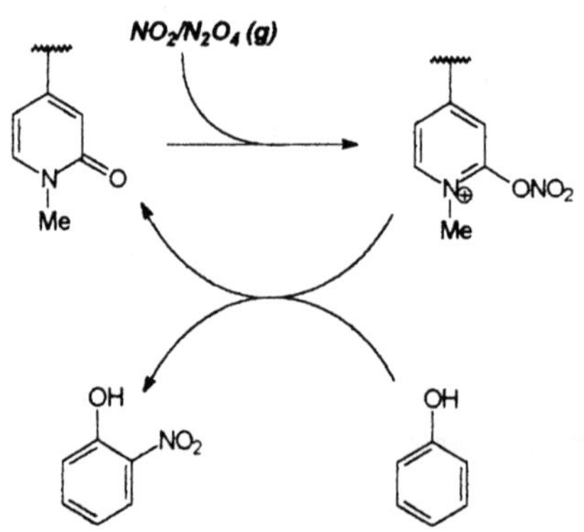


**Figure 8.** Polymer supported mononitration of phenol

We found that carboxy-substituted pyridinium containing surfactants could be converted into the corresponding nitro mixed anhydrides and that these intermediates nitrated phenol essentially quantitatively in the 2-position with no polynitration whatsoever. The most interesting embodiment of these results, however, was a polymer supported version in which the pyridine derivative was part of a styrene/divinylbenzene/4-vinylpyridine copolymer. Instead of using highly reactive nitric acid, the nitrating agent in this material could be supplied simply from gaseous $NO_2/N_2O_4$. Stirring the so-activated polymer with a solution of phenol and evaporation of the solvent afforded crystalline 2-nitrophenol in greater than 95% yield and 99% purity. We understand the selectivity as due to association of the electron rich phenol ring with the electron deficient pyridinium ring with the selectivity introduced by hydrogen bonding between the phenol hydroxyl group and the activated reagent which places the 2-position in direct line of attack. This system has not been extensively developed and one limitation is the need for hydrogen bonding; hydrogen bonding limits the range of solvents that can be used to relatively non-polar ones that will not compete with the reagent for the hydroxyl group. Also the polymer must be sufficiently swollen to allow access of the reactants to its active groups. So far, we have only been able to satisfy these requirements using chlorinated solvents, compounds that should obviously be avoided.

Although in the above example, the reactivity of the nitrating agent was deliberately stepped down from the normal power of nitric acid by design, it is commonly found that immobilised or supported reagents display reduced or modified activities compared with the same reagent im homogeneous solution (Clark et.al., 1992). They are thus a major focus for clean synthesis. In addition, the restriction of the reagent to the support means that it will not come significantly into the effluent stream. Accordingly, otherwise toxic or hazardous reagents such as chromates or arsenates can be used with improved cleanness. Strong bases can also be obtained in immobilised form just as the *Envirocats* provided Lewis acids. For further discussion, see Longley and Kirkwood (1994) and Blossey and Ward (1989).

Industry is, of course, well aware of the potential of immobilised reagents in processes. There is a good example from Merck of a process in which catalysis by a zeolite not only fulfills the immobilisation principle but in doing so makes it possible to accommodate the activating and leaving group principles also (Davis, 1993; Reuben and Witkoff, 1989) (figure 9).

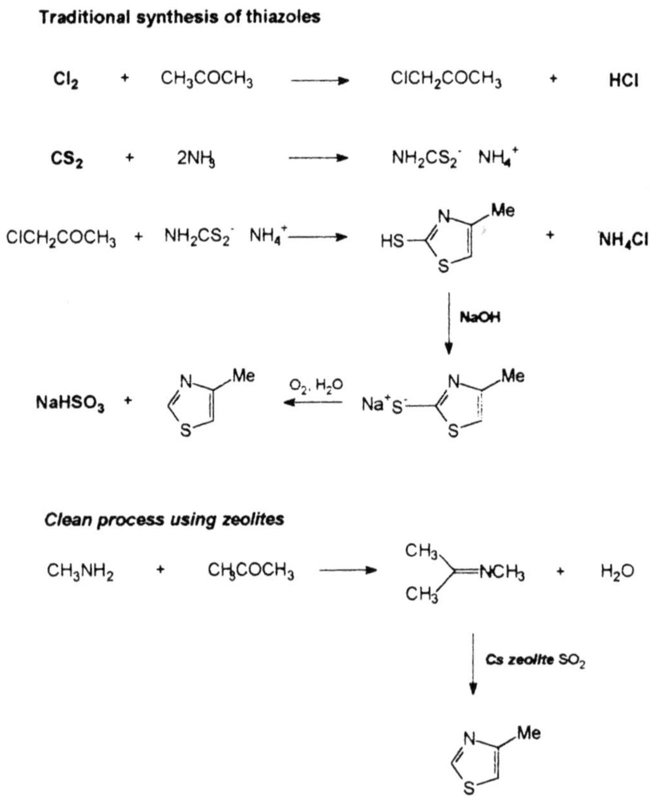


**Figure 9.** Syntheses of 4-methylthiazole

### 4. Is biocatalysis the answer?

Sometimes. Just as there are things that conventional chemistry is good at and things in which it has shortcomings, so biological catalysis is not the answer to all problems. It is not even new; recall the Weizmann process for acetone using *Clostridium acetobutylicum* using starches as the feedstock, a process that is now 80 years old. What is new, however, is that modern chemistry and molecular biology have made it possible to interact constructively with the organisms that produce the biocatalysts and with the catalysts themselves. The interest of chemists in biotransformations was stimulated more than 20 years ago (Suckling and Suckling, 1974), and its current position in the context of clean chemistry is worth further exploration.

Let us firstly consider feedstocks again because, as mentioned above, they control the chemistry that can be undertaken. The primary carbon-containing bulk natural products that can be considered for synthesis are carbohydrates and fatty acids. There is, of course, a substantial industry built upon each but these industries typically convert carbohydrates into carbohydrate-like products and fatty acids into hydrocarbon rich products. In the broader view of clean synthesis, organic chemistry requires typically C4 to C8 building blocks that it can organise flexibly into more complex structures as desired. There is therefore a problem with the currently available biological feedstocks: carbohydrates are too highly functionalised and fatty acids are too little functionalised for general synthetic use (figure 10). One bulk product that could resolve the problem is the so-called *Biopol* (Holmes et.al., 1982), a naturally occurring polymer of 3-hydroxyalkanoic acids produced by a number of bacteria including *Hydrogemonas eutropha* and *Alcaligenes eutrophus*. The bacterial polymerisation was also discovered more than 60 years ago but its commercial exploitation was only undertaken in the 1980s. *Biopol* can be fabricated into elastomers, films, and fibres and finds use as a biodegradeable polymer but, on hydrolysis, it affords the 3-hydroxyalkanoic acids themselves. These compounds are close to the synthetic requirement for generally convertible building blocks. They are not ideal, an aldehyde or a monoester would be more useful than a carboxylic acid, but as a source of carbon at the right sort of functionalisation level, they are significant. There is thus an opportunity for research to produce a plant or microorganism that would accumulate a relative of the hydroxyalkanoic acids. Metabolic engineering of organisms to create modified pathways leading to desirable products has already been shown to be feasible in fine chemicals conversions (cephalosporins (Conder et.al., 1994) and opiates (Bruce et.al., 1994)) so why not in the production of a basic feedstock chemical? Success would be a major advance in renewable resource provision for the chemical industry.

Turning from feedstocks to the biocatalysts themselves, principally enzymes, there have been enormous advances in the use of enzymes in organic synthesis in the last ten years (Wong and Whitesides, 1994). Perhaps the most notable recent feature has been the constructive partnership of selective chemical reactions with appropriate biotransformations; indeed the development of much selective modern organic chemistry has been stimulated by the selectivity of enzymic catalysis (Seebach, 1990). Biocatalysis in clean synthesis is, however, limited by those enzymes that nature

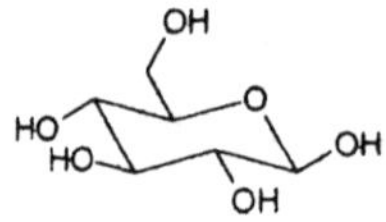


***Glucose***

***$C_6$ 6 functional groups***

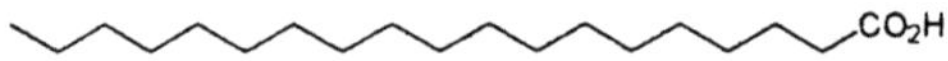


***Stearic acid***

***$C_{18}$ 1 functional group***

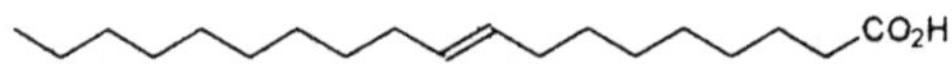


***Oleic Acid***

***$C_{18}$ 2 functional groups***

***Biopol and hydrolysis***

***$C_{4+5}$ 2 functional groups***

**Figure 10.** Prototypical feedstocks produced by plants or microorganisms

provides, and hence the chemistry accessible to biocatalysis is itself limited to a range of hydrolysis and reduction reactions. There is a continuing need for new catalysts. When such catalysts are found, it is interesting to note how new chemistry explodes around them.

During the 1980s, much work was devoted at ICI (now Zeneca) into the development of a benzene oxygenase from *Pseudomonas putida.* This enzyme is interesting in the context of clean synthesis because it introduces oxygen into the substrate thereby providing useful functionality and it does this in a *stereoselective* manner. Because nature is chiral, it is essential that compounds released into the environment as drugs or agrichemicals have been developed with the understanding that enantiomers of xenobiotic compounds will interact differently with nature be it in mammalian receptors, enzymes or in plants (figure 11). When things go wrong, the effects can be catastrophic. Much of the development of modern synthetic chemistry has therefore been to find efficient methods of controlling the stereochemical course of a reaction.

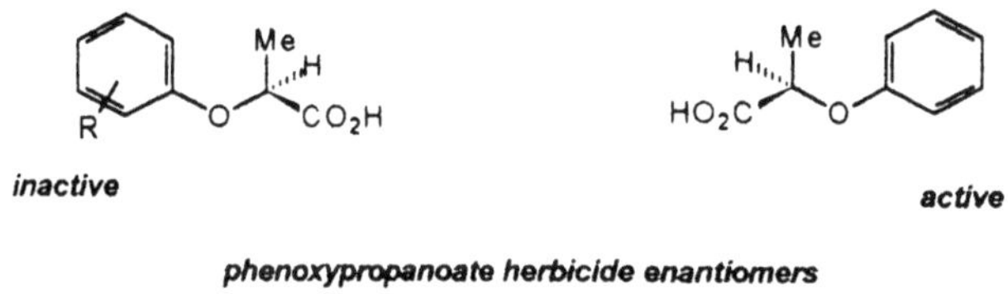


**Figure 11.** Chirality and environmental activity

Enzymes are chiral catalysts and are very good at producing compounds in high enantiomeric purity, as in benzene dioxygenase. The enzyme was first shown to be useful in preparative chemistry through the production of a strain that lacked the ability to cleave the oxidation product, *cis*-dihydrobenzene-1,2-diol, which then accumulates (figure 12). This compound itself has been used as the starting material for the synthesis of polymers (Ballard et.al., 1983) and of natural products (Taylor et.al., 1987). However, interest was greatly increased with the discovery that the enzyme will also accept substituted benzenes as substrates and the products of their oxidation, which are chiral, provide useful starting materials for specialised organic synthesis (Boyd et.al., 1993). It is also worth pointing out that this oxidation reaction is one for which no conventional organic reagent exists. This is one of the major niches for biological catalysis. More new biocatalysts are required.

There are several approaches to the discovery of new biocatalysts. Firstly, one can search in the communities of microorganisms that have evolved on dumps and landfills for organisms that are able to transform compounds of interest for chemical synthesis as was the case with benzene dioxygenase. Having identified an organism, strain selection using random mutation, a traditional approach, can provide useful catalysts. Alternatively, if an enzyme is highly characterised in terms of its molecular structure and its production in an organism, it is possible to modify its structure to improve its catalytic properties. This is not easy because our understanding of detailed interactions between functional groups in proteins and their connection to catalysis is notoriously patchy and progress is largely empirical. One of the most successful exploitations of this technology has been the work of Holbrook and his colleagues on lactate dehydrogenase (Casy et.al., 1992).

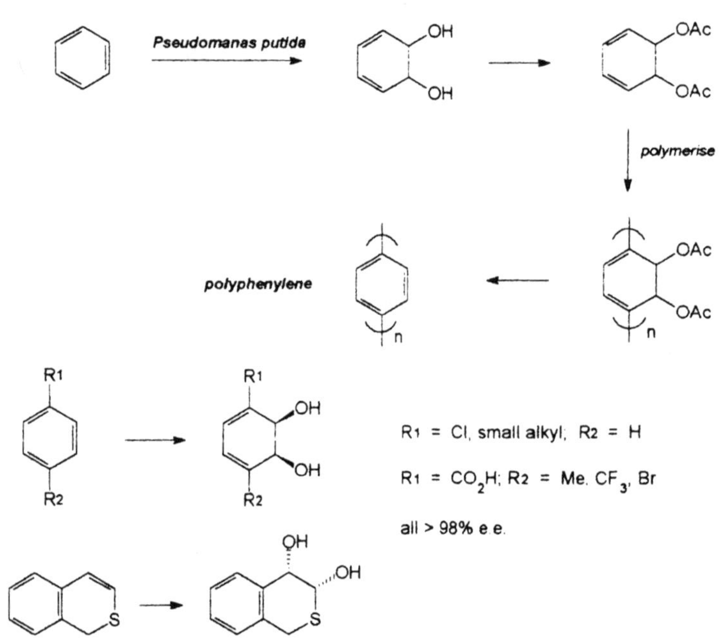


**Figure 12.** New chemistry stimulated by the discovery of a new enzyme for biotransformations.

It is also possible to modify the properties of enzymes using chemical reagents (Kaiser 1988; Kokubo et.al., 1987) but predicting the outcome of such modifications is difficult. Finally, mention should be made of the new field of catalytic antibodies (Lerner et.al. 1991). Antibodies raised to transition state analogues have been shown to be catalysts for many organic reactions.

In his work on lactate dehydrogenase, Holbrook recognised that the enzyme, although essentially completely stereospecific with respect to the reduction product, lactate, was strongly limited in terms of the substrates that it would accept (pyruvate and a small number of related α-keto acids). By identifying those residues responsible for forming the substrate binding pocket as distinct from those responsible for catalysis, he has been able to redesign the binding pocket so that it will accept a range of branched chain α-keto acids and reduce them rapidly with high stereoselectivity to the corresponding 2-hydroxyacids (figure 13). The essence of the chemical changes made were that a loop region close to the active site was made more hydrophobic and more flexible to increase the binding of larger alkyl chains and the introduction of greater plasticity into the helix on to which the loop folds.

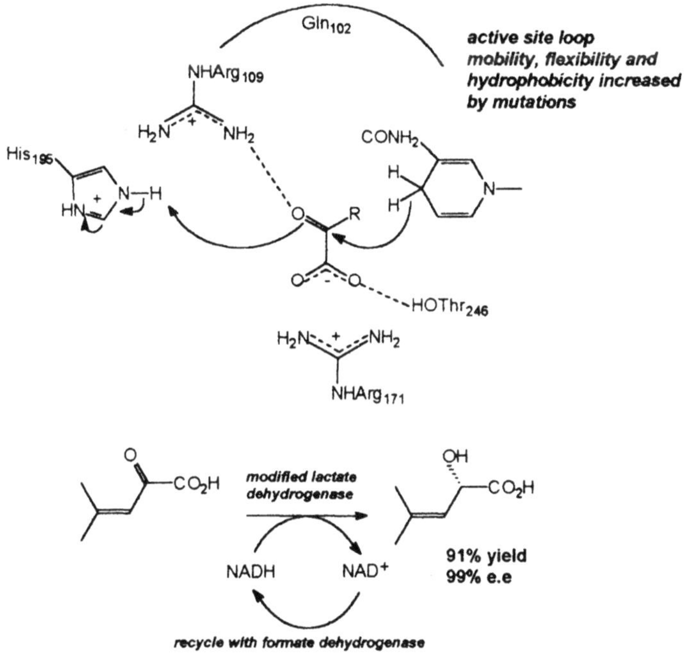


**Figure 13.** Modifications of lactate dehydrogenase by site-directed mutagenesis and their outcome.

A small number of catalytic antibodies have been developed with chemical synthesis in mind. It is obvious that, if one is going to go to the lengths of producing a new catalyst using antibody technology, it is important to choose reactions that are not catalysed by readily available enzymes. In the earlier discussion of the general prinicples of clean synthesis (figure 1) one example was Diels Alder cycloaddition which has been studied by several groups including our own (Suckling et.al., 1993, Gouverneur et.al. 1993). The target of our work was a multifunctional compound with several related chiral centres (figure 14); a number of relevant antibodies have been obtained one of which has the most unusual property of catalysing two reactions, a cycloaddition and a hydrolysis reaction. There are many details of the chemistry of this reaction to be established but it is interesting in the context of the production of new bio-catalysts that such protein with dual catalytic activity should be discovered.

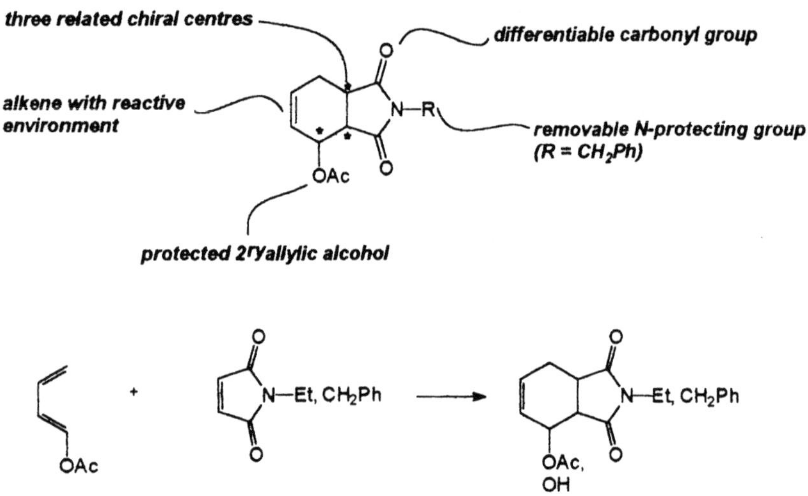


**Figure 14.** Catalytic antibodies designed for the production of synthetic building blocks.

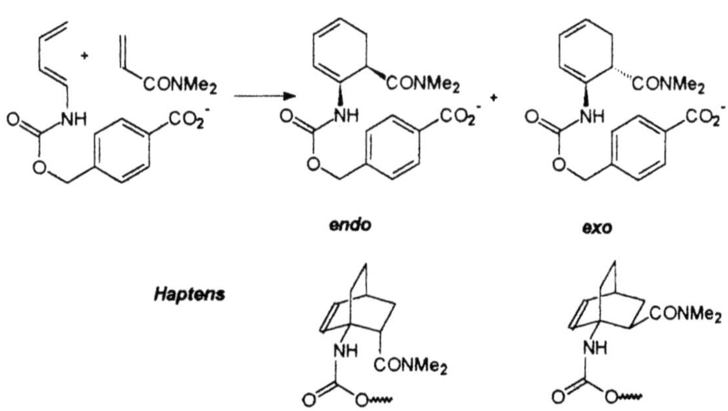


**Figure 15.** Control of Diels Alder cycloaddion by catalytic antibodies.

Lerner's group (Gouverneur et.al., 1993) have emphasised the potential that catalytic antibodies have in causing reactions to follow unfavoured paths by providing a confined environment at the active site (figure 15). Normally, Diels Alder reactions favour the formation of the so-called *endo* product but by choosing haptens with the required stereochemistry built in, they were able to demonstrate catalysis of two alternative paths by separate antibodies: the rate of *exo* isomer production was increased 18 fold over the uncatalysed reaction when the antibody derived from the *exo* hapten was used.

The Dieckmann reaction poses much more serious problems for catalysis. In order to build in the two steps, a hapten that would lead to a general base and to stabilisation of a tetrahedral transition state in the reaction was required. Although we have not been able to characterise the antibodies fully, we have been able to identify antibodies that catalyse at least two Dieckmann type cyclisations (figure 16) (Suckling and Stimson, 1994) and further work will be carried out.

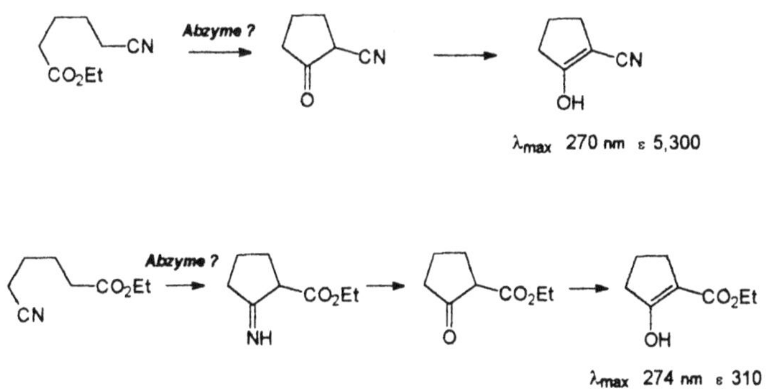


**Figure 16.** Reactions catalysed by a putative catalytic antibody for carbanion cyclisation.

This result is significant in two respects. Firstly, it shows that a complex reaction, not found in nature can be catalysed by a new protein. Secondly, the reaction itself using conventional chemistry would require the use of a strong base, a corrosive reagent that is relatively costly to prepare and use. The catalytic antibody avoids the need for this by carrying out the reaction in water. The increased use of water as a solvent for clean organic synthesis is an important development.

To emphasise that catalytic antibodies are genuinely applicable to synthesis and offer a new opportunity for the discovery of useful new catalysts as part of the drive towards clean synthesis, Lerner, a leader in the field (Reymond et.al., 1994) has shown that it is possible to prepare near gram quantities of products in high enantiomeric purity using catalytic antibodies (figure 17). It is interesting to note that a large amout of antibody was required for this work, a current limitation but one that can be overcome by appropriate molecular biological development.

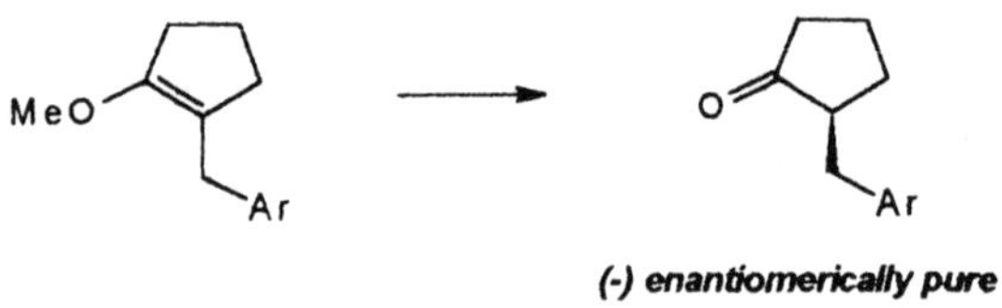


**Figure 17.** An enantiomerically pure compound produced by antibody catalysed hydrolysis.

The demand for a large quantity of biocatalyst in the last example highlights a problem with biocatalysis that is often overlooked in the glare of its great efficiency and selectivity. The production of a biological catalyst in pure form will leave behind much waste biomass. The work up of a fermentation will likewise create biological waste. There are often problems associated with down stream processing in order to extract products from dilute aqueous solutions. All of these factors influence the economic viability and environmental acceptability of a biocatalyst.

**5. Biomimetic chemistry, Hydroxylation, and Epoxidation**

If the biological catalysis approach is too limited, can chemistry learn from biology to devise clean reactions, especially new clean catalysts? Sometimes. One of the answers to this question comes in the field of oxidation chemistry. Oxygen is an ideal atom to introduce into an intermediate for synthesis because it can create highly acceptable activating and leaving groups. Research into hydroxylation to introduce alcohols, oxgenation to introduce epoxide, and further oxidation to carbonyl groups has therefore become a major focus in industrial and academic laboratories.

In nature, hydroxylation reactions are catalysed by cytochromes P-450, a superfamily of enzymes that is widely distributed. The catalytic activity of these enzymes in not high and they require a number of partner proteins in order to transform oxygen from the air into a substituent hydroxyl group in a product and water. Although these biological transformations themselves are intrinsically of interest for clean synthesis, the complexity of the enzymology has made progress slow. If model reactions could be made efficient, there would be good opportunities for clean synthesis. The active site in P-450s is a haem and in model reactions, this relatively labile ion porphyrin complex has been replaced by readily available *meso*-tetraphenylporphyrins.

Increasingly sophisticated synthesis has led to a porphyrin (figure 18) that, in the form of an iron complex, was able to oxidise cyclohexane to cyclohexanol and cyclohexanone with oxygen achieving turnovers (based upon the porphyrin) of 370 and 270 respectively (Bartoli at.al., 1994). This is one of the most impressive results so far with a P-450 model. The authors go further and point to 'one of the most spectacular results' using this complex is this essentially complete epoxidation of cyclooctene using dilute hydrogen peroxide as oxidant.

Alkane oxidation is one of the most difficult reactions to attempt. In organic synthesis, much effort has therefore been invested in the functionalisation of alkenes, in particular in their conversion into epoxides which are highly versatile clean intermediates. Several years ago, Sharpless discovered a highly enantiomerically selective epoxidation reaction for allylic alcohols (Gao et.al., 1987) which has since become a classic of organic synthesis. The reaction was, however, restricted to allylic alcohols. Recently, following the conceptual lead of metal complexes and atom transfer reactions, Jacobsen has developed a reagent capable of stereoselective transfer of oxygen to non-activated alkenes using a cheap oxygen source, sodium hypochlorite (Jacobsen et.al., 1991) (figure 19). This reagent can be scaled up to kilogram quantities and will undoubtedly contribute to clean synthesis by short cutting what would have required several steps to reach the same intermediate. Indeed, this can be identified as a principle of clean synthesis: *use the shortest practical route*. This is essentially a corrollary of the activating and leaving group principles: a shorter route will obviously reduce the requirement for activation and loss of leaving groups.

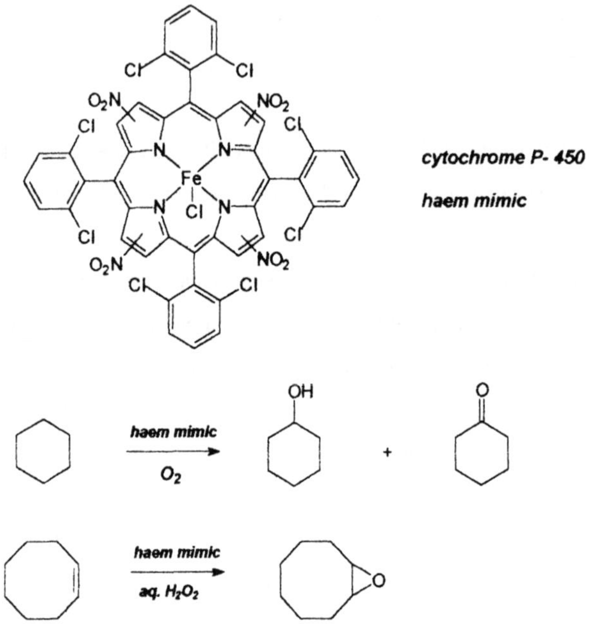


**Figure 18.** Reactions catalysed by an effective cytochrome P-450 mimic.

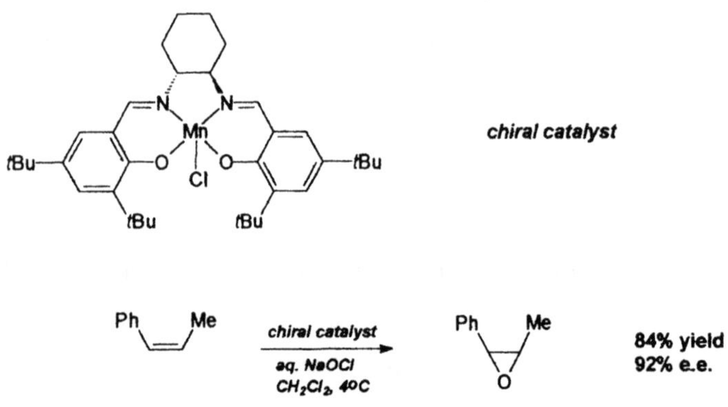


**Figure 19.** A new selective epoxidation catalyst.

All of the above oxidation reactions occur in homogeneous solution. On a large scale, it would be expected that the metal complexes would leak into the effluent stream. The need for supported reagents therefore returns. One of the most widely used polymeric supporting materials is polystyrene; however this polymer suffers seriously from the drawback that it is itself sensitive to oxidation at the benzylic positions. This leads to breakdown of the polymer and leakage of the metal into the effluent. Clean synthesis therefore would benefit from a more stable polymer capable of operating at modestly elevated temperatures. Recent research at Strathclyde (Miller and Sherrington, 1994) has shown that polybenzimidazole can satisfy this demand (figure 20). The epoxidation of propene, a notoriously difficult alkene to epoxidise because of ready oxidation of the methyl group, can be carried out effectively using molybdenum-based catalysts and *t*-butylhydroperoxide as oxidant. The leakage of molybdenum from the catalyst was especially low with this system.

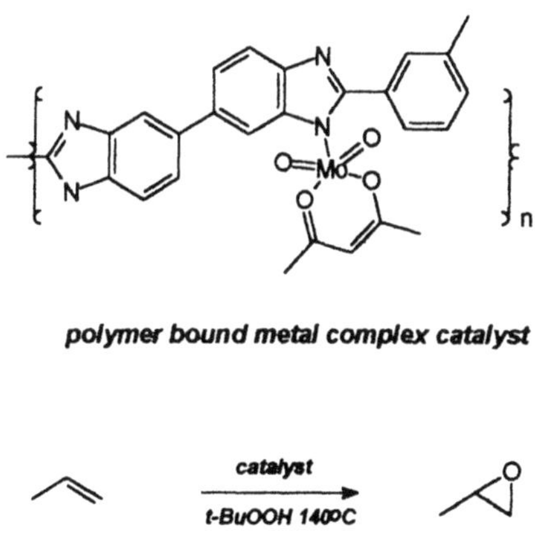


**Figure 20.** Oxidation over a molybdenum catalyst supported by a high performance polymer.

## 6. Conclusion

In this paper I have interwoven what I believe are useful concepts for the design and discovery of clean synthetic reactions with examples from current research in academic and industrial laboratories over a background of the basic chemical and biochemical problems. There are several principles that have emerged namely:

*Activating group*

*Leaving group*

*Keep as many atoms as you can*

*Immobilised/supported reagents*

*Avoid overkill*

*Use the shortest practical route*

Clean synthesis will be approached asymptotically. There will scarcely ever be a process without some disadvantages or a product without some drawbacks. Even the development of the chiral epoxidation of non-activated alkenes requires a reagent, sodium hypochlorite, a product of the chloralkali electrochemical industry, and chlorine does not remain in the product. However, this disadvantage is outweighed by the avoidance of many steps which together would have been less clean. Increasingly, reports are emerging of the development of older, established stoichiometric chemical reactions into modern, cleaner catalytic versions. For example, the well-known conversion of aldehydes into epoxides using sulphur ylides has recently been carried out catalytically using a coupled cyclic process in which a diazoalkane is transferred in the form of a rhodium carbene complex to form the required sulpur ylide (figure 21). Rhodium and sulphur are catalytic and the leaving group is nitrogen, admirable clean credentials.

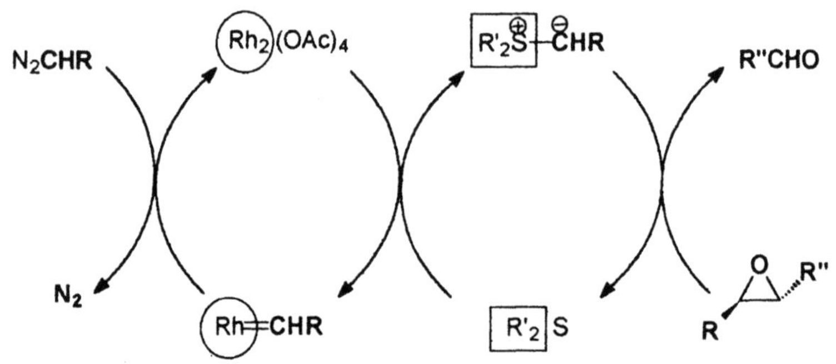


**Figure 21.** Catalytic modification of well-established stoichiometric reaction.

The choice of what process and what chemistry to use in industry is obviously not determined by academic elegance or by the current conventional wisdom. It should, however, include more thatn the local economics of constructing and running the plant in its total environment. This is the niche for clean chemistry. As explained above, we are not dealing with absolutes but with balanced judgments. I believe that there is much new clean chemistry waiting to be discovered to improve the environmentally sound choices to available. Therein lies the excitement of research. Chemistry has given society much; it has still more too offer.

**References**

Aggarwal VK, Abdel-Rahman H, Jones RVH, Lee HY, and Reid BD (1994) *J.Am.Chem.Soc.*, **116**, 5973-5974.

Ballard DHG, Courtis A, A.M. Shirley AM, and Taylor SJC (1983) *J.Chem.Soc.,Chem.Commun.*, 954-955.

Bartoli JF, Battioni P, DeFoer WR, and Mansuy D (1994) *J.Chem.Soc.,Chem.Commun.*, 23-24.

Blossey EC and Ford WT (1989) in 'Comprehensive Polymer Science', vol. 6, eds. Allen G, Bevington JC, and Eastmond GC, Pergamon Press, Oxford, pp. 81-114.

Boyd DR, Sharma ND, Hand MV, Groocock MR, Kerley NA, Dalton H, Chima J, and Sheldrake GN (1993) *J.Chem.Soc.,Chem.Commun.*, 974-976.

Boyd DR, Sharma ND, Boyle R, McMurray BT, Malone TF, Chima J, and Sheldrake GN (1993) *J.Chem.Soc.,Chem.Commun.*, 49-51.

Breslow, R and Campbell P (1969) *J.Am.Chem.Soc.*, **91**, 3085.

Bruce NC, French CE, Hailes AH, Lang MT, D.A. Rathbone DA, and Willey DL (1994) *Biotechnology '94, Applied Biocatalysis*, Institution of Chemical Engineers, Rugby 50-52; *Ann.NY.Acad.Sci.*, **721**, 85-100.

Casy G, Lee TV, Lovell H, Nichols BJ, Sessions RB, and J.J. Holbrook JJ (1992) *J.Chem.Soc.,Chem.Commun.*, 924-926.

Clark JH, Kybett APP, and Macquarrie DJ (1992) 'Supported Reagents, Preparation, Analysis and Applications', Verlag Chemie, New York.

Clark JH (1993) in 'Waste Minimisation: a Chemist's Approach', Martin K and Bastock TW (eds) Royal Society of Chemistry, Cambridge, 141-145

Clark JH, Kybett, AP, Landen, P Macquarrie, and Barlow SJ (1989) *J.Chem.Soc.,Chem.Commun*, 1353-1354.

Clark JH, Kybett, AP, Landen, P Macquarrie, DJ and Martin K (1989) *J.Chem.Soc.,Chem.Commun*, 1355-1357;

Conder M, Almeida N, Behrens S, Crawford L, Delawder S, Hoerner T, McAda P, Rambosek J, Reeves C, Schimme T, Stepan T, Stieber R, Tong I, and Vinci V (1994) *Biotechnology '94, Applied Biocatalysis*, Institution of Chemical Engineers, Rugby, 20-24.

Davis ME (1993) *Accts Chem.Res.*, **26**, 111-115;

Gao Y, Hanson RM, Klunder JM, Ko SY, Masamune H, and Sharpless KB (1987) *J.Am.Chem.Soc.*, **109**, 5765-5780.

Gouverneur VE, Houk KN, Depascualteresa B, Beno B, Janda, KD, and Lerner RA (1993) *Science*, **262**, 204-208.

Guy A, Lemaire M, Guette J-P (1982) *Tetrahedron*, **38**, 2339-2346;

Holmes PA, L.F. Wright LF, and Collins SH, EP 82 52 459.

Jacobsen EN, Zhang W, Muci AR, Ecker JR, and Deng L (1991) *J.Am.Chem.Soc.*, **113**, 7063-7064.

Jentzsch, W (1990) *Angew.Chem.Int. Edn.Engl.*, **29**, 1229-1334.

Kaiser ET, (1988) *Angew.Chem.Int.Edn.Engl.*, **21**, 913-922;

Kokubo T, Sassa S, and Kaiser ET (1987) *J.Am.Chem.Soc.*, **109**, 606-607.

Lerner RA, Benkovic SJ, and Schultz PG (1991) *Science*, **252**, 659-667.

Longley A and Kirkwood RC (1994) eds, Handbook of Clean Technology, Chapman and Hall, 1994, in press.

Malle, K-G (1993) in 'Waste Minimisation: a Chemist's Approach', Martin K and Bastock TW (eds) Royal Society of Chemistry, Cambridge, pp. 35-48.

Miller MM and Sherrington DC (1994) *J.Chem.Soc.,Chem.Commun.*, 55-56.

Pervez H, Onyiriuka SO, Rees L, Rooney JR, and Suckling CJ (1988) *Tetrahedron*, **44**, 4555-4568.

Reuben BG and Witkoff HA (1989) 'Pharmaceutical Chemicals Perspective', Wiley, New York.

Reymond JL, Reber JL, and Lerner RA, (1994) *Angew.Chem.Int.Edn.Engl.*, **33**, 475-477.

Seebach D (1990) *Angew.Chem.Int.Edn.Engl.*, **29**, 1320-1367.

Suckling CJ, Tedford MC, Bence LM, Irvine JI, and Stimson WH (1993) *J.Chem.Soc.,Perkin*

*Trans. I*, 1925-1929.

Suckling CJ, Halling, PJ Kirkwood, RC and Bell, G (1992) 'Clean Synthesis of Effect Chemicals', AFRC/SERC Clean Technology Unit, Swindon.

Suckling CJ and Suckling KE (1974) *Chem.Soc.Rev.*, **3**, 387-406.

Suckling CJ and Stimson WH (1994) *Biotechnology '94, Industrial Immunology*, Institution of Chemical Engineers, Rugby, 11-13.

Taylor SJC, Ribbons DW, Slawin AMZ, Widdowson DA, and Williams DJ (1987) *Tetrahedron Lett.*, **28**, 6391-6394.

Wong CH and Whitesides GM (1994) Enzymes in Organic Synthesis, Wiley, New York.

Wright, M (1993) in 'Waste Minimisation: a Chemist's Approach', Martin K and Bastock TW (eds) Royal Society of Chemistry, Cambridge pp. 20-34.

# EDUCATIONAL MATERIALS FOR WASTE MINIMIZATION AND POLLUTION PREVENTION EDUCATION AND TRAINING

**David Y. Boon**
**NATO/CCMS Fellow**
**Environmental Science and Technology**
**Front Range Community College**
**3645 West 112th Avenue**
**Westminster, Colorado 80030**

## INTRODUCTION

We must spur the development of a new generation of technologies that prevent pollution ... President Bill Clinton (NSTC, 1994). Schools, colleges and universities must be proactive in incorporating pollution prevention and clean production technology into their curriculum if we are to provide the future scientists and engineers necessary to carry out this vision.

Traditionally, pollution prevention education has been conducted in an "end of pipe" style. Pollution prevention and/or waste minimization assessments at a business or industry setting with subsequent training and retraining of their employees is needed in the United States as well as in other developed and developing countries. However, this type of training is "end of pipe" pollution prevention education. To provide for the greatest impact, pollution prevention education and training must be conducted at the academic institutions that prepared the student for the jobs of tomorrow. Emphasis has to be place on the entire educational system .... kindergarten through graduate levels programs. Providing exposure to pollution prevention principles within these educational systems provides a "source reduction" pollution prevention educational program.

NATO ASI Series, Partnership Sub-Series, 2. Environment – Vol. 2
Cleaner Technologies and Cleaner Products
for Sustainable Development
Edited by H. M. Freeman, Z. Puskas, and R. Olbina

**POLLUTION PREVENTION EDUCATION AND TRAINING NEEDS**

To be truly effective, pollution prevention, waste minimization, and clean technologies education needs to be introduced into the following target audiences:

- School Age Children (kindergarten - 12 grade)
- College/University Students
- Business/Industry
- Government and Non-government Organizations

Many kindergarten through 12th grade (K-12) teachers are incorporating environmental studies into their classroom by introducing the students to recycling, energy and natural resources conservation. Some innovative teachers have actually incorporated pollution prevention into their course materials but many more are needed. Fink (1994), through a U.S. EPA environmental education grant, is incorporating pollution prevention into an Environmental Literacy for Families (ELF) project which focuses on a third grade elementary school students and their parents.

Many community college, four-year colleges, and universities have incorporated pollution prevention, waste minimization and clean technologies into environmental and engineering curriculum. However, most are focusing on waste minimization and not on the "source reduction" concept of pollution prevention. Most programs (chemical engineering, environmental science and technologies, hazardous materials technologies) focus on how to reduce volume, toxicity or mobility of waste that has already been produced. There is little argument that this does not constitute pollution prevention but is more aptly called treatment technologies for waste reduction. Programs need to focus more on incorporating source reduction concepts and practices into their relevant courses.

The number of environmental laws and regulations that require components of pollution prevention continues to grow (Thurber and Sherman, 1994). If government regulations have the legal authority to require pollution prevention, then government regulators as well as individuals from the regulated community (business/industry) must be educated and knowledgeable in the principles and techniques of pollution prevention. Wigglesworth (1993) has published a practical guide for state and local government pollution prevention programs.

The regulated community can benefit the greatest by training all levels of employees (managers to line workers). The line workers are often the individuals that come up with simple and innovative solutions to waste production problems thus saving management the costs associated with waste management and disposal. In addition, workers are always striving to make the workplace safer and pollution prevention and clean technologies go a long way toward achieving this goal.

If we educate students in pollution prevention during their formal education/training programs it would be **POLLUTION PREVENTION SOURCE REDUCTION EDUCATION** ! If we wait until they are already employed and go to a company or business and educate them, it is **END-OF-PIPE POLLUTION PREVENTION EDUCATION.**

For instance, if we go to a business and train 30 of their employees in pollution prevention we will certainly impact that facility and likely reduce the volume of waste generated and produce a financial return for that business. However, if we train 30 students during their formal education/training those 30 students might end up at 30 different companies which would result in a greater total reduction on the volume of waste generated. We must provide for pollution prevention education at the educational/training source (i.e., K-12, vocational schools, colleges and universities).

**POLLUTION PREVENTION CURRICULUM**

Most college level pollution prevention educational materials and curriculum focuses on three types of delivery systems. First, pollution prevention and clean technology lessons and modules are infused into existing courses. This has been the most widely used method since it is easier, does not require long curriculum development time by the instructor, and does not displace other classes in the required curriculum. This has been common in university level engineering courses since the inclusion of an entire course dedicated to pollution prevention would be difficult due to the required number of courses and the accompanying accreditation requirements. The second method is to offer a full course (semester or quarter duration) dedicated entirely to pollution prevention. It is estimated that less than 10 college/university programs in the United States are teaching a course dedicated to source reduction pollution prevention (Foecke, 1994). However, well over 100 programs teach a course in waste minimization. A third method of pollution prevention education and training is to offer short-course or seminar style presentations. These are most commonly utilized by institutions and private companies for "fast track" training for employees of business, industry and government agencies.

In 1991, the University of Michigan was awarded funding for a National Pollution Prevention Center for Higher Education (NPPC, 1991). This center has developed pollution prevention resource compendiums for undergraduate and graduate engineering, business, and natural resources classes. These materials are for broad distribution to other colleges and universities nationwide. The Center has focused, to date, on six key areas of pollution prevention: accounting, business law, chemical engineering, environmental studies, industrial ecology, and industrial operations and management. Development is underway to develop resource materials for courses in chemistry and architecture. A summary of university based pollution prevention activities, courses, workshops, and pollution prevention/waste minimization assessment centers has been compiled (EPA, 1991; and Grulich, 1994).

Less information has been published on the pollution prevention activities of U.S. community colleges. In the United States, there are currently over 1200 community colleges, junior colleges and vocational colleges. The Partnership for Environmental Technology Education (PETE) has been established by the U.S. community colleges that offer environmental programs. The PETE member community colleges (200 to 500 institutions) are implementing pollution prevention and waste minimization courses into their environmental curriculum (Boon, 1993). The Partnership for Environmental Technology Education has been cited by the White House Office of Science and Technology Policy as a model education and training public-private partnership (NSTC, 1994). The impacts on pollution prevention education and training by such a partnership can be demonstrated by the success of one of its founding community college members, Front Range Community College.

The Front Range Community College (FRCC) Environmental Science and Technology Department offers both credit and non-credit courses in pollution prevention and waste minimization in its Hazardous Materials Technology program (HAZMATT). Although one entire course is dedicated to Pollution Prevention and Hazardous Waste Minimization, the ideas and concepts of pollution prevention are taught in a number of courses including:

- Introduction to Hazardous Material Management
- OSHA Health and Safety
- Environmental Laws and Policy
- Pollution Prevention and Hazardous Waste Minimization
- Environmental Sampling and Monitoring
- Treatment, Storage and Disposal of Hazardous Waste
- RCRA Regulatory Compliance.

This insures that all students in the program will be introduced to the concepts and principles of pollution prevention. This has a powerful effect on the numbers of future scientists, engineers and environmental managers knowledgeable in the principles of pollution prevention. Since the FRCC HAZMATT program began in the Fall of 1988, student class enrollment has exceeded 8000. The enrollment trends for the FRCC HAZMATT credit program are included in Table 1. In addition, over 400 students have taken the credit course SCI-211 (Pollution Prevention and Waste Minimization), which focuses on pollution prevention for an entire semester.

This course SCI-211, Pollution Prevention and Hazardous Waste Minimization, has traditional periodic exams but in addition requires the students to conduct the following projects for successful completion of their college credit class:

- **Conduct a Solid Waste Stream Assessment and Develop a Pollution Prevention Plan for Their Home.** Identify what's in their garbage and the percentages by weight for each waste type (i.e., paper, yard waste, glass, plastic, metal, etc). Develop a Pollution Prevention Plan for reducing the solid waste generated by their family.

- **Conduct a Hazardous Chemical Inventory and Develop a Pollution Prevention Plan for Their Home.** Students must inventory all hazardous chemicals located in the home and develop a Pollution Prevention Plan for reducing the number of hazardous chemicals in their home.

- **Conduct a Pollution Prevention Assessment for a Business/Industry Waste Stream.** The students must individually conduct a Pollution Prevention/Waste Minimization Assessment that includes both technical and economic feasibility analyses for a single waste stream from a local business or industry.

**TABLE 1**

**STUDENT ENROLLMENT**

**FRONT RANGE COMMUNITY COLLEGE**
**ENVIRONMENTAL SCIENCE AND TECHNOLOGY DEPARTMENT**
**HAZARDOUS MATERIALS TECHNOLOGY (HAZMATT) PROGRAM**

| YEAR | STUDENT ENROLLMENT | FTE's[1] |
|---|---|---|
| 1988 | 333 | 84 |
| 1989 | 1076 | 207 |
| 1990 | 973 | 214 |
| 1991 | 1371 | 258 |
| 1992 | 877 | 166 |
| 1993 | 1775 | 250 |
| 1994 | 1991 | 265 |
| TOTALS | 8395 | 1444 |

[1] = Full Time Equivalent (FTE) is computed by taking the number of total student credit hours for the course and dividing by 15 (the average course load for a full-time student).

The benefits of teaching this course in the credit program are numerous and include: 1) provides the student with the hands-on knowledge and skills in performing Pollution Prevention Assessments, 2) provides a mechanism for reducing waste at home, 3) provides a "free service" by conducting Pollution Prevention Assessments for local business/industry, 4) reduces waste generation from business/industry, and 5) provides a service to the local, state and federal government in conducting and implementing pollution prevention projects.

To date, the FRCC HAZMATT program through this single college credit course has conducted over 400 Pollution Prevention/Waste Minimization Assessments. These Pollution Prevention Assessments have included: Public Schools, Electronic Manufacturing Facilities, Laboratories, Mass Transit (Bus) Systems, Highway Departments, Fleet Maintenance Facilities, Automotive Repair Shops and Car Dealerships, Mining Operations, Downhill Ski Operations, Aircraft Maintenance Operations, Rocky Flats Nuclear Weapons Facility and the Rocky Mountain National Park.

As more and more schools, community colleges, four-year colleges and universities start providing pollution prevention in their various courses a rapid and diverse spread of pollution prevention education and training at the educational source will occur.

If more partnerships like PETE and nationally and international programs instilled the principles of pollution prevention in this manner, think of the enormous impacts on the local, national and international environment. In addition, the savings to our business/industry and governments from more efficient use of our natural resources would be immediate and long lasting.

**TEXTBOOKS**

Because pollution prevention is a relatively new concept in engineering and environmental programs, textbooks are scarce (EPA, 1991). Although no formal college level textbook has been developed to date, several professional series texts and guidance documents have been utilized for this purpose.

Most undergraduate programs in Environmental Science and Technology or Hazardous Materials Technology that offer courses in pollution prevention have primarily utilized two professional series texts (Freeman, 1990; Theodore and McGuinn, 1992). A timely and comprehensive professional series publications has recently become available, and will likely be utilized as a textbook for teaching pollution prevention (Freeman, 1994).

**POLLUTION PREVENTION RESOURCES (PUBLICATIONS, FACT SHEETS AND GUIDANCE DOCUMENTS)**

A large and rapidly growing body of case materials, fact sheets, resources materials and other industry specific information has been published and summarized. Most of the state environmental regulatory programs in the U.S. produce fact sheets and newsletters on pollution prevention. Summaries of these resources along with various training opportunities, technical assistant programs, publications, videos, and state and federal programs and contacts have been extensively reported (NPPC, 1991; EPA, 1992a and 1992b; Freeman et al, 1992; EPA 1993a; Wigglesworth, 1993; and Grulich, 1994). In addition, the U.S. Department of Energy produces a newsletter on advances in pollution prevention (DOE, 1994). The National Roundtable of State Pollution Prevention Programs is the most current and comprehensive source of information on state newsletters and fact sheets (NRSPPP, 1994).

Guidance documents and fact sheets are a valuable resource. The U.S. Environmental Protection Agency Pollution Prevention Research Branch publishes a list of documents available from the Center of Environmental Research Information (EPA 1993b). These documents include: Guidance Documents, Waste Minimization Opportunity Assessment Project Summaries, Project Summaries, and Environmental Research Briefs.

The United Nations Environmental Program (UNEP 1991) in cooperation with the United Nations Industrial Development Office (UNIDO) has developed an audit and waste reduction manual to assist in the identification of industrial emission and waste sources. In addition, the UNEP has published information on case studies on pollution prevention related to clean production from around the world (UNEP, 1993).

The Pollution Prevention Review is a quarterly publication that focuses entirely on pollution prevention articles that include case studies, regulatory and compliance reviews (PPR, 1994).

**VIDEO TAPES**

EPA (1993a) published a review of thirteen videos relevant to pollution prevention and waste minimization. These videos are suitable for use as supplements to a training course or workshop and as sources of background information on pollution prevention. Another good source of videos on pollution prevention for specific applications and case studies are the industry and businesses that generate waste and their various trade associations.

**INTERACTIVE POLLUTION PREVENTION TRAINING TOOLS**

Several innovative and interactive exercises for pollution prevention training have come from the Waste Reduction Institute for Training and Applications Research (WRITAR). WRITAR is a nonprofit organization dedicated to facilitating implementation of innovative strategies, techniques, and technologies that prevent pollution at the source. More recently, WRITAR has published two excellent "how to" guides: Participatory Approaches to Pollution Prevention Training - A Trainer's Toolbox (WRITAR, 1994a) and Training Evaluation - A Practitioner's Guide to Pollution Prevention Training Evaluation (WRITAR, 1994b). Several of the innovative exercises are briefly described below.

**Fun Factory** - The Fun Factory exercise was developed by the Minnesota Technical Assistance Program and WRITER in order to provide training to regulators who might have an opportunity to promote pollution prevention to the regulated community. This exercise has been repeatedly utilized by a number of academic institutions and trainers for a much wider audience including students, business/industry employees, and regulators. The exercise utilizes a childs toy (Play-Doh Fun Factory) to illustrate the extrusion molding industrial process and the waste types generated by this process. Waste from the process include quality control, process and clean-up wastes. Options for pollution prevention are incorporated into the group activity.

**Green Square** - The Green Square outlines an exercise which can heighten public policy staffs' awareness of how waste is generated from a surface finishing operation (painting) and how waste can be reduced. At the same time, this exercise provides participants with an sense of what it's like to work in an industrial setting and to use group problem solving techniques for pollution prevention.

**Cafe Reducto** - This interactive exercise was developed for an audience of small business owners and staff from local government offices. The Cafe Reducto coffee making exercise introduces the participants to a chemical extraction processes and the options for pollution prevention through a commonly understood process (i.e., coffee making).

**COMPUTER SOFTWARE AND PROGRAMS**

Computer software and databases have become increasingly popular for accessing pollution prevention information and resources. Several of the larger and publically available databases are described below.

PPIC - The Pollution Prevention Information Clearinghouse (PPIC) is a multimedia clearinghouse of technical, policy, programmatic, legislative, and financial information dedicated to promoting pollution prevention through efficient information transfer. PPIC is composed of four different elements: repository of technical information, PIES - an electronic information exchange, hotline for additional information, and outreach. For more information on PPIC contact: Jocelyn Woodman, Pollution Prevention Division, US EPA, 401 M. Street SW, Washington, D.C. 20460, (202) 382-4418.

PIES - The Pollution Prevention Information Exchange System (PIES) is a computerized information network for the Pollution Prevention Information Clearinghouse (PPIC). PIES helps organizations identify grant programs and technical pollution prevention documents, contacts experts in the field, and identify other pollution prevention resources. For more information on PIES contact: Myles Morse, US EPA, Office of Environmental Engineering and Technology Demonstration, 401 M. Street SW, Washington, D.C. 20460, (202) 475-7161.

SAGE - The Solvent Alternative Guide (SAGE) was designed to provide recommendations for solvent replacements in cleaning and degreasing operations. The system leads the user through a question-and-answer session and based on the responses provides a brief description of the solvent alternatives. For more information on SAGE contact: Chuck Darvin, Air and Energy Engineering Laboratory, U.S. EPA, Research Triangle Park, NC 27711 (919) 541-7633.

<u>SWICH</u> - The Solid Waste Information Clearinghouse (SWICH) was established by EPA to increase the availability of information on solid waste resource reduction, recycling, composting, and other education and training issues. For more information on SWICH contact: The Solid Waste Information Clearinghouse, Box 7219, Silver Spring, MD 20910, (800) 67-SWITCH.

<u>HSSDS</u> - The Hazardous Solvent Substitution Data System (HSSDS) was developed by the Idaho National Engineering Laboratory. The system ranks solvents that can be substituted for functions that the user predescribes (i.e., degrease stainless steel and cannot be a chlorinated solvent or aerosol). The advantages of this data system is that it has substantial analytical data backing its development and it is accessible on INTERNET. For more information on the HSSDS Contact: Kevin Twitchell, Idaho National Engineering Laboratory, P.O. Box 1624, Mailstop 1604, Idaho Falls, ID 83415-1604, (208) 526-6956.

**SUMMARY**

To provide for the most cost effective pollution prevention training and education it must be given throughout the educational system (K - graduate programs). Be a pollution prevention champion by providing quality pollution prevention education and training to our future managers, scientists and engineers.

Utilize these resources and materials. Create new and innovative ideas and ways of teaching pollution prevention to our students. Then share these ideas.

By creating partnerships, sharing ideas and curriculum, we will be able to spur the development of a new generation of technologies that prevents pollution.

Think of the environmental impacts this would have on our local, national and international communities. Think of the resultant savings to our business/industry and governments. Think of the immediate and long lasting impacts. "P2. Onward" (Freeman, 1994).

**Think of it !!**

**LITERATURE REFERENCES**

Boon (1993) Pollution Prevention Education and Training in U.S. Community Colleges. Proceedings of the Workshop on Pollution Prevention Techniques and Technologies pg 165-170. NATO/CCMS Pilot Study Pollution prevention Strategies for Sustainable Development. April 27-28, 1993; Graz, AUSTRIA.

DOE (1994) Pollution Prevention Advisor. Department of Energy Pollution Prevention Advisor Newsletter, 109 South Riverside Drive, Elizabethton, TN 37643.

EPA (1991). Pollution Prevention 1991: Progress on Reducing Industrial Pollutants. EPA 21P-3003. October 1991. US EPA, Office of Pollution Prevention, Washington, D.C. 20460.

EPA (1992a). Facility Pollution Prevention Guide. EPA/600/R-92/088. US EPA Risk Reduction Engineering Laboratory, 26 West Martin Luther King Drive, Cincinnati, OH 45268.

EPA (1992b). Pollution Prevention Case Studies Compendium. EPA/600/R-92/046. April 1992. US EPA Office of Research and Development, Washington, D.C. 20460.

EPA (1993a). 1993 Reference Guide to Pollution Prevention Resources. EPA/742/B-93-001, US EPA, Office of Pollution Prevention and Toxics, 401 M Street, S.W., Washington, D.C. 20460.

EPA (1993b). Recent Publications. EPA/600/F-93/009. US EPA, Risk Reduction Engineering Laboratory, Center for Environmental Research Information, 26 West Martin Luther King Drive, Cincinnati, Ohio 45268.

EPA (1993c). Pollution Prevention Compendium. US EPA, Pollution Prevention Research Branch, Risk Reduction Engineering Laboratory, Center for Environmental Research Information, 26 West Martin Luther King Drive, Cincinnati, Ohio 45268.

Fink, Mary (1994). Environmental Literacy for Families (ELF). Environmental Education Grant Profiles Fiscal Year 1994; page 10, U.S. Environmental Protection Agency, June 21, 1994.

Foecke, Terry. (1994) Personal communication (October 26, 1994). Terry Foecke, Executive Director, Waste Reduction Institute for Training and Applications Research, 1313 5th Street SE, Suite 327, Minneapolis, MN 55414.

Freeman, Harry M. (1990). Hazardous Waste Minimization. Harry M. Freeman (ed) McGraw Hill Publishing Company, New York, NY.

Freeman, Harry M. et al (1992). Industrial Pollution Prevention: A Critical Review. J. Air Waste Manag. Assoc. Vol. 42. No. 5, pg. 617-656.

Freeman, Harry M. (1994). Industrial Pollution Prevention Handbook. Harry M. Freeman (ed), McGraw Hill Publishers, New York, NY.

Grulich, Madeline M. (1994). Sources of Pollution Prevention Information. In Harry M. Freeman (ed.) Industrial Pollution Prevention Handbook. pgs. 885-910. McGraw Hill Publishers, New York, NY.

NPPC (1991) National Pollution Prevention Center for Higher Education. For more information contact: National Pollution Prevention Center for Higher Education, University of Michigan, Dana Building, 430 E. University, Ann Arbor, Michigan 48109-1115.

NSTC (1994). Technology for a Sustainable Future: A Framework for Action. July 1994., Produced by: The National Science and Technology Council, Environmental Technology Strategy Staff, Office of Science and Technology Policy, Room 443, Old Executive Office Building, Washington, D.C. 20500

NRSPPP (1994) The National Roundtable of State Pollution Prevention Programs, 218 D Street S.E., Washington, D.C. 20003.

PPR (1994) Pollution Prevention Review. Executive Enterprises Publications Co., Inc., 22 West 21st Street, New York, NY 10010-6990.

Theodore, Louis and Young C. McGuinn (1992). Pollution Prevention. Van Nostrand Reinhold Publishers, New York, NY.

Thurber, James and Peter Sherman (1994). Pollution Prevention Requirements in U.S. Environmental Laws. In: Freeman, Harry M. (ed) Industrial Pollution Prevention Handbook. pgs. 27-49. McGraw Hill Publishers, New York, NY.

UNEP (1991) Audit and Reduction Manual for Industrial Emissions and Wastes. UNEP. 1889 F Street, NW, Washington, D.C. 20006.

UNEP (1993) Clean Production Worldwide. United Nations Environmental Program, Industry and Environment, Program Activity Centre, Tour Mirabeau, 39-43 quai Andre Citroen, 75739 Paris Cedex 15.

Wigglesworth, David (1993). Pollution Prevention: A Practical Guide for State and Local Government., Lewis Publishers., CRC Press, Inc., 2000 Corporate Blvd., N.W., Boca Raton, Florida 33431.

WRITAR (1994a) Participatory Approaches to Pollution Prevention Training: A Trainers Tool Box. Waste Reduction Institute for Training and Applications Research, 1313 5th Street SE, Suite 327, Minneapolis, MN 55414.

WRITAR, 1994b Training Evaluation - A Practitioner's Guide to Pollution Prevention Training Evaluation (WRITAR, 1994b). Waste Reduction Institute for Training and Applications Research, 1313 5th Street SE, Suite 327, Minneapolis, MN 55414.

# ALL ABOUT POLLUTION PREVENTION

Harry M. Freeman
Risk Reduction Engineering Laboratory
U.S. Environmental Protection Agency
Cincinnati, Ohio 45268

## INTRODUCTION

"An ounce of prevention is worth a pound of cure." Trite, and something everybody knows? Maybe, but this proverb is coming to be quoted by individuals around the world as a rallying cry for programs to enhance the quality of the world's environment. The proverb is at the heart of the pollution prevention movement.

Pollution prevention is a term used to describe production technologies and strategies that result in eliminating or reducing waste streams. The United States Environmental Protection Agency defines pollution prevention as "the use of materials, processes, or practices that reduce or eliminate the creation of pollutants or wastes at the source. Pollution Prevention includes practices that reduce the use of hazardous materials, energy, water, or other resources and practices that protect natural resources through conservation or more efficient use." The idea underlying the promotion of pollution prevention is that it makes far more sense for a generator not to produce waste than to develop extensive treatment schemes to insure that the waste poses no threat the quality of the environment. The EPA definition of pollution prevention does not include off-site recycling as pollution prevention. It should be noted though that not including recycling in the definition is not meant to imply that it is not worthwhile or that it should not be actively encouraged. It is clearly preferable to many other waste management and disposal options and contributes to establishing the same sustainable society for which a pollution prevention program is designed.

The shorthand term for pollution prevention , "P2," is used at times in this paper in keeping

NATO ASI Series, Partnership Sub-Series, 2. Environment – Vol. 2
Cleaner Technologies and Cleaner Products
for Sustainable Development
Edited by H. M. Freeman, Z. Puskas, and R. Olbina

The shorthand term for pollution prevention , "P2," is used at times in this paper in keeping with the P2 spirit of conserving ink and paper. Pollution prevention or source reduction is the element at the top of the four part environmental management hierarchy shown in Figure 1. The idea is to begin at the top of the hierarchy in considering potential solutions for environmental programs, not at the bottom.

While "pollution prevention" is coming to be the most widely accepted term in the U.S. for such strategies and processes, there are other similar terms that have been used in the past and are to varying degrees still in use in the U.S. and elsewhere today. Among these are low and non-waste technologies, cleaner production, clean technologies, waste minimization, and waste avoidance.

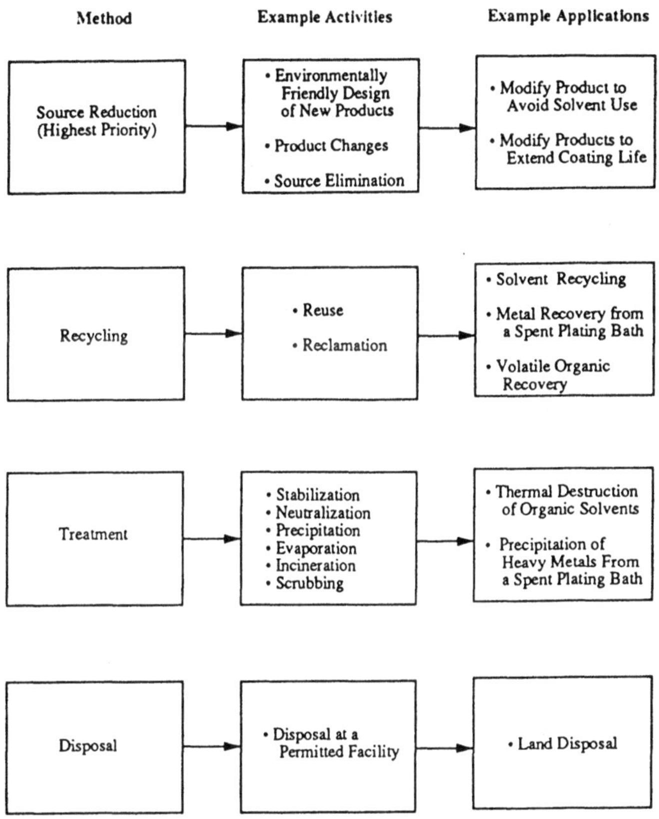


Source: Facility Pollution Prevention Guide (EPA/600/R-94/088)

Figure 1: Environmental Management Options Hierarchy

That pollution prevention and its related terms, sustainable industrial development, are ideas whose times have clearly come is illustrated by the following quotes:

> "Our long term strategy invests more in pollution prevention, energy efficiency and solar energy:. . . "
>
> President Bill Clinton
>
> April 1993

"I have four priorities for the Agency. First, pollution prevention... Pollution prevention is our best hope for the future of environmental protection: not just environmental regulation, but environmental protection."

Carol Browner
Administrator U.S. EPA
April 1993

BENEFITS OF POLLUTION PREVENTION

Answering the question, "Why should you undertake pollution prevention," in a manual to help the industrial generator make cost comparisons on the basis of costs and benefits of pollution prevention, the authors of the EPA's Pollution Prevention Benefits Manual state:

Pollution prevention can help you achieve the following:

- improve your firm's "bottom line"
- make compliance with environmental regulations easier
- demonstrate a proactive commitment to genuinely pursuing a pollution prevention program.

P2 is a compelling strategy for many reasons. If no pollution is generated, there are no pollutants to be managed. Thus, future problems are avoided, such as the problems which occur when previously accepted land disposal methods are discovered to be major sources of environmental contamination. Preventing pollution before it occurs also prevents situations that not only might endanger members of the community, but workers involved in the management of pollution.

One of the significant benefits of P2 is that it is often an economical approach. When wastes are reduced or eliminated, cost savings in materials result--more product is produced from the same starting materials. The close examination of manufacturing processes needed to plan a successful pollution prevention approach can produce a number of side-benefits as well, such as significant improvements in energy and water conservation, and improved, or more consistent, product quality.

P2 can also lead to large savings in regulatory and compliance costs, which are lowered as less pollution is produced. Frequently the dominant cost savings come from reduced future liability for the pollution. Ever since passage in the U.S. of the Federal Resource Conservation and Recovery Act, with its mandate that manufacturers have "cradle to grave" responsibility for the wastes that they generate, and enactment of the joint and several liability provisions of the Federal Comprehensive Environmental Response, Compensation, and Liability Act (better known as the Superfund), waste producers have been subject to the possibility of unlimited liability for any harm caused by their wastes. This liability includes even future problems caused by wastes managed using the best current practices. Because waste site cleanups can cost hundreds of millions of dollars each, these liabilities can dwarf all other costs associated with waste generation, which makes pollution prevention options even more compelling.

The environmental advantages of P2 approaches include improving effectiveness, minimizing uncertainty, avoiding cross-media transfers, and protecting resources.

## P2 TECHNIQUES

Two typical methods of source reduction are product changes and process changes. Both of these techniques can reduce the volume and toxicity of production wastes and of end-products during their life-cycle and at disposal.

Product changes in the composition or use of the intermediate or end products are performed by the manufacturer with the purpose of reducing waste from manufacture, use, or ultimate disposal of the products.

Process changes are concerned with how the product is made. They include input material changes, technology changes, and improved operating practices. All such changes reduce worker exposure to pollutants during the manufacturing process. Typically, improved operating practices can be implemented more quickly and at less expense than input material and technology changes. Figure 2 provides some examples of product and process changes.

## CLEANER TECHNOLOGIES

### Industrial Programs

Reflecting both an interest in saving money and avoiding increasingly stringent environmental end-of-the-pipe regulations, and responding to the concern on the part of the consuming public for more environmentally friendly activities, industries around the world have adopted pollution prevention with a vengeance and have initiated many broad programs. In the U.S., the Chemical Manufacturing Association reports a "quiet revolution that the chemical industry is conducting within its own operation to improve its performance. As part of the CMA Responsible Care Program, the industry has adopted a Waste and Release Reduction and Management Code that contains 10 management practices that provide a framework for reducing waste generation and releases to the environment. The

CMA notes "improved performance will take time, money, and hard work. As we move down this road, we invite others to pick up the challenge and join us." If you can believe the wealth of information being produced on all fronts, the challenge is being picked up very well indeed.

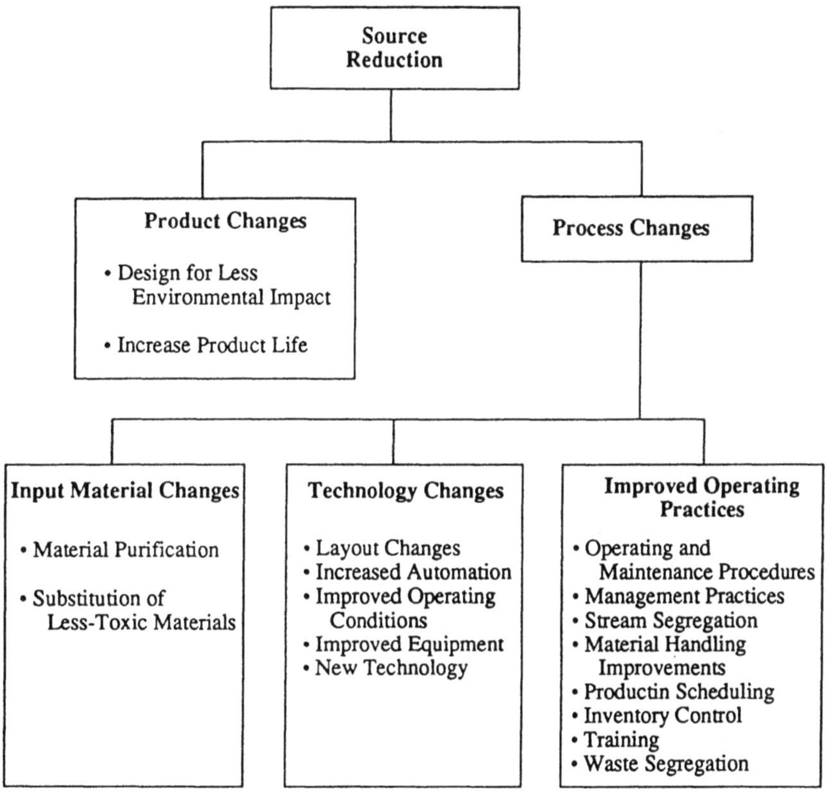


Figure 2. Source Reduction Methods

Pollution prevention planning is comprehensive and continual. A P2 program will affect many functional areas within a company. Therefore, it has much in common with the planning already done for other aspects of business operations.

Figure 3 outlines the major steps in the pollution prevention program.

An extensive discussion of the details of this approach is contained in the EPA manual, Facility Pollution Prevention Guide (EPA/600/R-92/088), available at not cost from CERI, US EPA, Cincinnati, Ohio 45268 (FAX 513-569-7566).

## Elements of a Successful Waste Minimization/Pollution Prevention Program

In the United States generators of hazardous waste streams, as a requirement to transport their wastes to treatment and disposal centers, are required to certify that they have in place a program to minimize their hazardous waste generation rates. The EPA has issued non-binding guidance as to what a good waste minimization program should contain. That guidance suggests that a good program might have the following elements:

A. Top Management Support. Top management should ensure that waste minimization is a company-wide effort. There are many ways to accomplish this goal. Some of the methods described below may be suitable for some firms and not others. However, some combination of these techniques should be used by every firm to demonstrate top management support.

- Make waste minimization a company policy. Put this policy in writing and distribute it to all departments. Make it each person's responsibility to identify opportunities for minimizing waste. Reinforce the policy in day to day operations, at meetings and other company functions.
- Set specific goals for reducing the volume or toxicity of wastestreams.
- Commit to implementing recommendations identified through assessments, evaluations or other means.
- Designate a waste minimization coordinator at each facility to ensure effective implementation of the program.
- Publicize success stories. It will trigger additional ideas.

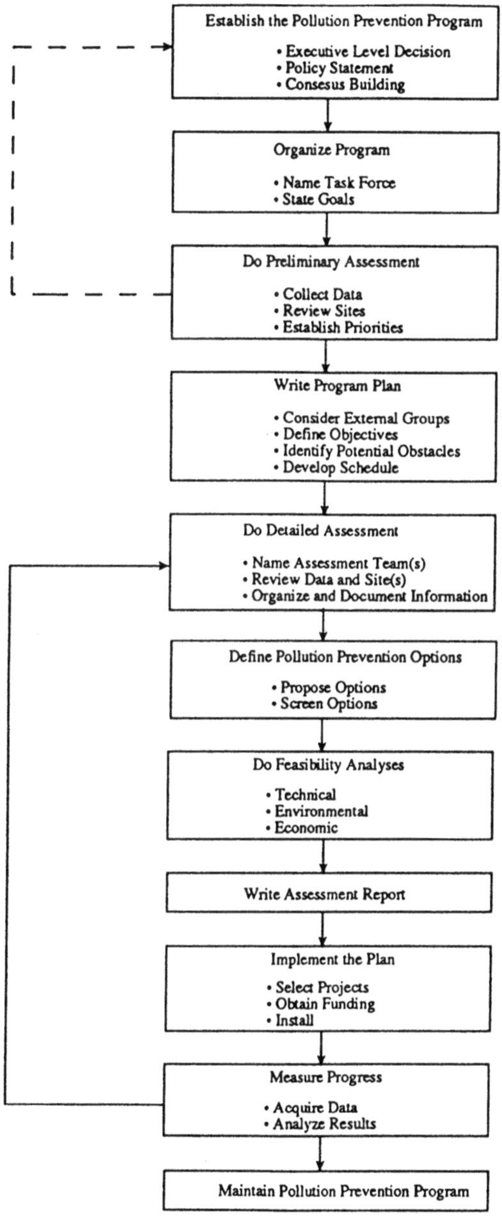


Figure 3. Pollution Prevention Program Overview

- Reward employees that identify cost-effective waste minimization opportunities.
- Train employees on aspects of waste minimization that relate to their job. Include all departments, such as those in product design, capital planning, production operations, and maintenance.

B. Characterization of Waste Generation. Maintain a waste accounting system to track the types, amounts and hazardous constituents of wastes and the dates they are generated.

C. Periodic Waste Minimization Assessments. Track materials that eventually wind up as waste, from the loading dock to the point at which they become a waste.

- Identify opportunities at all points in a process where materials can be prevented from becoming a waste (for example, by using less material, recycling materials in the process, finding substitutes, or making equipment changes.) Individual processes or facilities should be reviewed periodically. Larger companies may find it useful to establish a team of independent experts.
- Determine the true costs of the waste. Calculate the costs of the materials found in the waste stream based on the purchase price of those materials. Calculate the cost of managing the wastes that are generated, including costs for personnel, record keeping, transportation, liability insurance, pollution control equipment, treatment and disposal and others.

D. A Cost Allocation System. Departments and managers should be charged "fully-loaded" waste management costs for the wastes they generate, factoring in liability, compliance and oversight costs.

E. Encourage Technology Transfer. Seek or exchange technical information on waste minimization from other parts of your company, from other firms, trade associations, State and university technical assistance programs or professional consultants. Many techniques have been evaluated and documented that may be useful in your facility.

F. Program Evaluation. Conduct a periodic review of program effectiveness. Use these reviews to provide feedback and identify potential areas for improvement.

## P2 ASSESSMENTS

A major element in an industrial pollution prevention program is carrying out periodic assessments to identify opportunities for reducing wastes. These assessments originally called waste minimization assessments and later pollution prevention assessments have been the subject of much of the pollution prevention literature. A P2 assessment is a systematic planned procedure with the objective of identifying ways to reduce or eliminate waste. The assessment consists of a careful review of a plant's operations and waste streams, and the selection of specific areas to assess. Some typical questions that might be asked during a review are shown in Figure 4. After a specific waste stream or area is established as the assessment's focus, a number of options with the potential to minimize waste are developed and screened. Third, the technical and economic feasibility of the selected options are evaluated. Finally, the most promising options are selected for implementation.

Figure 4. Typical questions to ask during site reviews include:

- What is the composition of the waste streams and emissions generated in the company? What is their quantity?
- From which production processes or treatments do these waste streams and emissions originate?
- Which waste materials and emissions fall under environmental regulations?
- What raw materials and input materials in the company or production process generate these waste streams and emissions?
- How much of a specific raw or input material is found in each waste stream?
- What quantity of materials are lost in the form of volatile emissions?
- How efficient is the production process and the various steps of that process?
- Are any unnecessary waste materials or emissions produced by mixing materials--which could otherwise be reused with other waste materials?
- Which good housekeeping practices are already in force in the company to limit the generation of waste materials?
- What process controls are already in use to improve process efficiency?

There are many manuals that present guidance and suggestion to those interested in carrying out assessments. There are also many industry specific manuals and guidance documents that have been provided by Federal, State, and local waste reduction programs. A list of manuals available from the US EPA is shown in Figure 5.

Examples of Pollution Prevention Savings

Assessments carried out in 1992 at a manufacturer of base ball bats and golf clubs, a manufacturer of metal cutting wheels and components, and a manufacturer of outdoor signs produced the recommendations in Table 1.

It should be noted that several of the options have a payback period of 0, and that others are well within the range of return that would make the necessary investments realize profitable returns.

## CLEANER TECHNOLOGIES

Production processes that incorporate pollution prevention considerations to reduce either the quantity or toxicity of waste streams are referred to as cleaner technologies. There are many cleaner technologies either currently available or under development throughout the industrial sector. Some examples of available cleaner technology alternatives for common manufacturing processes are shown in Table 2. Examples of alternative cleaner technology processes under development are shown in Table 3.

Figure 5. Listing of Guides

| | |
|---|---|
| FACILITY POLLUTION PREVENTION GUIDE | EPA/600/R-92/088 |
| LIFE CYCLE DESIGN GUIDANCE MANUAL | EPA/600/R-92/226 |
| LIFE CYCLE ASSESSMENT: INVENTORY GUIDELINES AND PRINCIPLES | EPA/600/R-92/245 |
| GUIDES TO POLLUTION PREVENTION: | |
| THE PESTICIDE FORMULATING INDUSTRY | EPA/625/7-90/004 |
| THE PAINT MANUFACTURING INDUSTRY | EPA/625/7-90/005 |
| THE FABRICATED METAL PRODUCTS INDUSTRY | EPA/625/7-90/006 |
| THE PRINTED CIRCUIT BOARD MANUFACTURING INDUSTRY | EPA/625/7-90/007 |
| THE COMMERCIAL PRINTING INDUSTRY | EPA/625/7-90/008 |
| SELECTED HOSPITAL WASTE STREAMS | EPA/625/7-90/009 |
| RESEARCH AND EDUCATION INSTITUTIONS | EPA/625/7-90/010 |
| THE PHOTOPROCESSING INDUSTRY | EPA/625/7-91/012 |
| THE AUTO REPAIR INDUSTRY | EPA/625/7-91/013 |
| THE FIBERGLASS REINFORCED AND COMPOSITE PLASTICS INDUSTRIES | EPA/625/7-91/014 |
| MARINE MAINTENANCE AND REPAIR INDUSTRY | EPA/625/7-91/015 |
| THE AUTOMOTIVE REFINISHING INDUSTRY | EPA/625/7-91/016 |
| THE PHARMACEUTICAL INDUSTRY | EPA/625/7-91/017 |
| METAL CASTING AND HEAT TREATING INDUSTRY | EPA/625/R-92/009 |
| METAL FINISHING INDUSTRY | EPA/625/R-92/011 |
| MECHANICAL EQUIPMENT REPAIR | EPA/625/R-92/008 |
| NON-AGRICULTURAL PESTICIDE USERS | EPA/625/R-93/009 |
| WOOD PRESERVING INDUSTRY | EPA/625/R-93/014 |

Table 1. Example P2 Opportunities

| Waste Generated | Minimization Opportunity | Annual Waste Reduction | | Net Annual Savings | Implementation Cost | Payback Years |
|---|---|---|---|---|---|---|
| | | Quantity | Percent | | | |
| Scrap cardboard and paper | Cardboard and paper wastes are to be segregated and sold to a recycler. | 930,000 lb | 100 | 31,500 | 0 | 0.0 |
| Paint overspray and spent filter | Paint several golf club heads on a rack, instead of individually, for improved efficiency. | 148 gal | 33 | 3,365 | 900 | 0.3 |
| Paint overspray and spent filters | Paint golf club heads with an air-assisted spray for improved efficiency. | 202 gal | 80 | 4,589 | 5,600 | 1.2 |
| Methylene chloride | Use a template to maintain position of letters during over-night curing in place of methylene chloride | 330 gal | 100 | 1,980 | 200 | 0.1 |
| Spent scrubber water | Use water from the plant's wastewater treatment system in place of fresh water to scrub the chemical vapor deposition furnace exhaust gases. | 5,292,000 gal | 100 | 21,168 | 31,500 | 1.5 |
| Spent tricholoroethane and acetone from cleaning presses, maintenance cleaning, and cleaning cutting wheel disks | Install a batch distillation unit to recover the solvents for reuse. | 493 | 90 | 3,542 | 15,740 | 4.4 |
| Spent acidic degreaser solution | Preclean the tungsten carbide inserts with a hot water rinse before they enter the acidic degreaser tank. The quantity of degreaser solution needed will be reduced by about 50%. | 0 | 0 | 12,333 | 1,196 | 0.1 |

Table 2: Available Cleaner Technologies

| Technology Type | Pollution Prevention Benefits | Limitations |
|---|---|---|
| Alternatives to Chlorinated Solvents | | |
| Aqueous Cleaners | · No ozone depletion potential<br>· May not contain VOCs<br>· Many cleaners reported to be biodegradable | · Nonflammable and nonexplosive; relatively low health risks compared to solvents; consult Material Safety Data Sheet (MSDS) for individual cleaner<br>· Contaminant and/or spent cleaner may be difficult to remove from blind holes and crevices<br>· May require more floor space, especially if multi-stage cleaning is performed in line<br>· Often used at high temperatures (120 to 200°F)<br>· Metal may corrode if part not dried quickly; rust inhibitor may be used with cleaner and rinsewater<br>· Stress corrosion cracking can occur in some polymers |
| Semi-Aqueous Cleaners | · Some have low vapor pressure and so have low VOC emissions<br>· Terpenes work well at low temperatures, so less heat energy is required<br>· Some types of cleaners allow used solvent to be separated from the aqueous rinse for separate recycling or disposal | · NMP is a reproductive toxin that is transmitted dermally; handling requires protective gloves<br>· Glycol ethers have been found to increase the rate of miscarriage<br>· Mists of concentrated cleaners (especially terpenes) are highly flammable; hazard is overcome by process design or by using as water emulsions<br>· Limonene-based terpenes emit a strong citrus odor that may be objectionable<br>· Some semi-aqueous cleaners can cause swilling and cracking of polymers and elastomers<br>· Some esters evaporate too slowly to be used without including a rinse and/or dry process<br>· May be aquatic toxins |
| Petroleum Hydrocarbons | · Produce no wastewater<br>· Recyclable by distillation<br>· High grades have low odor and aromatic hydrocarbon content (low toxicity)<br>· High grades have reduced evaporative loss | · Flammable or combustible, some have very low flash points, so process equipment must be designed to mitigate explosion dangers<br>· Slower drying times than chlorinated solvents<br>· The cost of vapor recovery, if implemented, is relatively high |

| Technology Type | Pollution Prevention Benefits | Limitations |
|---|---|---|
| Coatings Removal | | |
| Plastic Media Blasting | · Eliminates VOCs and HAPs<br>· Uses nontoxic media<br>· Spent media are cleaned and reused several times for paint stripping<br>· Some spent thermoplastic media are recyclable to make plastic products | · Spent plastic media contain paint chips and so may be hazardous waste<br>· Requires workers to wear respiratory and eye protection equipment<br>· Blasting generates high noise levels<br>· May cause metal substrate damage<br>· More aggressive media types damage fiberglass or composite materials<br>· Contaminants in media cause stress risers in the substrate<br>· Uses flammable media |
| Molten Salt Coating Removal | · Eliminates VOCs and HAPs | · Generates by-product sludge that may be hazardous waste<br>· Will damage heat-sensitive materials such as heat-treated aluminum or magnets<br>· Must not be used for low-melting alloys<br>· Must not be used with pyrophoric metals<br>· May require offgas treatment, depending on local air permitting regulations<br>· Potential for generation of products of incomplete combustion |
| Cleaning and Degreasing | | |
| Completely Enclosed Vapor Cleaner | · Virtually eliminates solvent air emissions | · High initial capital cost<br>· Slower processing time<br>· Relatively higher energy requirement |
| Ultrasonic Cleaning | · Eliminates solvent use by making aqueous cleaners more effective | · Part must be immersible<br>· Testing must be done to obtain optimum solution and cavitation levels for each operation<br>· Thick oils and grease may absorb ultrasonic energy<br>· Energy required usually limits parts sizes<br>· Wastewater treatment required if aqueous cleaners are used |
| Low-Solids Fluxes | · Eliminates need for cleaning and therefore eliminates solvent use | · Conventional fluxes are more tolerant of minor variations in process parameters<br>· Possible startup or conversion difficulties<br>· Even minimal residues are unacceptable in many military specifications |

Table 3. Emerging Cleaner Technologies

| Technology Type | Pollution Prevention Benefits | Limitations |
|---|---|---|
| Alternatives to Chlorinated Solvents | | |
| Catalytic Wet Oxidation Cleaning | · Only $CO_2$ and water produced | · May damage/corrode some substrates |
| Absorbent Media Cleaning | · No water involved in cleaning<br>· Replaces solvents<br>· Media biodegradable and/or can be recycled | · Plant air on a shop vacuum needed in most cases<br>· Not useful on complex surfaces or detailed parts |
| Coatings Removal | | |
| Laser Heating | · Replaces solvents<br>· Produces a volume of ash residue smaller than the original coating volume | · Requires offgas collection and filtration for particulates<br>· Requires laser barrier wall to protect workers from lethal energy laser beam<br>· Requires air flow or other collection mechanism to prevent ash redeposition on the substrate<br>· Can generate products of incomplete combustion |
| Ice Crystal Blasting | · Replaces solvents<br>· Media are nontoxic<br>· Produces a small volume of coating chip waste | · Generates small volume of coating chips, which may be a hazardous waste<br>· Potential for worker injury from high-velocity ice pellet impact<br>· Requires workers to wear respiratory and eye protection equipment<br>· Requires workers to wear hearing protection |
| Flashlamp Heating | · Replaces solvents<br>· Produces a small volume of ash waste | · Requires offgas collection and filtration for particulates<br>· Can generate products of incomplete combustion<br>· Leaves oily residue on substrate |

| Technology Type | Pollution Prevention Benefits | Limitations |
|---|---|---|
| Cleaning and Degreasing | | |
| Vacuum Furnace | · Eliminates solvent use for cleaning | · Typical processes do not allow for oil recycling. If oil is not collected, it can degrade the diffusion pumps; so frequent cleaning would be necessary |
| Plasma Cleaning | · Eliminates solvent use for cleaning | · Requires a special cleaning chamber<br>· Relatively slow process |
| Replacement of Tin-Lead Solder Joints | · Eliminates solvent cleaning and hazardous fluxes | · May replace hazards of lead compounds with hazards of silver |

33/50 Program

The EPA's 33/50 Program is a nation-wide voluntary pollution prevention initiative that builds on the Agency's pollution prevention policies and programs. It aims, through voluntary source reduction, to encourage the reduction of 17 chemicals and chemical compounds used in manufacturing, from an aggregate of 1.4 billion pounds in 1988 down to 700 million pounds in 1995--a 50 percent reduction. Voluntary goals were set for a 33 percent reduction by 1992 and at least a 50 percent decline by 1995, as measured by the Toxics Release Inventory (TRI). The list of chemicals was drawn from TRI based on the following considerations: high production; high releases and off-site transfers; potential for pollution prevention; and potential for a wide range of health and environmental effects. Of the 3,000 companies emitting one or more of the 33/50 chemicals, EPA has contacted 600 with the largest TRI releases and transfers to ask that they voluntarily develop programs to reach the targeted reductions. One of the major approaches to achieve these reductions will likely involve source reduction and substitution of less toxic chemicals. EPA publicly reports on the status of company commitments each year, focusing on source reduction actions and chemical use substitutions.

EPA's Green Lights Program

In January 1991, the U.S. EPA announced its new Green Lights program. Green Lights is a program that encourages major U.S. corporations to install energy-efficient lighting wherever it is profitable, and only where it maintains or improves lighting quality. By using energy-efficient lighting technologies and designs, less energy and electricity are demanded, and less pollution is generated by power plants.

By encouraging corporations to install energy-efficient lighting, Green Lights has produced multiple national benefits. Corporations save energy and increase their profits, the air pollution caused by electricity generation is significantly reduced, and the country's energy needs will decrease. The program will also increase demand for quality energy-efficient lighting products and services, improve consumer knowledge about those products and services, and clarify the role lighting can play in protecting the environment.

The Green Lights program can set a precedent by demonstrating how a cooperative, nonregulatory partnership between government and industry can be effective in achieving national goals with minimal red tape. At a time when those goals include protecting the environment, saving energy, improving national energy security, and increasing profits and competitiveness, Green Lights can be an important model program for the present--and the future.

State Facility Planning Requirements

A new pollution prevention requirement appearing in the legislation of many states is to require that a "facility plan" be developed. Almost all of these facility planning statutes require industrial facilities to submit pollution prevention plans which must be updated periodically. Most facility planning statutes cover the facilities that are required to report federal Toxics Release Inventory (TRI) data. These facilities must use their TRI data when preparing state pollution prevention plans. Some of the statutes expand their coverage to include large- and small-quantity hazardous waste generators, or holders of specific types of environmental permits. Out-of-state generators may also fall under the scope of the planning statutes.

The chemicals covered by facility planning statutes vary. However, chemicals covered by many of the laws are also covered under the Toxic Releases Inventory. In addition, hazardous wastes covered by the Resource Conservation and Recovery Act and chemicals covered under the comprehensive Environmental Response, Compensation and Liability Act are sometimes covered by the statutes, and some state legislatures require coverage of additional sets of chemicals.

A state's programs are really the best indicator of how active it is in the pollution prevention arena. Although a law may be on the books, unless there are programs putting it into practice (and funding to support the programs), little progress can be made at the state-level. State pollution prevention programs show at least as much variety as state laws. Some programs are mature, independent and well-established within the state's environmental hierarchy, and administer a variety of initiatives dealing with pollution prevention. Other programs consist of little more than a coordinator who tries to pull together the pollution prevention aspects of the other state environmental programs and whose main job is education about the benefits of pollution prevention. Some states delegate their pollution prevention programs to outside groups, such as universities or other research centers, which they supply with state funding.

## Pollution Prevention in Other Countries

Pollution prevention is receiving widespread emphasis internationally within multinational organizations and within individual countries. The driving force behind this emphasis is the concept of sustainable development and the hold that this concept has over planning strategies and long-term solutions to global limits and north-south economic issues. The June 1992 United Nations Conference on Environmental Development (UNCED) in Brazil spent a great deal of time on pollution prevention.

The European Community has designed some of its rules and programs around pollution prevention. The Organization for Economic Cooperation and Development (OECD) has just completed a major assessment on pollution prevention. The United Nations Environmental Programme has a clean technologies program and the United Nations Industrial Development Organization (UNIDO) held major conferences on sustainable development. Joining this group of international bodies is the North Atlantic Treaty Organization (NATO) which has a non-military Committee on Challenges of Modern Society that has just begun a multi-year pilot study called "Pollution Prevention Strategies for Sustainable Development" in which 14 countries are involved in an information exchange program on pollution prevention policy, education and technology.

Individual countries have taken their own initiatives in developing pollution prevention programs. Canada has the Green Plan and The Netherlands has the NEPP (National Environmental Policy Plan). Denmark, and The Netherlands, are extensively studying life-cycle accounting applied to a host of consumer and commercial products.

A very interesting approach to using the regulatory system is being pursued by Germany to encourage P2. The 1986 Waste Act empowers the Federal government to bring its' influence to bear on waste generation prior to the production and use of products.

The options, if necessary to increase recycling and to simplify waste treatment, include:

- Subjecting certain products to mandatory labeling or separate handling
- Requiring the manufacturers to reclaim their products once they become waste
- Imposing bans or restrictions on marketing

Although these regulations are valid for all wastes the two main aims are

- To reduce the pollutant content of waste and thereby enable more recycling of these pollutant free waste
- To reduce the amount of household waste by reducing all kinds of packaging material

In carrying out the second of the above options, the German government has passed an ordinance that required, by January 1993, that commercial dealers have to accept for recycling all returned packaging in or near shops.

## WHERE DO WE GO FROM HERE

Two decades of environmental regulation in the United States have shown notable results in addressing air pollution, water pollution, and solid and hazardous waste disposal. Today, to quote a 1988 EPA report,

> "There is no question that the air in most of our cities is far cleaner and healthier than it was in the 1960's. Thousands of miles of rivers and streams and thousands of acres of lakes have been restored and protected for fishing and swimming. In addition, we have taken extraordinary steps to improve the management of hazardous wastes, toxic chemicals, and pesticides. Nevertheless, looking at the current roster of environmental problems, they appear to be as formidable as ever. They include global concerns such as climate change and ozone depletion; cross-media pollutants such as lead and heavy metals, solvents, and pesticides; small and disperse source concerns such as agricultural runoff and mobile sources; shortages of waste disposal capacity and massive waste cleanup bills; and a growing public concern over the basic state of the environment in which we live. Of course, our concerns in the United States differ little from the concerns of citizens of countries throughout the world.
>
> As we have achieved success in improving our environment, it has become increasingly clear that there is only so much that can be achieved through strategies based upon developing more effective pollution control technologies. There appears to be much more to be gained from strategies based on

eliminating polluted streams through improvements in the production processes themselves. . . . admittedly, this is a much longer term strategy, but the apparent benefits are impressive.

Zero discharge of wastes is not possible, but as we continue to enjoy the benefits of creative individuals around the world developing new approaches for encouraging cleaner products and production technologies, we can see that a world with much less waste and the problems arising from treating and disposing of that waste is certainly possible.

To quote a former head of the EPA's Office of Pollution Prevention, "As progress is made on the technological side of pollution prevention, there is a growing recognition of the need for prevention to become an integral part of our basic philosophy of environmental protection. Pollution prevention must become the strategy of first choice in addressing any environmental problem. Creating this new "pollution prevention ethic" requires a shift in the perspectives of those whose activities affect the environment. Without question, this is a massive undertaking, and one that will continue to challenge society in the years ahead."

# Green Technology by Design: A New Paradigm for Engineering Education for Sustainable Development

Charles Overby, Professor Emeritus
College of Engineering and Technology
Industrial and Systems Engineering Department
Ohio University
Athens, Ohio 45701 USA

---

*Design should not merely meet environmental regulations; environmental elegance should be part of the culture of engineering education and practice. Selection and design of manufacturing processes and products should incorporate environmental constraints and objectives at the outset, along with thermodynamic and economic factors.* [Ausuble 1989]

---

*ABSTRACT*

In this paper I explore some aspects of the challenge we face in changing the culture and perspective of engineering education so as to achieve a holistic and integrated systems approach to design that includes "design for resource conservation and environmental elegance" or "Green Technology By Design". These ideas are essential components of any philosophy or program for "sustainable development". My context for these ideas is the USA. Thus, specifics come from my own country, but the general themes, I believe, are applicable worldwide. Engineering education for the environment has grown up in the USA, in departments of environmental engineering, often housed in civil engineering, and with an "end-of-pipe", "pollution control" philosophy and mind-set that mirrors the pollution control, regulatory philosophy of government. Today we are beginning to recognize that while "pollution control" is very important and probably always will be, perhaps we also need to think more expansively than from just a "control" perspective to one of "prevention" -- from "pollution control" to "pollution prevention". It is this transition from a philosophy of "control" to one of "prevention" in the environmental education of engineers, that I address in this paper. I will also suggest a non-coercive model for achieving these needed changes -- namely the model of "quality by design" that has played such a major role in improving the quality of products and processes designed and produced by our society in recent years. Unless we in the "first world" are able to bring these ideas into reality, then I fear that our collective futures on

NATO ASI Series, Partnership Sub-Series, 2. Environment – Vol. 2
Cleaner Technologies and Cleaner Products
for Sustainable Development
Edited by H. M. Freeman, Z. Puskas, and R. Olbina

planet earth are likely doomed by "resource wars" and unbelievable entropic environmental degradation.

## *INTRODUCTION*

In the late 1970's it was my good fortune to spend a year of sabbatical leave as a staff person at the US. Congress, Office of Technology Assessment (OTA) working on resource conservation and environmental issues [OTA & Overby 1979a & b]. One of these publications, [Overby 1979a], was a report on the first national survey of remanufacturing in the USA with a focus on its resource conservation and pollution reduction potential in America. The experience from this research motivated me to write several additional papers at that time in which I began to ask if engineering educators ought to think of integrating "design for resource conservation and environmental elegance" into their courses [Overby 1979c & 1980]. Unfortunately few of these materials appeared in major technical publications. This writing produced a flurry of interest at the time, but then passed into oblivion in the Reagan era of limited concern with environmental issues -- until one day in Summer of 1989, I was invited to Grand Valley State University (GVSU) in Grand Rapids to explore the possibility of working with them and their small General Engineering Program on integrating environmental concerns into their curricula. It became my privilege to work as a consultant with GVSU for a couple of years in this domain on their Design for Recycling project.

## *THE IMPORTANCE OF DESIGN*

From the beginning, in discussing the integration of environmental factors into the regular engineering curricula, I have concentrated on engineering design courses as the place to focus our attention [Overby 1990 & 1991]. Please see a very recent paper by Keoleian and Menerey for an excellent review of the state of the art in using life cycle design concepts for sustainable development [Keoleian 1994]. Engineering curricula are already extremely over crowded, thus it would not be possible to introduce a special new "engineering and the environment" course for all engineering students. Rather, environmental factors must be integrated into regular ongoing engineering courses, and the key courses for this integration are those in engineering design.

Design courses are important because, as we have learned from our contemporary "quality revolution" -- "design" is where quality is created, and by the time the designer is through with his or her work -- some 80 to 90 percent of all of the life cycle costs of a product or system have been determined [Fabrycky 1991]. Once these design decisions are in place, not much can be done at the end of the pipe to fix things up.

Since design has such an impact on everything that happens after the design function is completed, it seems important that we consider the possibilities of integrating environmental factors into the design process so that, from the very beginning, these considerations receive the same treatment that more traditional engineering functional design criteria receive. By designing technology that not only satisfies its primary engineering functions, but which also minimizes resource consumption, and prevents or minimizes pollution, we would be making a shift from an "end-of-pipe" "pollution control" mind-set to a "pollution prevention" perspective.

*WHY IS THIS CHANGE SO DIFFICULT?*

Unfortunately, that which seems rather logical and obvious, and perhaps easy to accomplish, is often not necessarily the case. There are many reasons why it is not easy in engineering education to integrate environmental criteria into the front end of the engineering design process. Let me explore two -- (1) lack of interest by engineering faculty, and engineering education administrators, (Deans etc.), and (2) a commitment to the status quo which in engineering education is largely one of "pollution control", rather than "pollution prevention".

From the very first in my discussion with GVSU, I expressed concern as to our ability to interest the typical electrical, mechanical, or industrial engineering faculty member, or engineering education administrator in integrating environmental issues into regular design courses. This ought to be less true for chemical engineers since their work so directly and intimately relates "design" to pollution. It had been, and still is, my experience that "recycling" and other kinds of environmental matters, while of general interest to engineering faculty and administrators -- are not of sufficient "sex appeal" to motivate them to spend the time and effort to do the integrating. Furthermore, engineering faculty have so many other challenging and interesting things happening in their respective fields that there is little time for them to take on the additional challenge of integrating environmental matters into their design courses. Most engineering faculty, and administrators saw, and in my judgment, still continue to see -- environmental issues as the proper domain of specialists in the "environmental engineering" programs that have proliferated over the years in response to the "pollution control" philosophy driven by Federal and State governmental regulatory apparatus.

I was reinforced in my above assessments, when I spoke with the author of a paper titled "No Course Is An Island" -- a paper that explored the integration of environmental issues throughout the engineering curricula [Meade 1992]. Meade, in preparation for writing his paper, contacted many engineering faculty and administrators around the country, including me. When I again spoke with him after his article was published, he told me that very few of those contacted felt that environmental factors should be integrated into regular courses in the curricula. Most felt that the appropriate place for these "environmental" concerns was in the separate programs under the generic label "environmental engineering". Meade found in his survey, that most administrators (Deans etc.) and "older faculty" preferred the "old fashioned" route of "pollution control" that is implicit in the separate environmental engineering curricula. A few younger faculty did seem to feel that perhaps there was some merit in integrating environmental factors into their regular engineering courses. How did we come to this specialist, pollution control perspective in engineering education?

*SPECIALIST AND END-OF-PIPE CONTROL PERSPECTIVE*

Over the years since the late 1960's and early 1970's, as governmental involvement in environmental concerns took off and rapidly grew, engineering education responded by establishing excellent "specialist" departments of "environmental engineering" with many of them ending up in civil engineering as extensions of "sanitary engineering". The focus of these departments mirrored that of the approach used by government -- namely a regulatory, "end of pipe", "pollution control" philosophy and mind-set. There was a lot of pollution out there, and the object was to learn how to "control" it as it came out of the "ends of pipes". Thus was born the idea that concerns with the environment in engineering education were to be the domain of specialists with a pollution "control" perspective. This lets all other brands of engineers off the hook, so-to-speak. They need not concern themselves with environmental things because in Adam Smith's world of the division of labor, the environment was someone else's domain.

A similar situation existed in the "quality arena", before international competitive pressures forced American industry to recognize that "quality" was not just the domain of the specialist department called the "quality control" department. Rather, a major component of the "quality revolution" of which we have heard much these past few years, was the recognition that "quality" is the responsibility of all persons in the organization -- and especially is it the responsibility of design engineers. Thus, driven by competitive forces, American industry came to recognize that design is a key factor in the determination of the quality of products, processes, and systems. These insights led to a revolution in the practice of and education for "engineering design" , as illustrated in the National Research Council 1991 publication, *Improving Engineering Design: Designing for Competitive Advantage* [NRC 1991]. We thus now hear ideas of "concurrent engineering", "simultaneous engineering", "life cycle design", "expert systems and design", "quality by design", "defect prevention by design", "quality function deployment" (QFD), Taguchi's ideas of "robust design" etc. etc.. After we healthily escaped from the very constraining and limiting "conventional wisdom" that "quality" was the domain of the specialist "quality control group" -- the "quality revolution" took off.

*THE QUALITY REVOLUTION REFLECTS A PARADIGM SHIFT*

In the words of Thomas Kuhn in his unique and original work, *The Structure of Scientific Revolutions* [Kuhn 1970], the fields of "quality" and "design" had undergone a "paradigm shift" -- to a new and healthier perspective -- to a new world view.

*EARLY STATISTICAL QUALITY CONTROL IN THE USA*

When statistical quality control ideas were invented in the late 1920's and 1930's by Walter Shewhart and a few other early pioneers, and adopted into American industry, the "control" perspective was dominant and "quality control" came to be seen as the domain of specialists -- quality control specialists". In part this reflects the influence of Frederick W. Taylor's "scientific management" and division of labor mentality of that period in our history. Departments of "quality control" were established, and specialist "quality control" education was introduced, dominantly in industrial engineering disciplines. The idea was that "quality" was to become the domain of specialists who were trained to "control" the quality of product coming out of the factory door. This was an "end-of-pipe" philosophy like that which presently dominates much of the conventional wisdom about engineering environmental education -- a "pollution control", philosophy and "world view".

*DOWNSIDE OF THE CONTROL PERSPECTIVE*

One of the most insidious aspects of the "control" mentality, whether associated with the word "pollution" or the word "quality", is the way it constrains and restricts our creative perceptions as to the nature of our problem. For example, this limiting and bureaucratic view of "quality" as "quality control", caused us to accept almost as "God Given" that which came down the manufacturing pipe. Our job as "quality specialists" (and I have been one of these "quality specialists" for much of my professional life) -- was to simply take those "givens" and make sure that with the proper statistical tools, we "controlled" things so that product was manufactured to meet "specifications". Hardly ever did this kind of perspective and organizational ossification enable us take a more holistic healthier systems view -- to look upstream and ask important questions about the ultimate causes of poor quality, oftentimes the designs of products and processes themselves -- or to look downstream to more profoundly ask the consequences to users and customers of poor quality. Nor, did this perspective promote the insight that "quality" was everyone's job, not just that of the quality control specialist. Fortunately for us in the USA, a paradigm shift has taken place where we are now becoming refreshingly liberated to see "quality" and "design" in much healthier ways so as to enable us to become internationally competitive once again.

There is much pollution out there and the field of environmental engineering and its focus of "pollution control" is very important and will remain so. However, the "control problem" in "environmental engineering" is analogous to that of the "control problem" in the quality scenario described in the above paragraph. Environmental engineering is constrained and limited in its perspective by its pollution "control", "specialist" world view or operational paradigm. This is especially true in engineering education. In industry we now begin to see some focus on ideas of "green technology by design", but very little of this yet in engineering education.

As an academic, I regret that these more holistic systems views for both "quality" and "engineering design" have not emanated from the US engineering education world. In my view, these exciting changes have come mostly from industries coping with competition from the Japanese and other producers of superior quality products and systems.

*WHO SHOULD DO THE INTEGRATING?*

Engineers who have been trained in environmental engineering programs to clean up the mess have a "pollution control", "end-of-pipe" orientation. Furthermore, these engineers, are mostly civil engineering types, who are not the ones who design the cornucopia of products and the manufacturing processes that create these "goodies" -- and the environmental consequences of them. Therefore, it is really quite inappropriate to ask these engineers to be the ones to integrate environmental factors and pollution prevention ideas into engineering design courses. The majority of this integration must come from the mainline, chemical, electrical, mechanical, and industrial engineers who do design and create the products and manufacturing processes.

A very interesting change in attitude seems to have recently taken place in mainline academic engineering departments relative to "quality". It is now becoming accepted wisdom in these departments that design and quality are intimately related, whereas before the quality-design paradigm shift, quality was looked at as something that only those industrial engineers did -- a specialist function that should not concern other mainline engineering disciplines.

*ADOPT THE QUALITY--DESIGN PARADIGM*

As indicated above, it is not easy to get mainline engineering disciplines, or engineering college administrators to buy into this idea of -- integrating environmental factors and considerations into regular design courses. Engineering design textbooks are not yet written that way -- but perhaps things are changing here too. The field of engineering quality and design in practice and in education, is alive with new ideas and dynamism. It should not be necessary for us to have to create another paradigm shift for "green technology by design". All we really need in order to do "green technology by design", is to walk that new path and use the same ideas, tools, and techniques of "life cycle design", "concurrent engineering", "QFD", "Taguchi's robust design" etc. If we simply bring meaningful environmental questions, to the design process, at its very beginning, we will find ourselves doing "green technology by design", and -- I believe, feeling more ethically professional about our creative engineering work. Illustrative environmental questions might be -- "What happens to this PC computer, or VCR when its useful life is ended? Could I design it so that it could be more easily recycled when discarded? Could I specify different materials that would have a less damaging impact on the air, land or water, as this product is being manufactured, used or disposed of, or as the raw materials are extracted from the earth? Could I design this part so that it could be remanufactured more easily, and thus also save the embodied energy in it. How might I change this design so as to enhance its energy efficiency? etc., etc.. As you know, these questions and issues are now being quite intensely explored in Germany and other European countries these days [Kreft 1993].

*CHANGING PERCEPTIONS OF PRODUCT LIFE CYCLE RESPONSIBILITY*

To some extent, the things happening in Germany, the product "take back" laws etc., reflect new and changing attitudes toward the allocation of responsibility between government and the private sector for coping with the mounting levels of waste and environmental degradation arising out of unlimited growth in consumption of material goods. In the past, municipal and county governments have assumed responsibility for disposing of the wastes that start in our factory cornucopias as shiny new products and end up as post consumer solid waste. Governments assumed this responsibility arising from their so-called "police power" to ensure "public safety", "public health", "morality", "peace and quite", "law and order" etc.. With the "take back" laws, governments are now saying that perhaps someone else, the manufacturer or producer of the item, must now assume a larger share of the responsibility for disposing of it when it enters the solid waste stream.

If this idea of more solid waste responsibility for manufacturers and less for governments is contagious, and if it grows, then from an engineering design perspective we will see much more of that for which I am arguing in this paper. If a product manufacturer knows that the disposal cost of his product is now his or her responsibility, and if the cost to dispose of a product is high, then we might be sure that that product's design engineers will receive strong economic and direct verbal signals from producers to expand their horizons to include more of the life cycle of the product out to its disposal phase. Under the present arrangement where governments assume the responsibility for disposal of wastes, there is little incentive for the private sector to make "design for disposal" an important engineering design criterion.

*TAKE-BACK LAWS AND PRODUCT LIFE CYCLE COSTS*

I am sure that there are major problems with the introduction, implementation, and administration of laws like Germany's "take back" requirements. However, if somehow the transition to more manufacturer responsibility for their portion of the waste stream can be achieved, then we should have a healthier environment because, products costs would now reflect a larger share of the "full life cycle cost", including environmental costs, than is presently the case. It has always been difficult to include environmental costs in a product's cost calculus. "Take-Back" laws effectively do this for the environmental costs of "goodies" disposal without our having to engage in some kind of bureaucratic regulatory and/or abstract economic nonsense. The creator and producer of a "good" simply knows that the costs of disposal are hers or his, and as I learned in my beginning market economics class many years ago, that producer might well seek to minimize these costs. Cost minimization might well include instructions to engineers who design these products to expand their perspective to include more of the life cycle of a product in their deliberations on -- design, materials procurement, production, customer use, and ultimate disposal. With market economics operating as theory tells us it is supposed to, and with these environmental costs now included in the producers cost calculations, we will likely get a different and healthier mix of products -- products that will more likely be "green products by design."

*ANOTHER PRODUCT LIFE CYCLE COST EXPANSION IN THE USA*

I have focused on German "take back" laws, but it is interesting to note that in the USA there are some other things happening that relate to manufacturer responsibility for more of the life cycle costs of their products. I am referring to the State of Florida where the Governor recently brought in legislation that would enable the State of Florida to sue cigarette manufacturers for reimbursement of about $300 million a year in Medicaid expenses for smoking-related illnesses [BW 1994]. I think there is probably some concern that these kinds of ideas might catch on relative to other kinds of products that also create social and health costs that are presently not included in those products' cost calculus.

*SOME SIGNS OF CHANGE IN ENGINEERING ACADEMIA*

Fortunately a few engineering academicians are helping to show us the way. I have been in some discussions with some of the following about the need to convene a conference/workshop in these topics so as to propagate these important ideas more broadly within the engineering disciplines that design products and processes. If any of you are interested in this possibility -- please let me hear from you. Here are some names.

Shirley Fleischmann, School of Engineering, Grand Valley State University, Grand Rapids, MI [Fleischmann 1993]; Phil Barkan, Emeritus Professor, Mechanical Engineering, Design Division, Stanford University; Stanford, CA; Kosuke Ishii, Associate Professor, Mechanical Engineering, Ohio State University, Columbus, Ohio. Professor Ishii and his students have developed an approach that they call "Life Cycle Design". [Ishii 1993]; Nabil Nasr, Professor, Industrial and Manufacturing Engineering, and Paul Petersen, Dean of Engineering, Rochester Institute of Technology, Rochester, NY. [Owen 1993]; Nam Suh, Woodie Flowers, and graduate students David Wallace, and Li Shu, Mechanical Engineering, Massachusetts Institute of Technology CAD Lab., Cambridge, MA. [Wallace 1993]; D. Navin Chandra, Robotics Institute, Carnegie Mellon University, Pittsburgh, PA. [Navin 1994]; Jonathan Bulkley, Professor, & Center Director, and Greg Keoleian, Center Manager, The University of Michigan National Pollution Prevention Center for Higher Education (NPPC), Ann Arbor, MI [U of M 1994]. For a more elaborate listing of courses, including several of the above, please see the new *Directory of Pollution Prevention in Higher Education: Faculty and Programs 1994,* just released by the Michigan NPPC [Bloch 1994].

*A HOPE*

If we in engineering education are able to successfully carry out this integration, then in the 21st Century our students, working in enlightened institutions, will be better able to conceive of and create products and systems for a more sustainable future. Billions of "third world" people on earth, aspire to acquire all of the material "goodies" that we in the "first world" so eagerly consume. If they are to have even a small chance of joining us in our consumptions -- then we in the developed countries must demonstrate technology models, ***for ourselves*** and them, that are orders of magnitude more resource conserving and environmentally benign than anything presently in existence. Can we imagine the environmental and resource depletion impact of 1.25 billion Chinese people and a few more billions from India and Africa consuming as we do? This is exactly what these teeming billions would like. If we are unable to create this kind of resource conserving and environmentally elegant technology for the future, then I fear that we may exterminate ourselves in unbelievable entropic environmental degradation and resource wars. We have a good example of a recent resource war -- the 1991 Gulf "oil resource" war. Environmental elegance and resource conservation by design, "green technology by design", while not a panacea -- is an important small part of a sustainable future that must also include absolutely essential reductions in population growth .

*REFERENCES*

Ausuble, J. H., and Sladovich, H. E. (1989) *Technology and Environment*, National Academy of Engineering, National Academy Press, Washington, DC., 1989.

Bloch, Pamela, and Goodwin, Joanne (editors) (1994) *Directory of Pollution Prevention in Higher Education: Faculty and Programs 1994*, NPPC, University of Michigan, Dana Bldg., 430 E. Univ. Ave., Ann Arbor, MI 48109-1115

BW, "Tobacco: Does It Have A Future", *Business Week*, July 4, 1994, pp 24 -- 29

Fabrycky, W. J., and Blanchard, B. S. (1991) *Life-Cycle Cost And Economic Analysis*, Prentice Hall, 1991.

Fleischmann, Shirley, Project Director, (1993) *Teaching Environmentally Responsible Design*, An Engineering Curriculum Development Project of the School of Engineering, Grand Valley State University, Grand Rapids, MI,. 1993, Sponsored by the Padnos Foundation, and the State of Michigan Department of Natural Resources.

Ishii, Kosuke (1993) "Life Cycle Design: US vs Japan", Seminar at the U.S. -- Japan Technology Management Center, Stanford University, November 1993, personal communication.

Keoleian, Gregory, and Menerey, Dan (1994) "Sustainable Development by Design: Review of Life Cycle Design and Related Approaches", *Air & Waste*, May 1994, pp 644--668.

Kreft, Hansjurgen (1993) "Electronic Scrap: The Case for Required Recycling", *Siemans Review*, Vol. 60, July/August 1993.

Kuhn, Thomas S. (1970) *The Structure of Scientific Revolutions"*, Second Edition, Enlarged, University of Chicago Press, 1970.

Meade, Jeff (1992) "No Course is an Island", *ASEE Prism*, December 1992, Washington, DC.

Navin Chandra, D. (1994) (Editor), *Green Engineering: Sustainable Growth Strategies And Tools For Manufacturers And Engineers*, Academic Press, to be published in 1994.

NRC (1991) *Improving Engineering Design: Designing for Competitive Advantage*, National Research Council, National Academy Press, Washington, D.C., 1991.

OTA (1979) *Materials and Energy from Waste, and Beverage Container Deposit Legislation*, US Congress Office of Technology Assessment Report, Washington, DC., July 1979.

Overby, C. M. (1979a) "A Study of Issues and Policies Related to Recycling of Products", US Congress Office of Technology Assessment, Internal Report, Washington, DC., August 1979.

Overby, C. M. (1979b) "A Preliminary Technology Assessment of Product Recycling Remanufacturing) as a Conservation Option" ASME 1979 Winter Annual Meeting, Dec. 4, 1979, 79-WA/TS-8.

Overby, C. M. (1979c) "Product Design for Recyclability and Life Extension", *1979 ASEE Annual Conference Proceedings*, Washington, DC.

Overby, C. M. (1980) "Product Design for a Sustainable Future: A Matter of Ethics?", *1980 ASEE Annual Conference Proceedings*, Washington, DC.

Overby, C. M. (1990) "Design for the Entire Life Cycle: A New Paradigm?", *1990 ASEE Annual Conference Proceedings*, Washington, DC.

Overby, C. M. (1991) "Design for Environmental Elegance", *Proceedings*, Engineering Foundation/NSF Conference, "New Approaches to Undergraduate Engineering Education III", Banff, Alberta, Canada, July 28-August 2, 1991.

Owen, Jean V. (1993) (Senior Editor), "Environmentally Conscious Manufacturing", *Manufacturing Engineering*, October 1993.

U of M (1994) Course Modules For Integrating Environmental Concerns Into Engineering and Business Courses, National Pollution Prevention Center for Higher Education, University on Michigan, Ann Arbor, MI, personal communication, 1992, 93, & 94, as a member of the Program Advisory Committee.

Wallace, David; Suh, Nam; and Flowers, Woodie (1993) "An Information-based Model for Environmental Life-cycle Design Planning", Mechanical Engineering Dept., Massachusetts Institute of Technology, personal communication, Summer 1993.

# Evolving Concepts in Life Cycle Analyses

Professor Michael R. Overcash
Pollution Prevention Research Center
Department of Chemical Engineering
Box 7905
North Carolina State University
Raleigh, North Carolina 27695-7905
United States

## INTRODUCTION

Many environmental professionals, policy scientists, and government representatives are widely familiar with the concept of life cycle analysis (LCA). Most individuals know of the existence of the basic LCA concept of cradle-to-grave concerns and appreciate some of the common applications (such as paper versus plastic, aluminum versus tin versus plastic, etc). To some extent the history of life cycle analysis, stretching over thirty years in Europe and twenty five years in the U.S., gives it a mature characteristic as an environmental technique. There are also some characteristics of LCA which are unfavorable such as,

- almost any answer you wish can be obtained in such studies
- it is a tool used for non-environmental reasons (marketing, banning, etc)
- much of the use is as a defensive tool for products or materials

Efforts to eliminate these unfavorable characteristics have occurred. For example the Society for Environmental Toxicology and Chemistry (SETAC 1991) has subdivided LCA into three stages:

1) inventory analysis
2) impact analysis

NATO ASI Series, Partnership Sub-Series, 2. Environment – Vol. 2
Cleaner Technologies and Cleaner Products
for Sustainable Development
Edited by H. M. Freeman, Z. Puskas, and R. Olbina

3) improvement analysis.

Improvement efforts have also led to better and more uniform definition of boundaries for use in life cycle analyses. Development of computer aided systems for inventory and analysis calculations have also reduced the unfavorable characteristics of this environmental technique. Overall these improvements have been primarily the more through implementation of life cycle analysis, much as it has existed for the last decade.

The evolution of life cycle analyses is addressed more in the research and development efforts of this field. These, in some way, point to directions in future concepts related to life cycle assessments. A discussion in this paper addresses some of the evolutionary concepts for understanding the complexity of products, processes, and the environment. Such research is occurring in a relatively small, but dynamic group of organizations around the world, including the efforts at North Carolina State University. Four topics will be used to illustrate the research and evolving concepts in LCA

1) emission dependence on stages of manufacturing
2) partitioning decisions in life cycle inventories
3) improvement analyses directly from life cycle inventories
4) algorithms for life cycle impact analyses.

## EMISSION DEPENDENCE ON STAGES OF MANUFACTURING

The traditional LCI has provided an aggregate picture of the entire cradle-to-grave generation of individual chemical constituents. This is conventionally displayed as a series of comparative bar graphs for two competing products or processes in which the mass of numerous chemical species are shown. In an aggregate form contrasting two or more alternatives, this information serves as input to decision making. If there is a clear improvement due to one alternative, given that the technical specifications and the costs are similar, then the manufacturing or consumer choices are relatively unambiguous. New products or processes found in this way are thus more environmentally compatible. Of course, the comparisons are often unclear, with some attributes being better and others not as good. Thus decision-making, particularly for large capital expenditures, is not significantly improved based on the environmental profiles generated from a life cycle analysis.

Two research topics have focused on changes in decision-making through different descriptions of the emissions across the life cycle of a product, material, or processes. The first is referred to as cradle-to-factory gate. That is, information is separated in such a way as to depict the emissions at each successive step in the life cycle. While this uses the same information, the value of understanding the step-wise generation of wastes is substantial. The questions are what are the shapes of the emissions or waste generation curves as one proceeds through manufacturing, beginning with natural resources. Then as manufacturing changes are made, how are the emission curves improved. Are there limits on further improvement and if so then which processes or materials reach such limits first. In Figure 1 is an example of such a step-wise curve with the individual technologies given in Table 1 (Cohen Hubal and Overcash 1993). In this case one notes the limits of greater air pollution control sequences associated with the rapidly increasing amounts of solid wastes generated.

The second research topic focuses on understanding the relationships of hidden or off-site wastes and emissions to the total chemical losses for a given product or processes. Again the concept is to examine sequentially the generation of emissions, but to assess how large are those which might be outside the control of any given manufacturing organization. The shift to greater percentage hidden wastes would depict situations in

## Table I: AIR POLLUTION CONTROL SEQUENCES

| | SEQUENCE | CONTROL EFFICIENCY |
|---|---|---|
| 1 | FABRIC FILTER (FF) | 99.7% |
| 2 | ELECTROSTATIC PRECIPITATOR (ESP) | 99.9% |
| 3 | FF & ESP | 99.7%, 99.9% |
| 4 | DRY SORBENT INJECTION (DSI) & ESP | 70%, 99.9% |
| 5 | LIME SPRAY DRY (LSD) & ESP | 90%, 99.9% |
| 6 | WET FLUE GAS DESULFURIZATION (FGD) & ESP | 90%, 99.9% |
| 7 | SELECTIVE CATALYTIC REDUCTION (SCR), FGD & ESP | 80%, 90%, 99.9% |

# Figure 1: Total Waste Produced By APC Sequences Versus Total Mass Pollutants Controlled

**As APC sequences provide increasing levels of control, mass waste produced by the APC technologies increases dramatically. For several APC sequences total mass of waste produced is an order-of-magnitude greater than that of air pollutants controlled.**

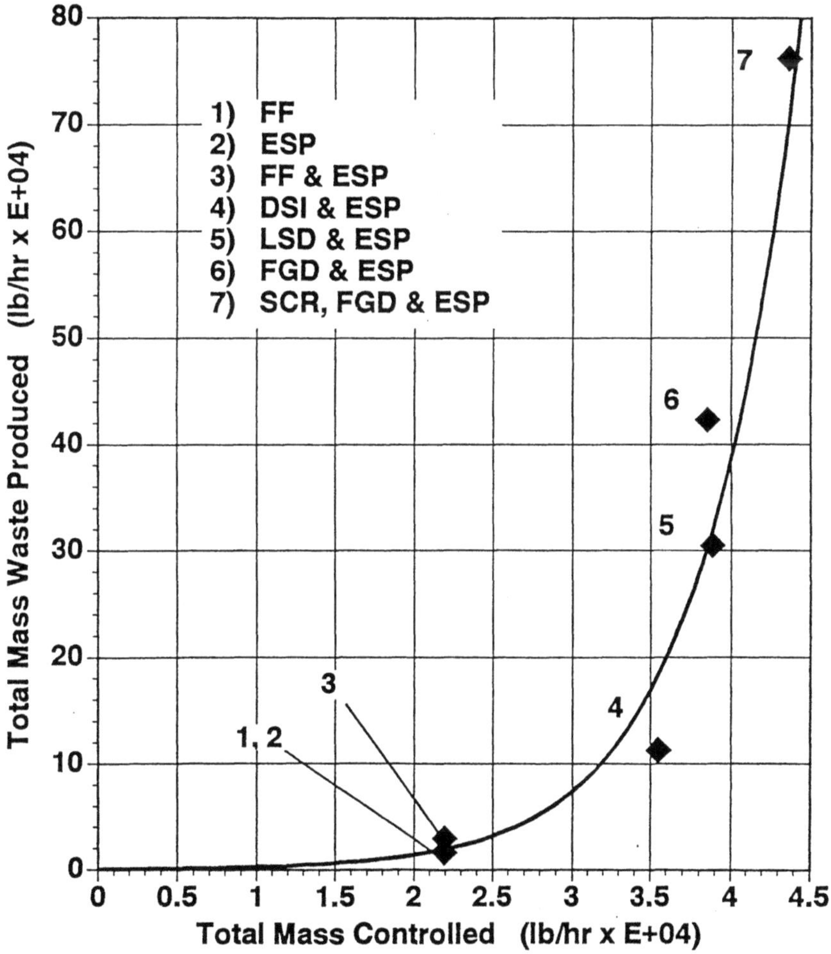

which different alternatives for pollution prevention are needed on a multi-industry basis. Figure 2 is the total emissions and the fraction found to be indirect for the series of air pollution control sequences given in Table 1. A similar type of analysis for chrome plating alternatives to approach zero aqueous discharge is given in Table 2 and Figure 3 (Cohen Hubal 1992).

The research issues are thus to understand the implications of numerous processing and utilization stages in the cradle-to-grave sequence on emissions and chemical losses, rather than only the aggregate of the same information. Such information can then be used to understand the relationship to improved processes or alternatives and a better theoretical understanding of the environmental implications of products and services in society.

## PARTIONING DECISIONS IN LIFE CYCLE INVENTORIES

There are some critical decisions in life cycle inventories which are necessary when dealing with the real world of manufacturing. In addition, there are principles of design or analysis for manufacturing processes which should be consistent with the approaches of life cycle inventory. Research is now focussing on these two types of questions to improve the scientific validity of LCA results.

First, there are inevitably multiple products from manufacturing plants and often from individual processes. Similarly there are characteristics such aş chemical losses, energy requirements, and wastes generated from these facilities or processes. How does one scientifically partition these waste-related characteristics to the individual products manufactured? These are usually involved in completely different sequences of cradle-to-grave calculations. Currently, empirical techniques for such partitioning include:

**Figure 2: Waste Produced By Air Pollution Control Sequences**

**Hidden waste constitutes 14-86% of total waste produced. When hidden wastes are included in analysis, waste consequences of APC treatment increase by 13,000-100,000 lb/hr.**

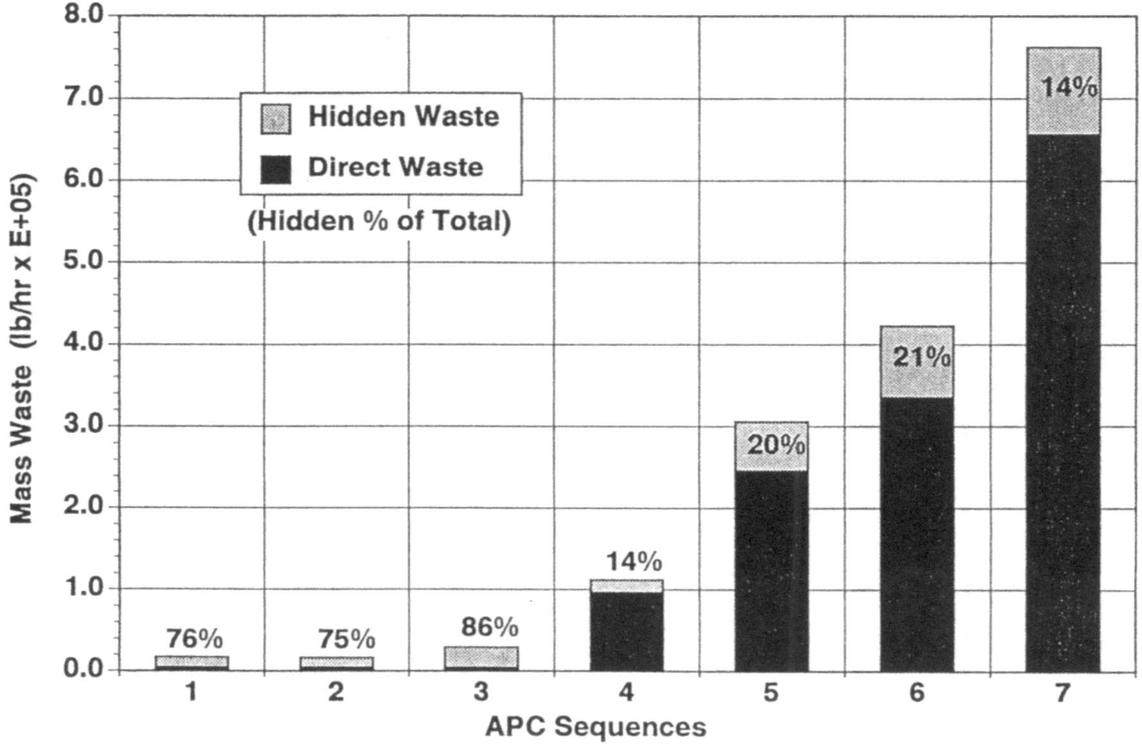

# Table 2 : CHROME PLATING RINSE TREATMENT/RECOVERY SCENARIOS

| | |
|---|---|
| BASE CASE: | ONE RINSE TANK WITH CONVENTIONAL TREATMENT |
| SCENARIO 1: | PROCESS IMPROVEMENTS WITH CONVENTIONAL TREATMENT |
| SCENARIO 2: | PROCESS IMPROVEMENTS WITH CLOSED-LOOP EVAPORATION |
| SCENARIO 3: | PROCESS IMPROVEMENTS WITH OPEN-LOOP EVAPORATION |
| SCENARIO 4: | PROCESS IMPROVEMENTS WITH ION EXCHANGE |

**Figure 3: Wastes Produced By Chrome Plating Rinse Treatment/Recovery Scenarios**

**Hidden waste ranges from 3-97% of total waste produced by zero-water discharge systems. Simple pollution reduction techniques may be as effective at reducing waste as high technology desalinization equipment used to achieve zero-water discharge.**

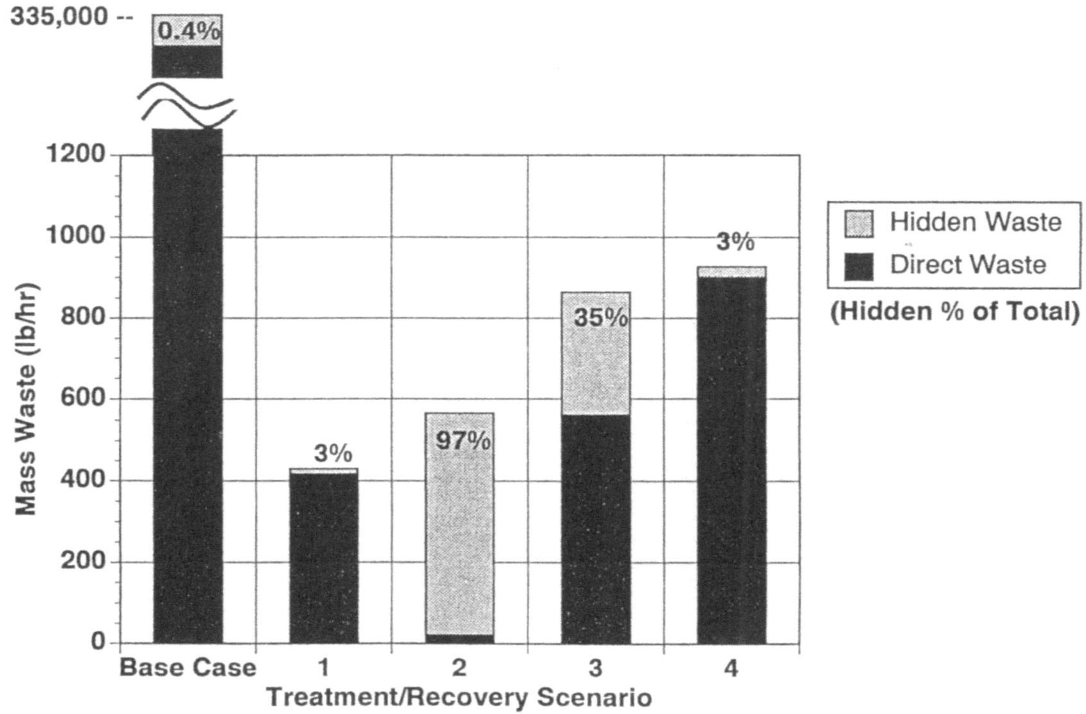

- mass of products
- molar bases of products
- heat of formation (a thermodynamic property)

  commercial value of products

The research edge is to study and compare these empirical alternatives to determine which accurately reflect the environmental implications of such products. Are there other process characteristics (such as the number of sequential processes leading to one of the multiple products) which better determine the assignability of emissions.

Secondly, in the actual determinations of life cycle emissions and wastes generation, the nature of the reported values is a topic for research and evolution in LCA. Currently many life cycle studies report the aggregate of chemicals which have regulatory importance (sulfur oxides, cadmium, carbon dioxide, ash, etc). However, these constituents which are used to represent the life cycle inventory rarely satisfy the constraints for a basic mass balance on products or processes. Thus the tools use in actual design and manufacturing are not consistent with those used in life cycle inventories. Until research is able to bridge these information differences, much of the lack of transparency of LCI and the usefulness for hard decisions will suffer. Some have proposed using a difference term to represent the need to satisfy issues of mass balance and to report that quantity (even though the chemicals are unknown) as a part of the life cycle inventory. Further research will be needed to make closure of mass balances a routine part of life cycle calculations.

## IMPROVEMENT ANALYSES DIRECTLY FROM LIFE CYCLE INVENTORIES

One recurring observation has been that life cycle analyses or impact analyses quickly introduce many qualitative factors into the life cycle inventory. The latter is reasonably agreed upon since basic engineering and mass balance principles are commonly used. However to determine the implications of the inventory information on environment, risk, or policy involves many judgements. For this reason, research is underway to go directly from the inventory stage to the improvement stage with little need to involve the issues of impact analysis.

In this evolving field, the questions are how to use the stage-by-stage information to identify opportunities for process improvement. In these cases less emphasis is placed on which chemicals or wastes to reduce, but on what options exist or could be developed to reduce overall emissions or wastes generated. Furthermore, the study of life cycle inventory improvement is needed to evaluate actual pollution prevention alternatives widely proposed for such issues as reduced chromium usage (Studt 1993), environmentally conscious manufacturing, biotechnology as a route to products, and concepts such as supercritical fluids in chemical processing.

Under this research field, the evolving concepts of life cycle assessments will lead to effective information for design and improvements in manufacturing. Current interest in chemicals selected during pharmaceutical R & D and in evolving electronic products illustrate the benefit in using life cycle inventory information directly in the concepts of improving manufacturing. This reduces the influence of large numbers of qualitative factors which are typically used in life cycle analyses and environmental policy.

# ALGORITHMS FOR LIFE CYCLE ANALYSES

The need to reduce the substantial information gathered during life cycle inventories to a few (or even a single) measures of environmental impact is a recurring dimension of this field. There is a belief in the policy arena that summary and overall environmental quality of products or processes is needed in order to invoke actions such as taxes, incentives, ecolabels, limitations, etc. Thus research must continue to broaden and improve these algorithms which generate summary measures such as ecofactors. A number of these algorithm concepts are included herein to illustrate the evolving and sometimes contradictory life cycle research on this subject.

Critical volumes is an algorithm concept developed in Switzerland. For materials discharged to air, water, and landfill, the mass emitted divided by the threshold concentration (mass per unit volume of a given media) yields a theoretical volume of that media which would be used to capacity. The threshold concentration is usually a regulated value. The sum of all critical volumes corresponding to the life cycle inventory of emissions from particular products or processes is thus the overall measure of environmental impact. A variation of this concept is referred to as ecological shortage. In this case the threshold concentration is replaced by concentrations believed to better reflect the sensitive portions of the ecosystem receiving the emissions. Generally these ecological threshold values are lower and hence larger critical volumes are obtained.

A third impact algorithm or approach is that of ecofactors. This approach (Buwal 1984) was also developed in Switzerland and is undergoing refinements based on new ideas . Within some geographical region one determines the flow of air and water resources and the available volume flow for solid waste management facilities. These critical flows in a region are divided by the actual flows generated by the particular product or industry under examination by life cycle analysis. This ratio is then divided by the critical flows to give an expression referred to as the ecofactor (with units of ecofactor per gram of chemical, such as sulfur dioxide, cadmium, chemical oxygen demand). Then the ecofactors for a given region are multiplied by the life cycle inventory results to provide an overall ecofactor for the given product or process. This approach provides some refinement for local conditions. The next evolution to this concept is being developed by scientists and economists in Switzerland (Braunschweig, 1994).

A fourth algorithm is evolving in Germany as a part of the advanced life cycle efforts of Prof. Eyerer (Univ. of Stuttgart, 1993). In this approach the economic implications of damages from chemicals discharged to the environment are envisioned to reflect a means of assessing impact. The sum of these costs for all constituents found to be attributed to a particular process, material, industry, or product is the measure of life cycle impact expressed in monetary terms. Such a monetarization algorithm would be useful in understanding investments and operating costs which offset or prevent such emissions to air, water, and land resources.

From a classical thermodynamic approach, a fifth algorithm is under development in Germany (Gilles, 1993). Research begins with defining a "state" function (much like Gibbs free energy) reflecting the quality function for environmental suitability. This function (W) is then related to the process variables and concentrations reflecting the product or processes under study. Then ecological modelling uses the environmental state function in available optimization techniques to establish the manufacturing conditions with the least impact on the environment. Having developed the theoretical basis for such an algorithm, the challenge remains to establish the appropriate function for environmental suitability.

Developments in Italy have produced a different algorithm to measure the aggregate impact of products. In this concept the emissions to air, land, water, raw material consumption, and the use of energy are determined for a given product (Scialdoni, 1993). This is done for each chemical constituent. Then the ratio of that emission, energy or raw material use is divided by that for the entire country or region from all sources. These fractions are then compared to that parameter with the lowest such fraction to provide a relative weighing factor. This is thus the relative importance attributed to chemical constituents and can then be used with the life cycle inventory to obtain a single environmental characteristic of each product.

A seventh approach is developing in The Netherlands (van Weenen, 1993). Their concept is that life cycle impact is heavily dependent on the perceptions of the public and policy decision-makers. Thus they begin with determining the characteristics of products, materials, or processes which are desirable from an environmental perspective to consumers and policy advocates. Such factors as repairability, litter potential, raw material scarcity, etc are developed. Then a fixed rating system (such as 1 - 10) is established. For the life cycle of each product the factors are given a score and then an overall average is determined. This average environmental rating can then be compared to alternatives reflecting the consumer and policy preferences adopted in this type of life cycle impact system.

## SUMMARY

Envisioning the cradle-to-grave assessment of environmental consequences of products, human activities, manufacturing processes, etc. as a multi-step concept, the principles which emerge are,

- inventory analysis
- impact analysis
- improvement analysis

The evolution of LCA as a tool for this cradle-to-grave assessment is occurring through steady, but modest research efforts around the world. New concepts for the principles or the stages of LCA are emerging in all three areas of LCA. The future of this field will thus reflect certain of these new research concepts.

## REFERENCES

Braunschweig, A. (1993) Schweizerische Vereingung für ökilogisch bewusste Unternehmungsführung, personal communications, Im stieg 7, 8134 Adlisnil, Switzerland.

BUWAL, (1984) Oekobilanz von Packstoffenstand, Swiss Federal Office of environment, Forests, and Landscape, Berne, Switzerland.

Cohen Hubal, E. (1992) Net waste reduction analysis applied to air pollution control technologies and zero waste discharge systems. MS thesis. Dept. of Chemical Engineering, North Carolina State Univ. 103p.

Cohen Hubal, E. and M. Overcash (1993) Net waste reduction analysis applied to air pollution control technologies, J. Air and Waste Mgmt Assoc.43:1449-1454

Eyerer, P. (1993) personal communications, University of Stuttgard, Institute für Kunststoffprufung and Kunststoffkunde.

Gilles, E. (1993) personal communications, University of Stuttgard, Institute für Systemdynamik und Regelungstechnik.

Scialdoni, R. (1993) personal communication, Agency for Energy, Environment, and New Technology, Rome, Italy.

SETAC (1991) A technical framework for LCA, Society for Environmental Toxicology and Chemistry, Washington, D.C., 134 p.

Studt, T. (1993) Chromium hazard drives replacement R&D. R&D Magazine, p.62

van Weenen, H. (1993) personal communication, University of Amsterdam, Amsterdam, The Netherlands.

# Prevention of Industrial Process Accidents for Sustainable Development

Adam S. Markowski
Faculty of Process and Environmental Engineering
Technical University of Łódź
ul. Wólczańska 175, 90-924 Łódź, Poland

## Introduction

The concept of sustainability described in the Brundtland report (World Commission 1987) relates to methods which do not injure the potential of nature to sustain future generation of humanity and other life forms. Approaching this concept the sustainable technology has been introduced. This type of technology can be achieved by hierarchical design approach which is characterized by the following features:

1. inherent efficient and clean (minimal use of energy and natural resources and minimal emissions to the environment: minimal environmental risks)
2. inherently safe (minimal safety risks; e.g. of explosion, loss of containment, leakage of toxics to the workplace and environment)
3. clean through recirculation of production wastes
4. clean through add-on methods (or *end-of-the pipe* technology)
5. clean-up techniques for improvement of the environment which have already been damaged by emissions.

The first three features are typical for the pollution prevention philosophy and they are a basis for sustainable development. This concept has to be taken into account at all stages of the production process called "total product life cycle".

The program of sustainable development cannot be completely achieved without the appropriate approach to the industrial process accidents. The plants which store, handle or produce hazardous chemicals due to accidents, may injure employees and the neighbouring community and result in major environmental and property damage (Marschal, 1989).

It is, however, evident that it is impossible to reduce these hazards to zero, since propensity to accidents is inherently built-in every industrial system. Therefore, authorities and industry have

NATO ASI Series, Partnership Sub-Series, 2. Environment – Vol. 2
Cleaner Technologies and Cleaner Products
for Sustainable Development
Edited by H. M. Freeman, Z. Puskas, and R. Olbina

had to face how to prevent effectively the industrial accidents, especially by well-founded decision-making process. This is why the concept of risk and method of its analysis have been introduced.

This paper presents the actual prevention approach to the industrial process accidents especially for major hazard industries like large-scale chemical and petrochemical plants, major storages and transportation facilities of hazardous materials, chemical warehouses, fertiliser work etc. There are three main concepts which can provide this approach:

1. Inherent safety design
2. Risk analysis
3. Risk management system.

This paper outlines the basic feature of each concept emphasizing the risk management as a main method to prevent accidents and their consequence to people, environment and property.

**Inherent safety design**

The chemical process industries are characterized by huge scale, capital-intensive nature, great complexity and large potential for different types of industrial accidents. The potential for risk depends on many different factors and it is often a combination of those which can lead to accidents or disasters. The most important factors concern the following: the scale of operations, physical and chemical properties of chemicals, types of operation and its complexity, human and equipment reliability and the effectiveness of management.

Industrial processes with a considerable amount of hazardous materials and/or energy may create a major hazard. A sudden loss of containment in production, storage and transportation systems, due to different types of process accidents causes the release of these materials to the workplace and environment sometimes leading to incidents such as large fires, explosions, generation of toxic smoke, dispersion of toxic vapour and ecological impact. As a result integrity of life, health, equipment and environment may be threatened.

Safety can be considered at different stages in the total process and life cycle. The earlier in the process it is considered, the less it costs to implement and the greater opportunity for so called inherent safety design. This concept can be understood as a method in which potential risks are removed and there is no chance for any types of errors (user friendly). The principle of this concept is summarized in Table 1.

Table 1. Principle of inherent safety

| Basic principal | Meaning |
|---|---|
| Intensification | Reduction of quantities of hazardous materials |
| Elimination or substitution | Avoid hazardous materials or use less hazardous alternatives |
| Attenuation | Application of hazardous materials under less hazardous process conditions |
| Simplification | Simplify the design equipment and safety systems |
| Separation | Separation of people from hazardous materials or hazardous situations |

Thus, inherently safe plants contain small inventories of hazardous materials, are simple, with few opportunities for error, easy to control and hard to assemble incorrectly. Consider the production of a particular product that requires a highly toxic chemical to be stored on site. The application of inherent safety may be realized by the elimination of that storage on site. This can be achieved in different ways, e.g. by change of technology where that dangerous chemical is used no more, or is not stored in big quantities, or is rapidly used as it is produced in another part of the plant.

The life of a plant can be divided into some stages: process development, process design, engineering design construction, operation and demolition. The inherent safety concept is especially important at the first two stages. Later, the safety aspects appear as extrinsic safety. It means that safety is built as *add-on-measures* or *end-of-the pipe* solution. Controls, alarms, interlocks, safety procedures etc. are the examples of extrinsic safety. At the operation stage there is still an opportunity for extrinsic safety but a very small chance for inherent safety. This is much more expensive than when it is introduced at the preliminary stage. Many successful inherent safety solutions named as friendly plants are presented by Kletz (1991).

The concept of inherent safety design can be, to some extent, implemented in qualitative hazard identification methods. A check-list or hazard ranking methods (e.g. NFPA or Dow Index) related to safety and environment can provide a useful insight in the plant from the principles of inherent safety and protect effectively against accidents.

Inherent safety concept may be especially efficient and useful in developing countries with restructuring economy like Poland. Our industry, due to a certain previous political system was designed uneffectively, equipment was over-dimensioned with huge consumption of raw materials and energy. As an example one of the big producers of polyurethane foams can be quoted where quantities of phosgene being used in processing are two-fold greater than in similar plants in western countries.The application of inherent safety may be one of the most efficient and cheapest methods to improve the standard of safety moving in direction of sustainable development ( Scharfe and Sewekow, 1991).

**Risk analysis in process industry**

In order to prevent industrial accidents and to optimize the process decision making concerning different aspects of industrial activities, the concept of risk and its analysis has been introduced.

Risk is a measure of the unsafety or the potential threat to the environment which is bound up with an activity or a system. The formal definition or risk is the combination of probability that occurs as a result of the undesired event.

There are three interrelated, however distinct categories of risk:

1. safety risks (S)
2. health risks (H)
3. environmental risk (E).

The interrelationship between the concept of "safety", "health" and "environmental" risks can be illustrated by the relation an exposure–risk presented in Fig. 1.

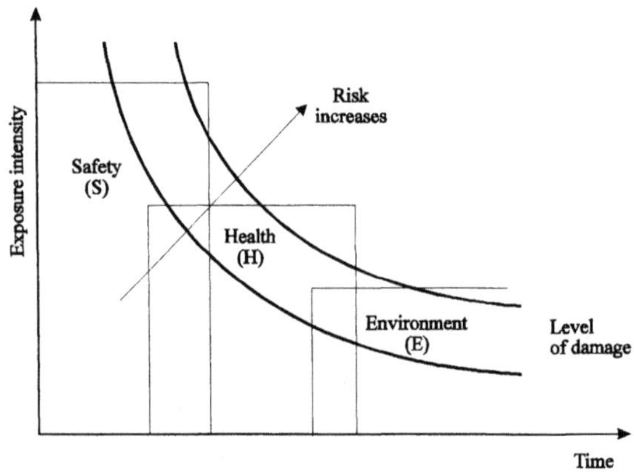


Fig. 1. SHE risks relationship

The "safety" risks refer to exposure situation in which the intensity level is high and time of exposure short. This is typical for accidental disruptions and catastrophic emission especially for major hazards (e.g. overpressure from an explosion, thermal radiation from an intense fire). The consequences used to be high and usually cover all types of losses, in particular casualties on the plant area and in the neighbourhood (acute effects), catastrophic damage to environment and property. The "environmental" risks represent regular pollution originating from daily operation of a complex industrial system with so called "long term effects", e.g. exposure to $SO_2$ from power plants. The "health" risks usually lie in between as, for example, those occurring in occupation situation.

Risk analysis refers to the method of risk identification and assessment. This method has been increasingly used since 1960s to assess the likelihood of serious accidents occurring in industrial installation classified as "major hazards" to describe the possible consequences, and to assign probability figures to them. The integral model which includes all categories of risk used in the Netherlands is shown in Fig. 2.

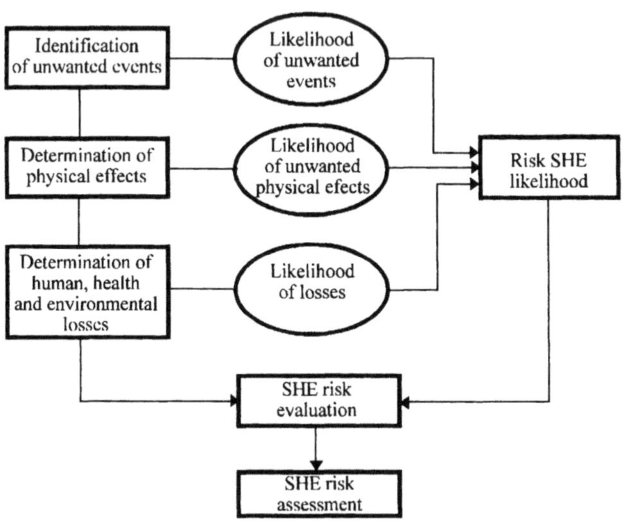


Fig. 2. SHE risks analysis integral model (Lemkovitz, 1993)

Quantitative methods of risk analysis (QRA) have been extensively presented in many publications (Guidelines, AIChE, 1989, Taylor, 1993). The results of QRA are presented in the form of frequency-distance diagrams or individual risk contour. Fig. 3 presents a risk contour for chlorine storage installation in one of Polish companies (Markowski, 1994).

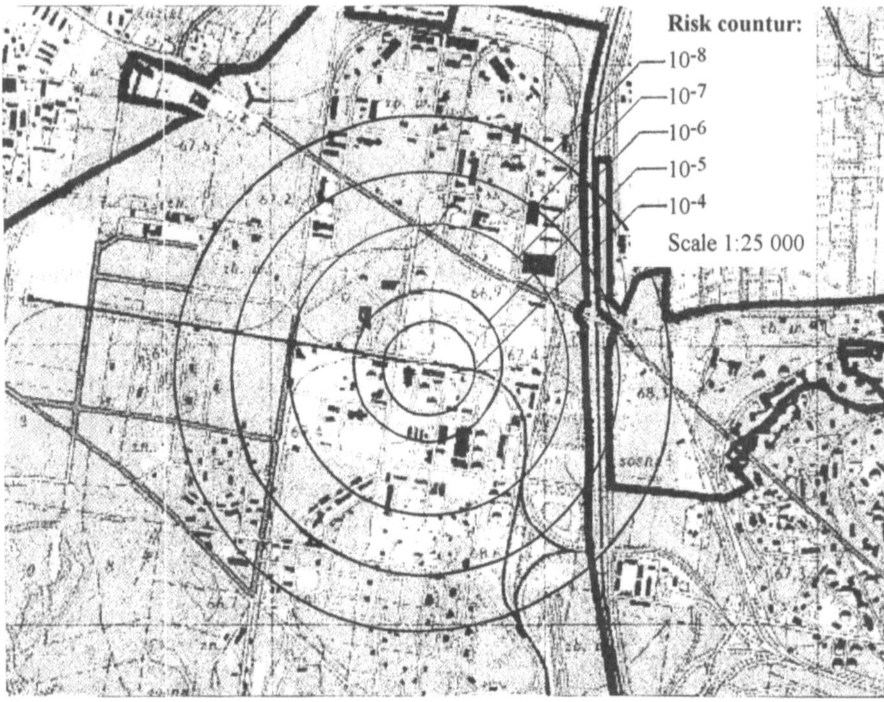


Fig. 3. Risk countur for chlorine storage and distribution at Chemical Works, Bydgoszcz

Although there is a lot of controversy and limitation on QRA (Roberts and Haynes, 1989; Ramsey, 1993) that methodology is successfully used as an input to decision-making process comprising the following major types of decisions:

1. Principal decision directed towards the choice of a certain activity or certain hazardous materials which should be developed or applied on large scale; important factor is the choice of raw and auxiliary materials, process technology and products.
2. Siting decision related to the choice of the location for activities in respect to the residential area and presence of special ecosystems; it includes also site and plant layout.
3. Routing decision connected with a choice of routes to be taken for transportation of hazardous substances.
4. Licensing decision directed towards meeting the goals of relevant regulation.

Beside that QRA is very useful for making priorities between alternative risk-reducing measures and allows for development of emergency preparedness plans and exercises. The use of QRA data in licensing and regulating processes has become much more common during the last years as a result of implementation of the post-Seveso directive (501/82/EC). In Poland there has not been yet such a regulation although we will accept it in the near future. It seems to depend, first of all, on the demand of society for safety and sustainable development and the relevant legislation which does not exist yet. Risk contours are to be used for physical planning and for emergency response planning.

Risk analysis represents a new safety philosophy. Accidents can be prevented by earlier identification of the potential causes and assessment of risks, their occurrence and consequences. Therefore, appropriate safety systems can be introduced before the accident may happen. This philosophy is called a "proactive" approach in comparison to the "reactive" traditional approach based on actions after accidental investigation. Both approaches are complementary although the major hazards in plants should be based on the risk analysis. Fig. 4 presents differences in proactive and reactive approach.

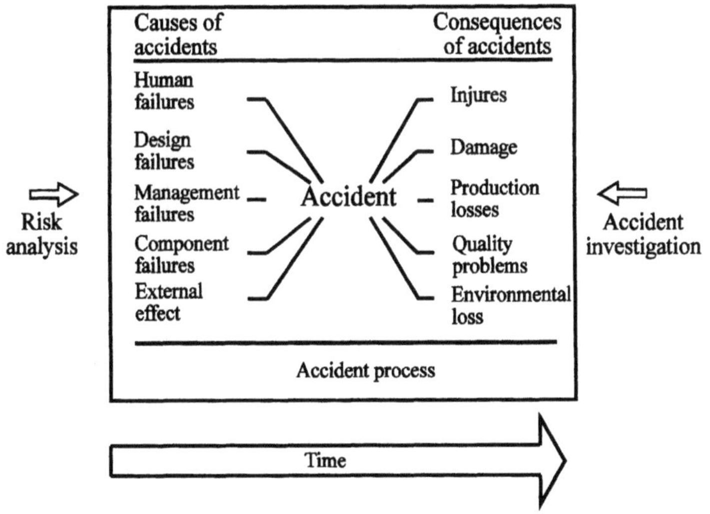


Fig. 4. Risk analysis and accident investigation approaches

**A general model of risk management**

Risk analysis applied to safety within the chemical process industry combined and integrated with modern management methods is extremely useful and necessary for reaching high level of safety and reliability. That is why recently, a rapid development of safety management methods has been observed. It consists in an introduction of structural management functions which usually cover: determination of policy and targets; organization; performance; control of performance and assumed targets.

Application of these functions to the safety system constitutes a safety management system which is based on 10 principles formulated by Petersen (1989). The system of safety management denotes a universal set of policies, programmes, procedures and practices to ensure efficient protection against accidents that might bring about losses. Risk management is an integral part of the safety management system and is a method for performing the safety programme. A general model of risk management is shown in Fig. 5.

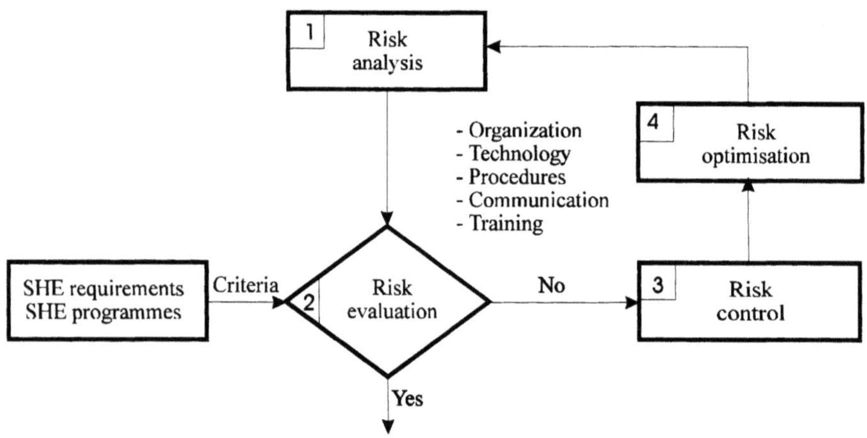


Fig. 5. Risk management cycle model

The first stage of the model is the risk analysis which covers

1. Description of the subject of study (installation, firm, eco-system), e.g. location, technology etc.
2. Identification of SHE risk aspects during normal operation and/or as a result of undesirable events (problem identification and description).
3. Quantitative calculation of SHE risk: effects and possible occurrence of hazards.

On the second stage, called risk assessment, we have

1. Determination of target criteria for risk evaluation.
2. SHE risk comparison of the result with a target criterion.

The third stage, called risk control, covers:

1. Determination of technology, procedure, organization, staff and social means of control (reduction).
2. Choice of control measures with preference given to these for hazard elimination, and then for the reduction of effects.
3. Realization of selected methods.

As a separate, the fourth stage can be taken:

4. Risk optimization and assessment and verification of efficiency of the control measures undertaken in relation to requirements (feed back) — closure of the cycle.

Integral parts of this model are organization which makes this model possible, technology for suggested control measures, procedures and standards as a guidance towards a safe operation, communication as a transfer of information concerning policy, standards, procedures to the employees and staff training.

Risk management is a closed cycle in which assumed or imposed requirements are satisfied. Risk analysis is used to obtain an input value of risk that occurs in particular fields of activity. This value, which is usually a resultant of technical standards and regulations applied and experience gained by a given company, is compared to SHE requirements determined by the acceptance criteria. In the field of safety and health these can be figures concerning exposure limits, toxic properties of products, risk criteria. As far as the environment is concerned these are the criteria of water eco-toxicity or ground water pollution criteria etc. Choice of these criteria should be in agreement with the principle of risk minimization. If the assessed risk is not in accordance with SHE requirements, it is necessary to undertake measures to control (reduce) the risk. These can be technical and organizational actions (procedures, permissions, inspection) as well as emergency response plans. Choice of the methods is based on the optimum substitution principle in which we select an optimal safety level from among different possible alternatives, or it is made on the basis of the risk cost benefit analysis (RCBA) in which — on the basis of economic factors only — we determine the minimum level of total costs.

Thus, the risk management can be called dynamic management. It aims at continual improvement of risk level as the starting point in this cycle is always higher. This provides a basic advantageous difference to a static approach in which a rigid system is set up on the basis of standards and regulations.

The key point in this model is risk management policy which concerns the target criteria for risk acceptability. Nowadays philosophy is based on two levels of risks (Ale, 1991). The upper one, the maximum acceptable level or tolerable risk level, should not be exceeded, irrespective of economic or social benefit that could result from the activity under consideration. The lower one (100-fold reduced with respect to upper one), the negligible level indicates the level below which it is not sensible to try to further reduce the risk, in view of the fact that man and the environment are already subject to other risks resulting from nature or society. Between these two levels there is a grey risk zone within which risk needs to be reduced according to the ALARP principle, requiring the application of at least the best practicable means (Kletz, 1992). The general model of risk management is an integral part of the safety management system. A company with a risk management system along the lines of the model will continually compare performance with targets, seek to correct flows and weakness in the system and search for improved ways to ensure achievement of objectives.

## Conclusions

1. Inherent safety design in earlier stages of the project can assure optimum standard of safety in design, construction, operation and maintenance and is a basis for prevention of industrial accidents. This concept may be especially efficient in developing countries as the most effective and simplest method to move in direction of sustainable development.
2. Prevention of industrial process accidents can be successfully achieved by the risk analysis which represents a new safety philosophy called "proactive approach". The accidents are identified before they happen which provides an appropriate basis for prevention. The decision-making process may be essentially supported by a risk analysis giving valuable input to sustainable development.
3. The general model of risk management as a dynamic cyclic process comprising risk analysis, risk evaluation and risk control presented in this work can be treated as the improved integrated methodology for prevention of industrial accidents which meets regulatory requirements as well as optimizes the safety, health and environmental risks.

## Acknowledgments

The author gratefully acknowledges the financial support given by the Organizing Committee for his participation in the NATO Advanced Research Workshop "Cleaner Technologies and Cleaner Production for Sustainable Development", Budapest, September 12-15, 1994. Special thanks are due to Mr Harry Freeman of EPA USA for his excellent leadership and truthful and valuable discussion at the workshop.

**References**

Ale BJM (1991) Risk analysis and risk policy in the Netherlands and the EEC. Journal of Loss Prevention in the Process Industries 4, 1:58-64

Guidelines for Chemical Process Quantitative Risk Analysis. (1989) CCPS AIChE, N.Y.

Taylor JR (1993) Risk Analysis for Process Plant, Pipelines and Transport. Chapman & Hall, London

Kletz T (1991) Plant Design for Safety – A User-Friendly Approach. Hemisphere, New York

Lemkovitz SM (1993) Systematic Hazard and Risk Study. Course MOSHE, v.5, TU Delft

Marschal VL (1989) Major Chemical Hazards. J. Wiley, N.Y.

Markowski AS (1994) Risk Analysis in Storage and Distribution of Chlorine. COPRO Łódź (unpublished, in Polish)

Petersen D (1989) Techniques of Safety Management. Aloray Inc., New York

Roberts LEJ, Haynes MR (1989) Limitation and the usefulness of risk assessment. Risk Analysis 9:4

Ramsay C (1993) Overview of QRA strengths and weakness. Risk 2000, IBC and DNV Technica Seminar, London

Scharfe G, Sewekow B (1991) Das Beispiel Bayer, Um Weltperspective II - Forum in Leverkusen

World Comission on Environment and Development (1987) Report to the United Nations, " Our Common Future ", Oxford University Press

## INDEX

### A


abrasives 201
absorbent preparation unit 300
absorption principles 299
accident investigation 477
activating group principles 374 377
actual design process 188
additional costs for investment and operation 115
advanced life cycle efforts 467
agitated immersion cleaning 202
agricultural
    production 55
    products 64
    waste 153
air
    cleaning technologies 133
    conditioning technology 91
    pollutant emissions 325
    pollution 69
    pollution control 329
    pollution control sequences 458
    pollution emission 310
    pollution problems 291
    stream 136
aldol reaction 375
alkali extraction 361
alkane oxidation 391
alternative energy sources 72
alternatives to chlorinated solvents 434
analytical methods 363
annual
    waste reduction 431
    volume of waste 112
antibody catalyzed hydrolysis 390
applied biocatalysis 396
aqueous
    cleaning processes 88
    based solvents 201
aromatic
    substitution reactions 380
    acylation waste problems 377
ash dam 314
automotive manufacturing 38


### B


Baltic Sea contamination 74
barriers to cleaner production implementation 48
base-line data 363
benzene alkylation 375

biocatalysis 383
biodegradability 236
biogas
    production 155
    program 155
biogeneous resources 32
biological
    method 204
    catalysis 372
biomass 25
biotechnological process 31
biotransformation 387
bleachery effluents 359
bleaching 335
    rinse water 338
    effectiveness 366
bleaching sequence 367
bottom ash 298
Budget of German Federal Ministry for Research and Technology 125
business/industries waste generation reduction 406

C

"Cafe Reducto" 409
Canada's Green Plan 439
carbanion cyclisation 389
carbon-containing bulk natural products 384
case studies 49 146
catalyst 133
catalytic
    antibodies 38
    conversion 138
    modification 394
    reaction 392
    wet oxidation cleaning 434
catalyzed prototypes 375
CEE countries 20
cellulosic fibers recovery 255
cement 309
    based solidification 169
    industry 165
    kiln 172
    manufacturing 312
    quality 176
ceramics manufacturing 320
changing raw materials 2
chelating agents 234
chemical
    education 159
    industry 52 141
    manufacture 371
    process quantitative risk analysis 481
    production 351
    reactions 128

reactivity 379
substitution 233 339
chemicals costs reduction 195
chlorinated
hydrocarbons 135
organic compounds 229
organic solvents 191 197
organic solvents alternatives 205
chlorination 362
strategies 370
chlorine dioxide substitution 369 366 364
chlorofluorocarbons substitution 87
chrome plating rinse treatment/recovery scenarios 462 463
classical thermodynamic approach 467
clean production promotion 21
clean
air 139
chemistry 395
production 1 28 32 43 414
synthesis 371 393
synthesis leaving group principle 376
synthesis principles 373
technologies 13 81 98 107 151 165 185 259 396 400 422 429
technology research and development 125
cleaner
energy production 279
production concept 47
production practice 210
production rules 347
production technologies 441
products 441
technologies implementation 270
types 201
cleaning
effectiveness 200
media 200
method 199
cleaning and degreasing 435
climate change 440
clinker 166
clinker manufacturing 167
closed
loop evaporation 462
water circulation 89
co-product allocation 60
coal preparation 313
collaborative projects 105
college credit class 404
combustion chambers 298
company performance 267
completely enclosed vapor cleaner 433
comprehensive polymer science 395
computer
programs 410

software 410
concrete 309
Confederation of Norway Business and Industry (NHO) 193
consensus building 425
construction materials 303
contaminated land 78
contrasting routes 376
control efficiency 458
conventional treatment 462
cooling conditions 357
copper slags 351
cotton
consumption 342
flowering 333
production 342 347
textile industry 331
yarn production 335 344
"cradle to grave" 421
cross
border pollution control 218
flow microfiltration technology 97
media pollutants 440
cumulative emission factors 326
cyclone operation 175
Czech
Cleaner Production Center (CPC) 47
Norwegian Cleaner Production Project 51 260
Republic 43

**D**

damaged land by military exercising 79
Danish
Environmental Agency 76
experience 58
data collection 57 177
decomposition mechanism 139
degreasing 204
delignification 368
demonstration of clean processes 35
design
for entire life cycle 454
for the environment 8
desizing 334
destruction and removal efficiency 171
diesel fuel consumption 310
Directory of Pollution Prevention in Higher Education 452
dispersible waste 272
dredged material 112
drinking water quality 75
dry cleaning industry 37
dust
control 314
emissions 111

Dutch Ecodesign Program 185
dye
consumption 277
house 271
house process 344
house production process 273
staff recovery 255
dyeing 334 336
process 273

**E**

Eastern
Europe 17
European countries 160 217 251
Eco
design MET-matrix 188
factors 466
labelling programs 22
Tex Standards 346
ecologically sound solution 186
economic
benefits 51
difficulties 79
instruments 21
operation 97
recession 263
transition 215
economically feasible solutions 186
economics in transition 17
economy 66
restructuring 217
transformation process 261
education 399 412
education program 9
effluent
properties 365
quality 364
electric
circuit improvements 189
furnace 134
electricity consumption 183
electricity-originated emissions 325
electroplating 82
sludge 82
electrostatic
precipitator 172
spray painting 85
emission 168 174 457
control 340
factors 296
employee training 401
end-of-pipe 445

technologies 13 102 115
energy
conservation 332
consideration 168
costs reduction 195
consumption 72 93 160 226 262 338
efficiency 58
efficiency improvement 216
production 52 70
products 64
requirements 292
saving 187
engineering
curriculum 400
education 443 448
design 447
design courses 449
Envirocat
catalytic reaction 388
reuse & recovery 388
environmental
activity 386
analytical chemistry 163
aspects 190
auditing training 182
awareness 214
centers 181
control 16
costs 306
crisis 26
education 9 157 159 236
educators 5
effects 54 61
engineering 158 181 446
engineering courses 161
engineering institutes 182
impacts 113 307 332 466
impact assessment 246
impact reduction 225
implications 460 464
issues 214 444
legislation 111 245 266
management 220
management hierarchy 418
management system 222
monitoring system (SEM) 222
policy 65 69 99 109
pollution 67
pollution control improvement 254
problems 6 244
program 213
protection 162 245 256
research 126
risk 474

science and technology department 405
situation in Estonia 65
strategies 15
technique 455
technology 100 107 126
Environmental
Law Act No. II - Hungary 18
Law 244
Protection Agency of the USA 170
Protection - Germany 108
Protection Act - Slovenia 18
Protection Technology 108
environmentally
friendly design 419
pure compound 390
enzymes 386
equipment and technical support 218
executive level decision 425
experimental setup 286
extrinsic waste 144

F

federal facilities 37
Federal
Immissions Control Act 101
Ministry of Research and Technology 123
Water Act 101
fertilizers 349
FGD trial runs 299
final product design 189
financial support 205
finishing costs 62
fixation strength 345
fixed rating system 468
flue
gas 298
gas purification facilities 110
Flue Gas Desulphurization (FGD) 297
flux
glazing solutions 320
system development 88
fly ash 71 298
food
production 153
products 53
forests 68
formic acid substitution 340
fossil fuels 279
fresh catalyst 137
fuel use 307
fugitive emissions 311
"Fun Factory" 409

G

garment manufacturing 344
gas bypass 175
gasification
    molten carbonate fuel cell 281
    unit 283
general framework 269
geographical origin 60
German Federal Ministry for Research and Technology (BMFT) 81
Germany's "take-back" law 451
glass manufacturing 322 323
global environmental chain analyses 190
good housekeeping 2 275
government
    financial support 109
    policy 50
    responsibility 450
graduate courses 46
gravel 309
Great Lakes
    Pollution Prevention Centre 34
    St. Lawrence Pollution Prevention Initiative 33
"Green Square" 409
green technology by design 443 448 453
gross land utilization 305
groundwater
    contamination 75
    protection 76
guidance documents 408
gypsum dewatering system 300

H

halogen-free processing methods 90
halogenated hydrocarbons substitution 87
hard to treat waste 272
hazardous
    organics 176
    substances 249
    waste 42 221 272
Hazardous Solvent Substitution Data System (HSSDS) 411
health risk 474
healthy food production 216
heat
    losses 183
    recovery 275
heavy metals 71 229
hidden waste 463
hierarchical approach 142
high
    energy conversion efficiency 283
    frequency application 233
    oxidation rates 355

volume waste 272
higher quality 98
homogeneous solution 381
housing area 327
hydrogen sulfide removal 280
hydroxylation 391

I

impact analysis 468
improved
operating practices 423
processes 460
rinsing technology 83
improvement analysis 468
improving engineering design 454
increase environmental standards 106
industrial
accident prevention 480
activity cycles 374
effluent 238
immunology 397
management 180
pollution prevention 414
process accidents 471
products 64
synthesis 372
technology type 432
industry 113
specific manuals 428
information 460
communication 50
exchange 194
support 20
transfer 211
infrastructure facilities improvement 223
inherent
safety design 480 472
safety principle 473
input
materials 104
material change 423
insufficient education 3
Integrated Gasification Combined Cycle (IGCC) 280 282
integrating environmental factors 445
intrinsic waste 144
intrinsically 'clean' reaction 373
inventory analysis 468
iron manufacturing 316

K

key elements 11
kraft bag production 315

L

Lake Superior 39
land use 305
disturbances 306
lead smelting 92
leaving group 372
legislative provisions 48
less stable complexes 234
licensing decision 475
life cycle 53
analysis (LCA) 337 455 464 469
analysis concept 468
impact analyses 456
inventories 456
screening 59
studies 63 303
lignite
characteristics 294
fuel 293
lime
manufacturing 318
stone quarry 318
limited bleaching 365
liquid
fuels 173
waste incineration 170
list of target substances 36
Lithuanian
experience 179
industry 184
loading
time 308
capacity 308
"Low Waste Technology" 131
low
emission processes and products 103
temperature plasma treatment 235
waste technology 83

M

magnetic
iron oxide 352
oxide production 357
properties 356
management system 267
managerial skills 264
manufacturing technologies 328
marginal methods 64
mass
balance principles 465
exchange network 148
material

cycle closing 187
flow reduction 27
transport 307
membrane distillation 84
metal
complexes 393
dissolver plant 130
finishing 38
parts cleaning 206
parts degreasing 191
metalworking operations 197
methodology 59
cleaner production implementation 45
microorganisms 385
middle management 12
military environment 78
minimizing emission/pollution 121
mobility of information 29
model compound 135
molecular biology 383
molten
alloy 3354
salt coating removal 433
Montreal Protocol 192
multi
cycle sulfidation 288
product process 61
municipal wastewater treatment plants 74


**N**


national
environment education center 223
environmental policy act (NEPA) 4
pollution standards 215
National Higher Education Pollution Prevention Center (NPPC) 402
natural
gas 293
product synthetic chemistry 379
resources 66 122 328
naturally occurring polymer - biopol 384
net waste reduction analysis 469
Netherlands's National Environmental Policy Act 439
new
energy sources 26
chemistry 387
nitrating agent reactivity 381
nitric acid
over-all reaction 129
production 127
nitrogen
leaching 62
oxides emissions 296
nitrous

gas emission 127
oxides 128
North Sea Deceleration 192
Norwegian
Environmental Protection Agency 193
Society of Chartered Engineers (NSCE) 44
NOx
gas absorption 129
gasometer 130

O

oil shale
deposits 67
processing 70
on-site recycling 149
organic constituents 171
Organization for Economic Cooperation and Development (OECD) 438
overall flowchart 56
oxidative extraction 362 367
oxygen-reinforced extraction 368
ozone depletion 440

P

packaging 348
paint manufacturing 321
paper/pulp industry 120
particulate
emission 314
monitoring instruments 301
Partnership for Environmental Technology Education (PETE) 403
periodic waste minimization assessment 426
permitting process 177
pesticides 349
PHARE
environmental sector program 219
program 213
phenol mononitration 380
phosphates 230
plant design for safety 481
plasma cleaning 435
plastic
building material manufacturing 319
foaming 90
plating operations 198
policy
instruments 266
statement 425
Polish
industry 207
National Environmental Policy 219
pollutant
emissions 324

handling 16
sources elimination 93
pollution
categories 241
control 268 445
pollution control technology 302
load reduction 277
loads 227
pollution prevention 4 179 184 243 248 260 311 331 356 399 407 408 412 445
agreements 40
assessment 406 427
benefits 420 432
centers 240
compendium 413
concept 256
curriculum 402
education 403
education in the USA 3
guide 350
ideas 449
pays 209
plans 36
program 152 424
promotion 33
research 257
savings examples 428
seminars 12
strategies 337
techniques 257 422
training 39 415
Pollution Prevention
Advisory Group 5
Information Clearinghouse (PPIC) 410
Information Exchange System (PIES) 410
Hazardous Waste Minimization course 404
polyvinyl alcohol 96
portioning phase 353
Portuguese industry 243
postgraduate
education 158
specialist course 162
potential
dry grinding 124
savings 276
Pressurized Fluidized Bed Combustion (PFBC) 281
preventive
environmental protection 19
policy 208
solutions 6
primary raw material 119
principal decision 475
printing 336 344
problem solving 11
process

classification 147
design 473
development 473
improvement 142 465
optimization 86
related investments 114
waste 143
product 104
change 419
integrated environmental protection 117
life cycle costs 451
modification 28
related investments 114
service intensity 29
system 54
production
integrated environmental protection 103 117
process 104 121 429
program
evaluation 427
integration 40 42
project
creation 34
organization 194
partners 14
PREPARE 19
prototypical feedstocks 385
public education 10
pulp
bleaching 360
bleaching effluents 360
consistency 361
quality 369
pulverized coal combustion system 301
"Pure Technology" 131
putative catalytic antibody 389
pyro-processing 166

**Q**

quality
control 447
concerns 268
enhancement 187
quantitative methods 476
quartz tube 284

**R**

radioactive waste 77
raw material
consumption 262 450
consumption reduction 246
conversion 24

preparation 313 317 323
requirement 14
reactive dyes fixation 235
recovery 252 237
recycle 7
Recycle Economy 118
recycling 237 247
economy 123
reduce 7
reduction procedure 356
reference guide 413
refrigeration 91
regeneration reaction 287
regional environmental programs 220
regular operation 89
renewable
natural resources 68
raw materials 253
resources 30
research and development 99
residue material recyclability 86
resource conservation 444
Resource Conversation Recovery Act (RCRA) 421
reuse 7
risk
analysis 472 474 477
analysis integral model 476
assessment 479
control 478
management 478
management system 472
optimization 478 479
Romanian perspective 254
rotary kiln 312
routing decision 475
rye bread life cycle 56

S

safety risk 474
sand & gravel production 319
sanitary engineering 446
saturation temperature 134
scanning electron microscopy 137
scare materials replacement 30
scouring 228
screening results 57
secondary
diploma course 161
materials 122
selected catalyst 136
selective epoxidation catalyst 392
separation 203
sewage sludge 112

shift of pollutants 102
simplified procedure 63
single
    closed reactor 92
    phase arc furnace 353
sintering 285
    zone 174
siting decision 475
sizing
    agents 94
    agent recovery 95
slag utilization process 352
Slovak industry 270
Slovakia 261
Slovenian industry 271
small and medium size companies 106
socialist market system 263
solar energy 253
solid
    waste 77
    waste formation 169
Solid Waste Information Clearinghouse (SWICH) 411
solvent
    based cleaners 201
    containing paint 84
    emission reduction 85
    soluble sizing agent 96
Solvent Alternative Guide (SAGE) 410
source
    conservation 452
    elimination 419
    reduction 230 247 252
special waste 105
spray washing equipment 202
standard
    chemistry 376
    reference 63
state facility planning requirements 437
state environmental
    funds 265
    regulatory programs 407
steam-cooled system 282
steel 309
    manufacturing 305 317
    production emissions 326
stenum-working group 1
stoichiometric reaction 394
strategy of cleaner production implementation 49
student
    education 10
    education/training 401
    enrollment 405
sulfate formation 287
sulfidation efficiency 285 286 289

sulfur
 dioxide emissions 110 295
 oxides scrubbing system 300
surface
 quality control 199
 treatment 198
surfactants 230
sustainability 31
sustainable development 15 157 251 259 443 471
 definition 27
 implementation 269
synthetic
 building blocks 372 388
 dyestuffs 345
systematic
 procedure 150
 studies 211
systems approach 196

T

teaching environmentally responsible design 453
Technical University of Budapest 163
technically educated professionals 180
technology change 423
textile
 chemical use minimization 232
 chemicals 226 228 241
 chemicals manufacturers 239
 chemicals substitution 242
 effluents 232
 effluent treatment 238
 industry 94 120 225
 processes 242
 raw material quality improvement 240
 wet process 272
 wet processors 239
thermal
 cleaning 203
 power plants 292 299 302
top management 12
 support 424
total
 air pollutant emission 327
 chemical losses 457
 mass pollutants control 459
 waste generation 459
toxic substances 35
 elimination 41
Toxic Release Inventory 437
traditional synthesis 382
training 412 399
 & education 264
 programs 8 46

evaluation 415
transport 348
trichloroethylene 138
tubular furnace 284
Turkish
agricultural areas 343
Air Quality Protection Regulation 295
coal reserves 294
Gross National Income 304
National Science Foundation 370
pulp & paper plant 359
Scientific and Technical Research Council 289
Statistical Yearbook 329
thermal power plants 291

U

U.S. EPA 417
U.S. EPA's
Green Lights Program 436
Guides to Pollution Prevention 430
Pollution Prevention Research Branch 408
Pollution Prevention Benefits Manual 420
30/50 Program 436
Facility Pollution Prevention Guide 430
ultrafiltration 95
ultrasonic cleaning 202
United Nations Industrial Development Organization (UNIDO) 438
University of Hannover 124
usable dwelling floor space 304
utility
waste 143
waste minimization 146

V

varnish manufacturing 321
vegetable and animal products 55
video & audio tapes 352
volatile organics 173
voluntary
agreements 22
pollution prevention agreements 41

W

waste 116
chlorine gas 315
constituents 461
disposal 354
disposal costs reduction 195
gas 116
generation 274
generation characterization 426
generation reduction 248

management 100
projects 45
processing 119 154
reduction 141 276
reduction alternatives 145
reduction programs 209
sources 145
treatment 461
waste minimization 14 23 44 100 150 151 152 206 249 271 339 397
options 274
opportunities 431
projects 208 210
techniques 147
Waste Management and Product Recycle Act 118
wastewaster 116
generation 322
management 221
minimization 148
treatment 73 154 341 350
water
consumption 226
contamination source 227
pollution 73 333
reuse 231
use minimization 231
weaving capacity 343
weight loss 355
Western European countries 160
wet-process clinker manufacturing 167
wood product manufacturing 324
world commission 23


**Z**


zeolites based clean process 382
zero discharge 441
zinc ferrite regeneration 288

# Springer-Verlag and the Environment

MIX
Papier aus verantwortungsvollen Quellen
Paper from responsible sources
FSC® C105338

If you have any concerns about our products,
you can contact us on
**ProductSafety@springernature.com**

In case Publisher is established outside the EU,
the EU authorized representative is:
**Springer Nature Customer Service Center GmbH**
**Europaplatz 3, 69115 Heidelberg, Germany**

Printed by Libri Plureos GmbH
in Hamburg, Germany